CW00747245

Military Maverick

In precious memory of my RAF father Ernest, who
served throughout both world wars and lived to be
ninety, and my mother Gladys, who died when I was ten.

Military Maverick

Selected Letters and War Diary of 'Chink' Dorman-Smith

Foreword by
Frederic Raphael

Edited by
Lavinia Greacen

Pen & Sword
MILITARY

First published in Great Britain in 2024 by
Pen & Sword Military
An imprint of Pen & Sword Books Limited
Yorkshire – Philadelphia

ISBN 978 1 03610 227 2

Typeset by Mac Style
Printed in the UK by CPI Group (UK) Ltd, Croydon, CR0 4YY.

Pen & Sword Books Limited incorporates the imprints of After
the Battle, Atlas, Archaeology, Aviation, Discovery, Family History,
Fiction, History, Maritime, Military, Military Classics, Politics,
Select, Transport, True Crime, Air World, Frontline Publishing, Leo
Cooper, Remember When, Seaforth Publishing, The Praetorian Press,
Wharncliffe Local History, Wharncliffe Transport, Wharncliffe True
Crime and White Owl.

For a complete list of Pen & Sword titles please contact

PEN & SWORD BOOKS LIMITED
47 Church Street, Barnsley, South Yorkshire, S70 2AS, England
E-mail: enquiries@pen-and-sword.co.uk
Website: www.pen-and-sword.co.uk
or
PEN AND SWORD BOOKS
1950 Lawrence Rd, Havertown, PA 19083, USA
E-mail: uspen-and-sword@casematepublishers.com
Website: www.penandswordbooks.com

'I would have liked to have been you, with my luck and your brains. But first I would have had to have studied under you at Camberley. This makes it a little complicated. If we could have mixed our lousy talents it might have made a fairly sinister and dangerous Condottiere of some kind. And a better writer than me.'

Ernest Hemingway to Chink, 7 November 1954

'It has always seemed to me that most people seeing the muddles of war forget the muddles of peace and the general inefficiency of the human race in ordering its affairs. War is a wasteful, boring, muddled affair; and people of fine intelligence either resign themselves to it or fret badly, especially if they are near the heart of things and can see matters which ought to be done, or done better, and cannot contrive to get them set right.'

Field Marshal Archibald Wavell, correspondence 1943

Contents

Foreword

It is curious, if not sinister, how certain people fall into the oubliettes of history. In view of the obsessive refighting which World War II, in particular, has engendered, it might be assumed that someone described by Liddell Hart as 'one of the most brilliant soldiers that the British army has produced in modern times' would be, if not a household name, at least frequently flagged in the literature. Yet who would put Eric 'Chink' Dorman-Smith in the first 50 generals of distinction in what may, if we are very lucky (or very unlucky), be the last great land war fought by the British army?

What comes out in the wash still depends on how the machine is loaded. It is Chink's case that he was ill-used during his active life and deliberately, even scandalously, blackguarded after it. Churchill, Alanbrooke and Montgomery – to name but a triumvirate – were variously at pains to brighten their own lustre by tarnishing anyone who might otherwise outshine them.

Brilliance is not a quality which the British are, in general or in generals, disposed to tolerate, unless it is mitigated by self-deprecation or jolly-good-chapmanship. Prudent careerists are advised that it is not enough to suffer fools gladly; one must also regularly congratulate them on their intelligence. 'Chink' not only failed to make the right friends, he was also conspicuous (almost diligent) in acquiring the wrong enemies. He made the cardinal error of imagining that, in collegiate life, it could be profitable to be both right and outspoken.

Had Chink been better connected, he might have done without the good opinions of those he despised; had he been more diplomatic, they might have volunteered them. His greatest misfortune was that his insolence had so thin a skin that, having courted ostracism, he suffered hellishly from its consequences. Had Auchinleck not aggravated Churchill with the accuracy of his caution, Chink's star might have risen with that of a commander whose respect for him was close to deference.

As soon as Montgomery arrived, determined to date the creation from his own advent, the reputations of those who had preceded him were, inevitably, on the bonfire, especially if their battle-plans were necessarily to be incorporated into his own.

The hyphen in Dorman-Smith acquires unnerving significance as one reads his story. He was a split personality in almost every regard, except sexually. At a touching moment during the most desperate period of the desert war, Auchinleck was so keyed up by the anguish of command that he could not drop off to sleep except by holding Chink's hand. But for the rest Chink was a ladies' man, gallant and a little unscrupulous, in a profession where 'that sort of thing' tended to be the cad's domain. His fall from grace was accelerated by his wartime infidelity to the wife he left in India, although it was less his passions than their lack of concealment which affronted. Absence of humbug is the last straw.

Chink's hyphenate character began with his Anglo-Irish origins. Having been baptised a Catholic, he was to see his younger brothers recruited to the Church of England. He became in some sense an only child who happened to have siblings. As a boy, he was at once excited and alienated by the Orangemen's drums. The final twist to his story, when he changed his name to O'Gowan and became a military advisor to the IRA, is perhaps implicit in the circumstances of his childhood.

He was not only brave (his Great War MC was almost less than he deserved), he was also bookish. The library was as congenial to him as the battle. Hemingway, whom he met by chance in Milan in November 1918, treated him as the genuine hero the author would have chosen to be. If Chink-clones recur in Hemingway's work, from Fiesta all the way through to the 'posthumous' *The Garden of Eden*, it is typical of his rotten luck that Chink figures most prominently as Cantwell in *Across the River and Into the Trees*, widely agreed to be Hemingway's worst novel.

These letters, and the War Diary in particular, convey the doubleness of Chink – his charm and his odiousness, his will to prevail and his aptitude for self-destruction – in such a way that we gain a parallax view of him. His bad luck was as much a function of his intelligence as was the superbly lucid plan by which he defeated the Italians in 1940 and helped the Auk to fight Rommel to a standstill at the first battle of Alamein. If his enemies had been only on the other side, he would have risen from his temporary rank of Major-General to the knighted generalship he deserved, instead of dwindling into dejected colonelcy and, finally, into closet treason (not that he would have seen his Irish nationalism in any such light). Failure to recognise merit is not limited to the army.

His own words deserve wide attention. They are at once shiningly honest and unapologetically partisan; they illuminate the nature both of courage and of command. I cannot think of any account of soldiering, by a man or a woman, which more convincingly conveys its bloody allure.

Frederic Raphael

Preface

Eric 'Chink' Dorman-Smith (1895–1969) was the subject of my earlier biography *Chink*. That book portrays his life from the standpoint of my own view, having been given access to his papers, and through the opinions of his contemporaries and those who met him in retirement. He raised strong emotions in people throughout his life – both for and against – and it is poignant to recall that range of lively conversations, as I do, and to know that the majority of my interviewees are no longer alive.

Military Maverick is in a different league from a biography. From beginning to end it contains Chink's own riveting descriptions of people and events, exactly as written in the heat of action. It admits readers to the hidden inner life of the man himself, and every page is a direct communication across time.

Gossipy, honest and compelling, his gift for the witty phrase brings alive his first-hand experiences in the army, which range from the horrors of trench life in World War I to brief fulfilment on the desert sands of Alamein at a critical moment in World War II, before renewed claustrophobia at Anzio in 1944. Written on the spot, his record highlights his complicated and passionate love affairs, his working relationships with colleagues over the years, and his postwar disillusion and despair. It shines a new light on his long friendship with Ernest Hemingway whose lifelong hero he became, so much so that Chink's principles of chivalry and honour strongly influence the world-famous novels. As a result, his impact on the course of the twentieth century was to be two-fold: in both the literary and the military worlds.

Chink's private unvarnished words, switching from enthusiasm to anguish, have the force of autobiography. Memoirs are written selectively, with at least one eye on reputation and history. This account is so much better. As we look over his shoulder, we read the genuine story of the First Battle of Alamein as it unfolds. We follow Chink's exhilaration, his despair and his self-reproach, and we can glimpse the arrogance, as well as the strands of narcissism and self-pity that raised so many hackles. The outwitting of Rommel during that supremely dangerous month of July 1942 is still controversial, but historians can comprehend the value of the tactical advice and emotional support he gave Auchinleck in the emergency, which together they averted.

'I know what I owe to you', Auchinleck would write privately to him in 1967, 'and realise very clearly, as I have always done, that without your wise and indomitable thinking always at my side and in my head, we could never have saved Egypt – and India and all the rest.' Today their achievement is seen as a pivotal turning point for the Allies in the war.

This is a gripping and intimate one-sided conversation with a fascinating man, and he has an extraordinary story to tell.

Lavinia Greacen

Apologia

This collection of Chink's letters is authentic; he did indeed write down each word here as it entered his mind. They are his choice at the time, and not mine. For historical value, and for readers to experience a genuine frisson of eavesdropping across time, their transmission must be accurate. So his descriptions of the enemy during the war in North Africa as Huns (Rommel's troops) or Wops (the Italian army) have been retained, at the risk of giving offence to readers. Other occasional terms include Yanks (the American army) and Japs (the Japanese army). Everyday language for soldiers in the Second World War is often unacceptable today. If seeing those words in print is a shock, I apologise in advance.

An Outline of Chink's Life

He was born on 24 July 1895 at the Palladian mansion Bellamont Forest, near Cootehill in County Cavan, in the days when Ireland was still part of the British Empire, and christened Eric Edward Dorman Dorman-Smith. His Catholic grandfather, initially a poverty-stricken emigrant, had made the family money, and his father, Edward Dorman-Smith, was a shy *nouveau riche* landlord whose brief army service in the Boer War had been rewarded by appointment as Justice of the Peace for the county. Amy, his Protestant young mother from England, resented the Catholic ceremony so much at her first baby's christening that her own religion was imposed on Eric's two younger brothers.

A solitary and energetic child, he gloried in having the freedom of the large estate in his parents' frequent absence. Authority lay with his local nurse – today she would be called a nanny – and a procession of tutors whom he could usually outwit. He would gravitate to the servants' hall where they made much of him, and he liked listening to their stories. Being a voracious reader from an early age, he was rarely bored; lying on the rug by the library fire with a book chosen from the well-stocked shelves was his idea of bliss. Out of doors his companion was his dog, a tough springer spaniel.

Eric did not go to school, but he was allowed occasional visits from the only boy of his own age judged suitable, the local doctor's son. His brothers were too young to be good company, and from the start he felt set apart. 'I am the heir,' he corrected them sternly one day. Admiration focused on his father's military friends, whom he could observe from a distance. Edward's reputation as a crack shot stemming from his service in South Africa was well known, and officers stationed in Ireland were invited to regular shoots at Bellamont. In boyhood, indelibly, the notion of a future career was planted.

The Irish idyll could not last, however. At twelve Eric was sent to Lambrook, a 'prep' school, near Maidenhead and, envisaging a future dominated by the cross-channel education of all three sons, his parents closed up Bellamont Forest and took a house by the Thames. At school Eric's accent was promptly ridiculed, and, never having played any team games, he was scorned for knowing none of the rules, isolating him further. By public school at Uppingham in 1910, though, he had externally adjusted. Inwardly, he remained a loner. His closest friendship at Uppingham would be with Brian Horrocks, a contemporary with the same ambition of attending the Royal Military College, Sandhurst, for which the entrance exam was stiff for the 172 annual places. In the event both seventeen-year-olds proved successful. Dorman-Smith came 69th in order of merit, assured of entry at once; Horrocks managed to scrape in, getting the 171st place.

At Sandhurst, Eric found his feet at last. He relished the atmosphere and teaching, and after two terms was allowed to leave early in February 1914 to join the Northumberland Fusiliers, his choice of regiment. Tall, thin and immaculately dressed, in the regiment his air of nervous energy won him the nickname 'Chink' after their mascot, a Chinkara antelope. He was thrilled, never having had a nickname before, and he liked being referred to in the army as Chink, or more forcefully as Dorman-Smith; Eric was kept for private use. His pride in belonging to 1st Battalion, the Fifth Fusiliers – fifth in line when regiments had first been numbered – was absolute. The Old and Bold, as the Northumberland Fusiliers were also known, had fought with renown under every great British commander and at every famous battle since their foundation in 1684. Happiness was a novel sensation.

Peace was to last a mere six months. It is at the start of his wartime experiences that these letters begin, written to his parents in spare moments. Chink was to be badly wounded three times, awarded an early Military Cross, and mentioned three times in dispatches, but no medals or citations reveal the impact on his nervous system, disguised in letters home. The cataclysms of loss increased through Mons, Messines, Armentières and Ypres; once, within a few hours, 486 out of 645 other ranks of 1st Battalion became casualties and two wounded officers were left, one of whom was Chink. Outwardly he kept going, until shellshock from the death by sniper-fire of his most admired regimental hero, Lindsay Barrett, was confirmed at the Medical Board in London.

When he returned to the front in 1917, after a year's sick-leave spent teaching trench warfare to raw recruits in England and frowning at night over the implications of casualty lists on such a massive scale from the Somme, Chink's opinions about cause and effect had hardened. 'I saw how senseless it was to use men against metal in frontal attacks', he would write later. 'If there was such a thing as military science, the war had shown little evidence of it.' He was coming to believe that reverence for regimental tradition was holding back development, and he already possessed an oblique Anglo-Irish view of the British establishment.

Smartly turned-out with polished boots, neatly ironed uniform, three prominent wound stripes and a row of top medals medals that drew curious attention to the youthful face in such contrast to his acting rank of major, it is understandable that Chink was to fire Ernest Hemingway's imagination when they met on Armistice day in Milan in 1918. They both happened to be in the Anglo-American Club when that news came through, at which Chink turned to the seated young American nearby with bandaged leg and the Croce de Guerra. "So that is that," he said casually. "Have a drink!"

Hemingway, who was in quest of a role model, would turn out not to be the soldier Chink had taken him for, but a Red Cross volunteer whose decoration was for rescuing a fellow colleague under fire while distributing chocolate. Chink struck him at once as the quintessential Kipling hero he sought, and Chink, in turn, was to gain a disciple who viewed tradition from a fresh and illuminating perspective. They found each other stimulating, and their letters here – necessarily from Chink's side alone – demonstrate the friendship that rapidly ensued, and would overcome a twenty-year gap to last a lifetime.

In the 1920s Chink's army leaves were spent with Hemingway. He approved at once of the early marriage to Hadley; she fitted in as the sister he had never had. He skied with them in Switzerland, stayed with them in Paris, went bullfighting in Pamplona and joined their decidedly non-military circle, mingling easily with James Joyce, Ezra Pound, Ford Madox Ford, Gertrude Stein, John Dos Passos, Nancy Cunard and Scott Fitzgerald. In 1924 he stood as godfather to their baby son, 'Bumby', and Hemingway's first book, *In Our Time*, was dedicated to him.

Meanwhile, a snapshot of his parallel army life shows that he climbed rapidly, rising from adjutant of his regiment in 1919 to an instructorship at Sandhurst by 1924. In 1927 he set his sights on the high-achievers goal of The Staff College at Camberley, and confidence swelled with success. In the tactics entrance paper he was awarded 1000 marks out of 1000, a record that still stands, and the student with such a flying start was widely expected to go right to the top. In class it was an unpopular tag, though, and his relish for argument and flagrant contempt for orthodoxy and those who went 'by the book' soon gained him a reputation for arrogance. Teamwork counted even more at close quarters, and the opinions of contemporaries were being formed.

In 1931 Chink was made brigade-major to like-minded 'Archie' Wavell at the forward-looking 6th (Experimental) Infantry Brigade, a harmonious working relationship, and within three years he was moved to the War Office in London's Whitehall, with the rank of brevet lieutenant-colonel. It was a sound upward trajectory for an officer hoping to command. With the potential now to advance military development, his anger at unnecessary loss of life mounted, setting him at odds there with the Cavalry, so loyal to the horse, and with traditionalists like the more senior Alan Brooke of the Royal Artillery.

Believing that fifty per cent of British casualties in war were the result of penny-pinching by the Treasury, he took to tipping his hat derisively to the war memorial of the Cenotaph in passing. It was no accident, he would explain to his companion, that it was located in Whitehall. And he took pleasure in ignoring the directive forbidding any contact with the *Times'* controversial military journalist Basil Liddell Hart, whose tactical 'Theory of the Indirect Approach' matched his own analysis, and got in touch. Regular lunches together enabled them to let off steam about lost potential and discuss radically new ideas, which built a genuine friendship based on mutual respect. In this century the two men are recognised as the outstanding military thinkers of their age.

Chink might take pride in not being a clubbable man, but the social skills of hospitable Estelle, his wife since 1927, continued to smooth his path. In 1937 he was promoted to Colonel of his regiment, stationed out in Cairo, and a year later was raised even higher to become Director of Military Training for India, a plum post given only to men judged fit to reach the top of the ladder. India was the largest repository of British soldiers in the world, making Chink responsible for the training of half a million men.

It is between that auspicious period – where in the small world at senior level of Delhi and Simla he found many ideas in common with Claude Auchinleck, ten years older and ahead in rank – and Chink's 1940 appointment as Commandant of the new Haifa Staff College in Palestine, to which Estelle was not allowed to accompany him, that his career began to falter. What happened?

The answer is in these pages. While old friends like Brian Horrocks would continue to soar, Chink was to be reduced in rank from Major General to Colonel, and he left the army he loved in 1944, before the war was over. *Across the River and into the Trees*, a sombre novel about soldiering and disillusion, would be Hemingway's lament upon learning the truth.

Postwar, after a brief flirtation with British politics, Chink returned to Ireland with his young wartime lover Eve, and they had two children; divorces enabled them to marry in 1949. He inherited run-down Bellamont and death duties, moved his family there, lobbied extensively for removal of the border, and openly supported the IRA. In the minds of most men who had once soldiered with him, he was now beyond the pale.

That opinion would hurt him as it filtered back, as did the emerging public view about El Alamein. Montgomery's battle in November 1942 was credited as a unique achievement in the Desert War, yet four months earlier Auchinleck had stopped Rommel's headlong offensive at a far more perilous and critical moment. In his sixties Chink embarked on his final mission. He knew what had happened, because he had witnessed the entire crisis at first hand during that pivotal summer. Instead of retiring quietly, he devoted his remaining years to correcting the record, until his death from cancer on 11 May 1969.

Introduction to the War Diaries

It should be explained that the War Diary section is compiled from Chink's vast array of contemporary letters to his lover, Eve, whom he would marry upon his divorce from Estelle after the war. It is different, but therefore essentially similar, to the genesis of the War Diaries of Field Marshal Lord Alanbrooke.[1]

Editing Chink's great quantity of wartime letters soon revealed that they evolved from passionate love letter to regular journal, back and forth, and winnowing the grain from the chaff, as he himself put it in a different context, led to so many ellipses initially that those had to be omitted.

Like Alanbrooke, Chink's War Diary was written 'out of love and loneliness', as the introduction to Alanbrooke's War Diaries begins. Unlike Alanbrooke, who wrote a daily letter to his adored wife Benita and kept his late-night diary comments for a series of bargain notebooks to be sent to her alone, Chink's letters to Eve were his sole outlet.

For both men, under such different circumstances in the war, the emotional release on paper to their soulmate of 'frustration, depression, betrayal [and] doubt' would grow into a 'necessary therapy.'

For military historians and the general reader today, that selfsame compulsion is our gain.

Letters Home from the Front: 1914–1918

War was declared on 28 July 1914, and 1st Battalion, Northumberland Fusiliers marched to Portsmouth Railway Station to join the British Expeditionary Force within two weeks, on 13 August. Chink disembarked in France with his fellow officers and men at 3am next day. In the darkness they faced an exhausting march that took its toll on parade-ground smartness, and the next train took them to the front. Eighteen-year-old Chink looked about on arrival and saw adventure at every turn. On 22 August, at Mariette Bridge in Mons, his first opportunity came to put theory into practice. Time for writing home was limited and overt affection unimportant.

To his parents   4 December 1914

I am writing nearly every day. Of course, I can't write in the trenches, still I get your letters sent up, which is nice. Today the company is billeted in a convent [and] one of the sisters is an English woman [who] calls us all 'her children'. The poor creatures were shelled out of Ypres, which by the way is a blackened ruin and absolutely deserted.

The countryside round the two firing lines is the most mournful thing you ever saw. Dumb skeletons of horses, dead horses, broken ploughs and farm waggons, here and there a cow lying where the hail of shrapnel struck it, and all along the shattered hedges and houses lie little patches of red and blue, side by side with the field grey of the Prussian. Only the crack of a German sniper's Mauser or the scream of a big shell overhead relieves the dread silence of a lost world. A few pigs, rejoicing in their new-found liberty, gallop round the shattered farm buildings or scamper to their old prison when a shell bursts near.

Unless peace is declared pretty quickly we shall spend Christmas in the trenches. The men have got news of a peace conference being held and are asking me if it's true, but I've heard nothing. Everything pretty quiet. I hope 'Germano' has had enough and is preparing to clear out – or perhaps preparing for one more effort to break through.

Eric

Telegrams to Mr and Mrs Dorman-Smith   *12 December 1914*
*Regret to inform you that Lieut. E.E. Dorman Smith was slighty wounded on 9
December. Secy, War Office'*

The wound was more severe than he was prepared to admit, and led to three
months' sick-leave in England.

To his parents   12 December 1914
It's only a slight affair, a graze along or rather just below the back of my right
knee caused by a sniper who was enfilading our trench at long range. He got in
some jolly pretty shooting, bagging two men and an officer (myself) in 2 shots,
myself just as I had finished tying up the other two.

I got down here yesterday and we are jolly comfortable. The Duchess,
disguised as a hospital nurse, looks us up occasionally and usually manages to
wound my feelings by saying I ought to be at school or 'Did you bring your
nanny with you?' I'm told my wound was providential as my left foot might
have got rather bad if I'd stayed on up country. It's swollen and puffy with very
little feeling and the doctor is busy giving it iodine baths. The trenches are
very wet indeed.

I'm able to hobble about with the aid of a stick [and] don't know how long
I'll stay here. It all depends on how the wounded come in. Lindsay[1] asked to
be remembered to you. He's very fit and has so far survived the campaign. It's
jolly bad luck the way I'm always stopping German bullets, but we're better off
here than in the trenches. They are miserable at present.

Eric

To his parents   English Channel
                              18 March 1915
The best crossing I've had yet. I went to sleep on my sofa just outside
Southampton water and didn't wake up until we were just off [Le] Havre.
We're waiting till the tide rises enough to take us up-river to Rouen, where I
pick up my draft. You had best start sending out a small weekly hamper with
cigarettes included – it helps the company mess so much. I crossed with a fellow
I hadn't seen since prep school and we recognised each other immediately. Isn't
it a small world? I'm awfully pleased with my new fur-lined coat and soft cap.
It makes sleeping in one's clothes so much easier.

Our machine gun officer has been wounded again so I may get the guns if
I'm lucky. It would mean a lot to me to be MG [machine gun] officer, a horse
to ride, living with HQ where I hear all the news and am usually with Lindsay
and am an independent command. I ought to rake in at least a Mention in
Dispatches this time.

Eric

To his parents   Flanders
                                                    2 April 1915

Our servants bring up our letters when they take our grub in the evenings.
I've swopped trench since I last wrote [and] it's a great change for the better.
We've been spending the last 48 hours in a château in support and are now
back in the front line. Things are quiet – too quiet; there's a lull before the
storm feeling about it.

This morning I was looking through my periscope at the Teutonic trenches
when I spotted a loophole being uncovered by the light flashing through it. I
took the looker down so as not to get it smashed and pushed my cap up on the
end of a stick but the blighter never fired.

We live like dukes in the trenches these days. For breakfast we had bacon
and eggs and rolls and jam and tea. For lunch we're going to have salted herring
as a first course and bangers and bacon to follow. Tea and toast, if we can
manage on a brazier, will form a light repast at 5 o'clock. The Germans have
just opened fire and are bursting the shells half a mile back among support
dugouts on the canal bank. The bank's about 10 feet thick, so much good may
it do them.

The German trenches run round our little redoubt (in a spinney which gives
some cover) on three sides of a circle. It's really a bad position but comfortable,
with a commodious dugout which I hope is shrapnel proof. I've not shaved for
a week now nor washed my neck, and except for a sore throat caught through
sleeping indoors am extraordinarily fit.

Eric

To his parents   4 April 1915

I was pleased to see in last night's papers that we carried the mound and
600 yards of German trenches. These journalists are marvellous! I'm still in
the same trenches, and rapidly growing roots. When I come home again I'll
build a dugout for myself in the back garden and only wash once in every three
days. I'm suffering slightly from the effects of overeating. The only thing to do.

Eric

To his parents   Sandbag Hall
                                                      Bomb'd Street
                                      Telegraphic address: Whizbang
                                                  St George's Day 1915

Irritating gases are Allemande's latest speciality – beastly smell, just like strong
mustard and cress, and very irritating to the eyes. All the men were weeping
but I was asleep so almost escaped. Poor old Wipers [Ypres] is for it again;
17 inch shells are nasty townfellows.

We're still in the trenches. I've read one book since Monday and got halfway through *Lorna Doone*, so am quite busy in my spare time. We gave the Bosche[2] a thin time last night with rifle grenades, and the company on our right report that they were bursting just right. Please don't worry, for I'm awfully careful these days and the rashness has gone out of me.

Eric

To his parents   30 April 1915

The OC Y Coy has gone sick so I've been pushed up into the front line to do his job. I came up last night and spent all today making a sketch of my trench and trying to dodge the German snipers. The whole trench has been so neglected that I don't know where to start. I've just sent in the deuce of a report to the Brigade. Don't you believe the 'Hill of Death' tales, except that the Germans do the dying. The 7th are in the trench next to me and I've been paying them a visit today. All old friends.

Eric

To his parents   Flanders, St Eloi
                                                               5 May 1915

These trenches are so good now that we're starting gardens to pass the time. A certain amount of shellfire this afternoon and once I was certain I smelt the gas but there was a strong wind and it was much diffused. They've just started twisting rifle grenades practically at the door of the palatial dugout shared by John Lawson[3] and myself. The first burst hurt no-one but the 5th wounded a sergeant and corporal, our best bomb throwers, and an officer standing near. None dangerously.

I wish they'd give us a show, only the brigade seems to be reserved for a much more difficult job. It's really easy to go on once you've seen the enemy's trenches shelled a bit and got your blood up. Casualties aren't noticed then. The loss of whole platoons in an attack isn't half as bad on the nerves as a few men shot through the head in trench work. Besides, it's a real test of first-rate troops.

I'm still fit, though rather short of sleep. A thunderstorm is expected so we've been busy digging irrigation works behind the trenches. We are all provided with aspirators, but no chance yet of testing their efficiency. We've good dugouts, furnished with chairs and tables. One home from home even has a spring mattress and is decorated with fresh flowers every morning. All the sandbag walls are full of field-mice, tame as no-one hurts them. They come out and run about the floor in the evenings. I wonder when this show will end? It's hopeless trying to guess, for the strategy is so colossal. But the sooner the better.

Eric

To his parents   14 May 1915

Trench life has its own routine. We turn in about 2am and I don't usually turn up again till 12 noon, when I breakfast. We read or play patience or write till tea time at 5.30 pm. At dusk, about 8pm, the servants arrive with our mail, papers, grub, and we have dinner. Afterwards I go for a walk to stretch my legs around the front line and visit my various pals, taking a drink and cigarette off each. This gin crawl usually lasts for an hour or so and is the cheeriest time of day. Everyone has a yarn and is jolly pleased to see a visitor.

I nearly died of laughter last night. Lawson, my brother subaltern but a month older, was coming with me over a particularly broken bit of ground. There were lots of spare bullets about, and whenever a starlight went up we were exposed to the savage enemy so I didn't want to hang about. John pretended he knew the ground and was just saying 'I know there's a six foot hole just here,' when he vanished. I was so powerless from laughing that I only just missed falling in myself. The only person who didn't see the joke was John – it was full of tins.

I wish they'd give us a short rest. We stand by the whole time, so I can't even go for a ride. I've written to Hawkes[4] for a new pair of breeches. My old ones are only fit for the trenches, but they've done jolly well, poor things.

Eric

To his parents   29 May 1915

We took over a new bit of the line last night – poor trenches made by the French and mainly consisting of a good parapet and a rabbit warren of small dugouts behind. Officers' dugout like a pigsty. The second bombardment of Ypres has absolutely reduced it to ruins. It was still on fire when I went through last night and smelt terribly. We occupy a reserve line 1000 yards from the firing line and 1500 from the Hun. It still smelt of gas from the last attack: sickly sweet and cloying, with a hint of chloride at the back, but not to be afraid of if you put your respirator on quickly and get well out of the bottom of the trench. We wear respirators for half an hour a day to breathe the solution well into our lungs to coat them with a protective layer.

Whizbanging has been the order of the day. We've had two rations: total bag 2 men wounded, both slight thank goodness. Allemande's plugging away at one of our aeroplanes – gives his dispositions away to hear the whole front line suddenly blaze away into the air, so we never do it. The blighter's going back over their lines again and he's fair stirred up a hornet's nest, plucky devil. I must go and watch.

There's a big Château within 50 yards so we've furnished our dugout with sofas and chairs and have some topping china we found in the trench. Topping weather now, though still cold at night, but I get on well with a Burberry only, though they're poor coats for active service. I've not seen the paper for two days but our *Comic Cuts* keep us credibly informed as to the progress of the war.

Eric

On 15 June 1915 1st Battalion took part in an attack to capture German trenches in the St Eloi sector of Ypres, near Bellevaarde Wood. Planning was complex, and ultimately proved disastrous. Chink was to be wounded, but his resourceful action despite that would result in the award of an early Military Cross. That honour had greater impact on his parents than the third War Office telegram sent on 21 June 1915. 'Regret to inform you … again … remains on duty with his regiment.'

To his parents   28 June 1915
We're shifting to a more stormy and gaseous quarter, [but] the more we hear of the German infantry, the less we're inclined to think of them. A lot came past here early this morning as prisoners. The poor 2nd battalion has had another cutting up, second within 3 weeks, but behaved very gallantly. All the youngsters who were at East Boldon[5] with me have either been killed or wounded. We move at 3am today. I am OC 2 troops.

Eric

To his parents   24 July 1915
It's been a funny 20th birthday. The queerest I'll ever spend, I suppose. The tuck hasn't turned up, [but] with any luck the servant will bring it up tonight. Still rather moist hereabouts and hard to keep one's feet dry. The trench is built on clay and holds the wet, which makes drainage difficult.

I'd quite a birthday party in my dugout this afternoon. 5 people to tea – not so bad. We've just been hammering the opposite trench with our new 18 pounder – good! Makes the deuce of a detonation. Our tenth day, and we're going to do a fortnight at least. However, I'm not complaining, and all the men seem very cheery too.

Eric

War Office communications to Mr and Mrs Dorman Smith:
8 August 1915 'Lt. E. Dorman Smith, Northumberland Fusiliers, admitted Fourth London General Hospital, Denmark Hill on 6th August.'
<u>*Later same day*</u> *'Capt DS going on very satisfactorily.'*

Suffering from shellshock (generally known today as PTSD) Chink went missing a week later, having discharged himself prematurely from hospital. His father responded at once but there was no sign of his son at their favourite meeting place, Berners Hotel,[6] so he took a cab to the old Maidenhead house, long locked-and-shuttered. Eventually, in the deepening dusk, he came across him sitting on an iron garden seat overlooking the Thames, utterly remote. This time the sick-leave was prolonged. But Chink's insistence on returning to the front, in spite of his invalid's role as instructor ('Bombs and Grenades') at the Army School of Trench Warfare, achieved his purpose.

To his parents   February 1916
I am the only combatant officer who sailed with the Battalion in 1914. The quartermaster is still here, but he barely counts. We're still busy training. Very cold though, as a heavy fall of snow today. Thank goodness we aren't in trenches.
    Eric

To his parents   28 February 1916
There's a new proposal out about giving a badge for every time one's been wounded. Suggestions were asked for, and ours was a small bar of gold braid for each wound, to be worn on the right breast. No news of our moving anywhere from the [Ypres] Salient. The Bosche are having a hell of a time at Verdun, tremendous casualties – worst in the war.
    Eric

To his parents   12 March 1916
Major Lindsay is in town with his regiment and I dined with him last night and will be lunching at his mess today. It's very nice seeing him again. Lin says I could easily get a GSO3 job – divisional intelligence – so I'll probably try for one.
    I hear we're returning to the line I was in last year. Don't be worried about my clothes. I'm warm enough and have everything I want.
    Eric

Lindsay Barrett, already a temporary lieutenant-colonel at 25, was the senior soldier whom Chink had modelled himself upon since joining the regiment. He continued to rely on Barrett's instinctive grasp of command, and agreed with his immediate criticism about the way the war was being run. The sense of relief, however, was to be short-lived. On return from front-line trench duty, he learned that within three days of their meeting Barrett had been killed by a sniper, shot in the head.

To his parents   Flanders
                                                                         27 March 1916

That divisional job is practically in existence, hastened by the fact that the CO wanted to send me home again because my nerves were just as bad now as last August. Lin's death has been a most awful shock. I protested, after getting out here again with such difficulty, and explained to the brigade major that if I went I'd be stuck on the side list for months. He knows me of old and was very sympathetic. The idea is, but it's not official yet, that I get attached to the division for a month and then get a job somewhere. One thing's certain. Trench work has finished as far as I am concerned.

    Eric

In April, with his application for a Staff job going forward, he crossed to London for the presentation of his Military Cross by King George V at Buckingham Palace. On his return to the regiment, however, it was obvious that he still needed treatment, so he soon found himself in front of the Medical Board in London once again. 'Nervous disability resulting from shellshock at St Eloi in March' was confirmed, and lengthy home rest prescribed. Casualty lists from the Somme multiplied, and his experience made him more valuable at home. He was sent to teach trench warfare to fresh recruits once more, and in frustration grew closer to his father, man to man.

To his father   School of Instruction
                                                                                    Bedford
                                                                             16 July 1917

I'm extremely bored and counting on steps to procure my release from this *durance-vile*.[7] Hey for the merry land of France again.

    I'm pretty busy here during the week and can say that I've fairly got the Trench Warfare Department on its legs. It will soon be as up to date as the best of the tacticians shows. But it's been a great struggle. I had to learn the whole subject before I started, and don't pretend to know it thoroughly yet. The peace of this place is horrible. I exist in a sort of monastic calm, only enlivened by occasional purple patches which are my descents on London town.

Eric

To his father   31st Infantry Base Depot
British Expeditionary Force
27 July 1917

Rather sudden, wasn't it? The War Office shot me over here in about 24 hours, and am now marking time here for a few days, as don't know my battalion. I hope my kit arrived OK as it was packed in rather a hurry,

By Jove, it's a relief having left Bedford. I really thought I was stuck there permanently. At least there's a chance of getting something worth having this side of the water. An uneventful crossing. We left Folkstone at 12.30 and I'd hardly finished lunch before we'd arrived. No sign of U boats or even an escort. I'm now close to where I was in hospital in December 1914, but the country has changed enormously. It's one mass of camps – training, convalescent – and until you learn your way about a hopeless jungle. They put us through a little course of instruction and tomorrow I arm myself *cap à pied* and run about bombing and sticking dummies. Still, it does hush one up, which is what I especially want.

Eric

To his father   Undated 1917
Wednesday

The powers that be are undecided about what to do with this remarkable addition to the Brigade Strength. I drove out from St. Ouen in the Battn. Mess cart and called on the OC 10th Bd who told me he'd talked to the brigadier about me, who wants me attached to his HQs to have a look at me. I've been there ever since. Did one or two jobs of work for him, understudying the staff captain – a lad with only a year more experience. The old man tackled me today, saying he's going to send me to the 12th Bttn, The Durham Light Infantry. He explained they needed 'waking up', and I told him I'd be delighted to work anywhere.

So the war starts for us again fairly shortly, and I expect to be in my element studying maps and writing orders. If I can make good and am spared, it ought to do a lot for me; even if I'm not with the Fifth, it means good work. We're in charming country, oh so different from when we were last here with the 1st bttn. in February, when I rejoined. Today I rode over the way to our old billets, now a brigade HQs. What a lot has happened since.

The brigade run a jolly good mess. If you can buzz out any game [i.e. wild animals] that's going they say it will keep well enough, and I'd like to make the general a present for his hospitality.

Eric

To his father   British Officers' Club
                                                              No. 31 Infantry Base Depot
                                                                        3 August 1917

Dear Pater.

It's been raining relentlessly for the last 4 days, [and] I live in a tent. Moisture permeates everything, and my flea bag feels like climbing into a sponge when I turn in. When I think of the fuss the mater makes over damp sheets I smile sardonically.

One thing about a new army unit, one has more chance of distinction. Decorations are easier to get and acting promotion too. The only trouble is the junior officers. I'm fairly broad-minded but they're rather terrible. They aren't even the beginning of soldiers as we know the word, and have poor powers of command. Never open their mouths on the march and seem afraid of the men. From all telling they are very gallant fellows, though.

The saving of this place is the very excellent officers' club. I spend most of my time off parade here. As many meals as possible here, too, besides I meet a certain number of folk I know and have already been greeted warmly.

Eric

To his father   23 August 1917

Alas, I move tomorrow to a spot I know too well, where I left the 1st Battalion over a year ago. After that our destination is rumour, but everyone seems to think we'll have to tackle a pretty big thing that so far has not been properly dealt with. It would be a tremendous feather in our hat if we succeeded. The Bosche must be pretty well done in too and the French success at Verdun is more than opportune. It may not be altogether fun being a British soldier, but by Heavens I'd loathe to be a German at present.

I don't know whether I want to keep this present job of 2nd in command. They're all very nice to me and it's a great compliment and I think I could do it alright, but I've superseded men much older than myself who've had almost as much service and experience. Unless they make me an acting major the situation isn't easy, especially as I know that the CO would sooner choose his own. If I stay I'll have to insist on acting rank, but what I want is a staff job. We'll wait and see.

Eric

To his father   12 September 1917

For the time being [I'm] commanding a company in our 10th btn, as the Company Commander has gone down the line. Busy training them for possible operations. I say, it's a bit thick about the Russians – Civil War. But

I'm all for a soldier getting at the top of this business – after all Russia stands or falls by the army.

I hope the grouse are succumbing to the gun, as the snipe used to. This country is full of partridge, but of course we can't shoot them.

Eric

To his father   15 September 1917

Alas the grouse arrived – I say 'alas' for their presence was so undesirable that they found their last resting place in Flemish soil as soon as they were unpacked, but several ardent fishermen nearly went off their heads when they saw the maggots.

The Bosche has developed a nasty habit of bombing the back areas, so the result, with only a canvas hut as cover, is that one lives in a concentrated essence of Hun bombs, bullets (both Bosche and ours), dud Archie shells[8] and long range gun fire. Rest is not all it might be.

Eric

To his father   29 October 1917

You mayn't expect to see me for a long time, I'm afraid. Leave is absolutely off. Sad but true. Still, it can't be helped [and] the delay will make it the more valuable. I'm now Major Dorman-Smith, terrific, isn't it? I'm tickled to death when sentries present arms and aged subalterns salute. I've been lucky enough to get my other ranks confirmed but am unlikely to keep my crown unless someone dishes out a brevet at the close of operations – that's too good to be true.

At last mine enemies are scattered. Lazy devils whom I fought against have been relegated to the limbo of failures and such a clearance was badly needed. It's astonishing to see how our muddy scarecrows smarten up after a short rest period. Wouldn't recognise 'em at all at the end of two days rest.

We're mighty chilly in the line nowadays, [so] I've bought a chic waterproof from a French shop here. It looks and is extremely smart, but whether it's waterproof I can't say till I test it.

Eric

Serving in battle-scarred France or in apparently sleepier Italy, where he and his battalion were rumoured to be posted, his determination to raise standards was ingrained.

To his father   Piave Front, Italy
                                                             15 December 1917

I'm still fit, in spite of the very unpleasant tummy complaint[9] that strikes each mess in turn. So far I've not been in the line, though of course in pretty close support. However, I fear Christmas will be spent in a more unpleasant proximity than at present. Still commanding a company. Bother, they've started shelling so I'll postpone this till it's over.

I'm told that staff's the best job to have, as unless one's likely to get a brigade, a battalion is an unsatisfactory existence. You go on till you're killed. Things are pretty quiet here although the organ-grinders[10] say they're fighting like hell. They've never had to handle anything like the bombardments at Ypres. Night operations this evening, with most objectionable ground to work over. Busy training the whole time, and really the battalion is damn smart.

Eric

To his parents   Piave Front, Italy
                                                                                            21 May 1918

I'm beginning to forget what you all look like, so long ago is it since we last met. It's over a year now surely, and looks like being much longer. I'm a little tired now. It's a long time without a break of any sort. Also I'm in rags. No tails to my shirts, collars reversed, cuffs worn – and as far as my riding breeches go, I'm absolutely disreputable. It's so hot that khaki drill is a necessity, and I don't know how to procure any. Worst of it is my clothes are so bad that I can't go on Italian leave.

I'm in the line at the moment, but it's mighty quiet. Hard to know there's a war on. Still, it's topping weather and *very* bracing up on top here.

Eric

To his parents   27 May 1918

You know one gets into a kind of coma when away from people for so long. England's got very visionary in consequence. Our positions are jolly strong and morale of the troops excellent – full of fight. It's mighty hot on the plains and bad enough up here when the sun's out. Everyone below is in drill. It's alright for us living in the shade of the pine forests, but the Boche must grill in the open plain.

Fitness and single-mindedness were no protection from highly contagious gastroenteritis, and Chink, too, was about to fall victim. Hospital treatment in Genoa led to a convalescent's posting as Commandant of the British Troops in Milan, known to be a soft R and R billet. His few duties were light, and preoccupied with military lessons from the war he was about to play a role in twentieth century literature.

Peacetime:
1919–1939

O n Armistice Day 1918 Chink was idling in Milan's snug Anglo-American Club when the news was announced on the wireless. He caught the eye of the only other person present, a wounded young Red Cross officer sporting the Croce de Guerra. 'So that is that,' Chink addressed Ernest Hemingway impulsively. 'Have a drink!' No intimation of the long friendship that would mould both their private and their professional careers.

For Chink, the effect of Hemingway's irreverent American outlook emphasised the dangers of conventional British respect for traditionalism when it came to military evolution. 'Hem was sufficiently remote from my own stereotyped world', he would write later, 'for me to be able to see that world through his eyes.' Analysis would increase as a result, and impatience shorten.

For Hemingway, Chink came across as a real-life Kipling hero, and the idea of war and death as the ultimate test of character would influence all his novels. 'I was very ignorant at nineteen and had read little,' he, too, would write in turn. 'And I remember the sudden happiness and the feeling of having a permanent protecting talisman when a young British officer first wrote out for me, 'By my troth, I care not: a man can die but once; we owe God a death ... and let it go which way it will, he that dies this year is quit for the next.'

The heady atmosphere of postwar Milan fostered intimacy and confession, and Chink's tone became more mannered in contrast. Self-mockingly he appointed Hemingway to be his *A de C*[1] and loftily addressed him as Popplethwaite; the name was just right, they agreed, for an ambitious *A de C*, and in due course this became shortened to Pop. Chink liked Hemingway instinctively, freeing his own charm and sense of fun; they were not only equally idealistic but intellectually well-matched. Affinity rapidly deepened into friendship in the brief time – four or five weeks at most – they spent together before moving on, and when Chink had to rejoin his regiment in January they resolved to keep in touch.

During the following two years of postwar anticlimax in England, Chink's efficiency was recognised by promotion to Adjutant, chief assistant to the commanding officer. In June 1921 the battalion was included when British

Army reinforcements were needed in Ireland to contain Sinn Féin unrest, and 1st Battalion, the Northumberland Fusiliers were despatched to the Curragh 5th Division, to patrol the martial-law county of Kilkenny from their HQ in Carlow town.

Chink viewed it professionally from the soldier's angle, not the political; in 1916 the Somme had had his attention, not the Irish Rising. The Government of Ireland Act of 1920 envisaging separate parliaments north and south within the United Kingdom of Great Britain and Ireland had led to Sinn Féin boycotting the Dublin elections and setting up their own, Dáil Eireann. The British Army were to be military policemen in a deadlocked situation, with Law and Order left to the mercenary Black and Tans and the Auxiliaries, both rumoured to be drunken or sadistic.

In Carlow barracks extreme boredom alternated with extreme tension in the beginning, but as soon as a Truce was announced in July the soldiers found themselves in an invidious position. Off duty they had to go about unarmed, while Sinn Féin members carried revolvers openly and jeered. Chink made sure from the start, as Adjutant, that his own battalion behaved itself.

One of his tasks was to write up 1st Battalion notes for the regimental magazine, *St George's Gazette*, and a new voice emerges: that of – almost – the stereotypical British army officer with a *Punch* cartoonist's view of Abroad, especially Ireland. A cross-section of his wide readership shared that attitude, despite – or, more likely, as a result of – serving around the world, and he approached his role dutifully, tongue in cheek. His characteristic ridicule, though, could bloom into satire; even criticism. After all, this was a chance to educate, not simply to report and entertain.

First Battalion Notes by 'Chink'   Carlow Barracks
                                                        County Carlow
                                                        30 July 1921

Within four short weeks we've experienced the whole scale of emotion which the Irish situation can produce. It's difficult to say whether we're performers in a harlequinade or a tragedy. At present the wry humour of 'This Agreement'[2], the spectacle of the chieftain of the Irish murderers[3] hailed as an honoured guest by the *canaille* of the Metropolis[4] must surely command the laughter of gods and men. It's difficult to disentangle the events of this paradoxical month into their proper order but the battalion has survived our period of incubation, and takes its place in the line. 70 years ago we entered Lucknow on its final relief,[5] so today we are suppressing an almost similar rebellion.

The IRA in this county confine their activities mainly to the blocking of roads and destruction of bridges, and as usual the spirit of burlesque appears triumphant. The curtain rises on a pleasant village basking in the July sun.

Exhausted peasantry toy with the sunbaked soil, the porter shop does a brisk trade. The only other brisk thing is the parish priest hurrying to a Sinn Féin Council. Suddenly peace is shattered by the rattle of approaching lorries. The porter house empties *via* the back door. The priest vanishes into a neighbouring house. The labourers find they have urgent private affairs some distance away and depart. Conscience makes cowards of us all.

The lorries stop in mid-street and hot canvas-clad soldiery leap out and dash as if their lives depend on their agility to the nearest points of tactical importance. Shortly they appear accompanied by the priest, the publican, the village idiot, two or three slow-moving labourers, and a cyclist or two. The idiot is discarded and the remainder of the bag ushered into a lorry. The convoy departs and all is peace, broken only by the loud blasphemies of an aged crone who curses the Crown and the Republic indiscriminately.

Until negotiations with Sinn Féin broke out, life at Carlow was not without excitement. We had been engaged in drives in the west of the county and most operations, as far as they went, were successful. About 50 doubtful characters found a night's rest under lock and key in the barracks and made an undesired journey to the Internment Camp[6] at the Curragh next day.

At present we lead a curious existence. 'There is a lot of freedom about,' to quote a local Sinn Féiner who observed an officer strolling down the main street on the night the curfew restrictions were removed, but it is not the true brand. It will be many years before the Army forgets the fate of the many officers and men who fell into Sinn Féin clutches, and those brutal murders on the very day of the agreement.[7]

In the midst of war we are in peace. Nobody who doesn't actually participate in our amusements seems to realise that war can flourish quite nicely even in the British Isles. There is no need to go to the Continent for it.

First Battalion Notes by 'Chink'   Carlow Barracks
                                                                     26 August 1921

We have been busily engaged in eating mud. It's a curious fact that one becomes almost used to that unpalatable food. Firstly, most of the patriots that we have been at pains to catch and intern are out and about again. Secondly, those who had joined the Dispersion in the local hills and bogs and dared not show their noses now swagger up and down their villages. Thirdly, we have to deal with Liaison Officers, such medlers of mock heroics and wrong-headedness, such mirrors of murderous patriotism. It's something to discover a spark of humour in the murk of this degrading business.

I suppose even to illogical rogues the fact that the British Army would keep the terms of an agreement, which they themselves had already violated in sixty different ways, seemed incredible, but it is so. We do it, and as a reward platoons of flat-footed rebels dared to drill within sight of barracks. Of course, such a breach of the agreement could not be tolerated so a series of reports rang out on the typewriter.

Dáil Eireann still sits in the Mansion House in Dublin, admiring hero-worshippers stand outside and applaud the murderers[8] as they arrive. The more murders, the more applause. It's all very democratic. In the *Daily Mail* we read that Mr Michael Collins admitted to a contributor that he runs more danger from the flatterers who surround him than ever he did from Brutal British Bullets. Sinn Féin uniforms are openly worn, Sinn Féin Courts deal with cases sufficiently important. As a kindly concession to a beaten foe, Resident Magistrates are graciously permitted to try cases which deal with the trespass of chickens and other important matters. Our future is uncertain. It's impossible to anticipate the turn of the wheel. Farce and disappointment are for the moment predominant, tragedy may at any time re-appear.

It should be realised that if this agreement breaks down Sinn Féin will start with a new lease of life. Re-organised, re-rested, re-armed, filled with confidence and contempt, and once again officered by its original leaders, it will be a force seriously to be reckoned with. The harvest is more or less in.

First Battalion Notes by 'Chink'   Carlow Barracks
                                                      December 1921

Our whole happiness now depends on what that successful literary product the Treaty[9] will do to us. It may decide to let us go home or it may decide to blow us up. The suspense is terrible. Nobody who saw our haggard worn faces would recognize the careless happy youths who marched from Guadaloupe Barracks[10] a short six months ago. If we're permitted to withdraw from the unequal contest, then will we appreciate the true ecstasy of the *puissant* army of the Duke of York, who, 'Marched up the hill with twenty thousand men' only to come down again when he and his forces had ceased to admire the view. So far so good, but one never knows….

To Ernest Hemingway   The Military Barracks
                                         Carlow
                                         13 January 1922

Popplethwaite – I'm as excited as a maiden at her first wedding – I'm all out of control. There is nobody in this Distressed Continent or any other that I would sooner get a letter from. Wonderful – marvellous. It takes me miles away from this miserable Free State where nothing is free, not even the drinks, and which at one time was the best domicile for a poor man in the world.

No Sir, the Fifth of Foot are still in Carlow amazing the peaceful inhabitants – I'm the Devil of a tyrant myself and by dint of practising a magisterial frown I flatter myself that I can silence a Sinn Féin band with one stern look. However what has all this to do with the fact that my one and only *A.de.C* is back in Yurrup[11] – and a successful plier of the editorial plume. Marvellous – comes of course of my education in Milan.

I too got into print, by accident a copy of my regimental journal fell into an unscrupulous loyalist's hands and my opinions on Ireland were launched onto an unsuspecting world as 'What the Soldier Thinks' – as if soldiers ever think. The army council was angry and said nothing. Sinn Féin was very angry and went after me with threats. But Nowt happened.

I am still adjutant of the aforesaid Fifth. My leave since December 1920 has amounted to 5 nights at intervals away from Barracks or camp. I've worked on an average 9 hours a day and I get no pay worth talking of.... We are hanging around waiting to quit[12] and the inhabitants, who have always had a sneaking fondness for the Military, are busy organising a railway strike to prevent the exit of the Hated saxon invader. You should come over here and study 'Paradox as a paying proposition'....

Remember I prythee my Hemmingstein that I am absolutely poverty stricken as befits an Irishman. I have no means but debt and live on the credit I obtain by producing my card with my high society home address [Bellamont Forest] on it at appropriate intervals.... I've got to quit writing, old friend. People come hopping in and firing questions at me – and when I say 'Go away, I'm writing to Poppelthwaite,' they get angry. I'm supposed to employ tact as an adjutant and not to have any private life that counts. I haven't – not till you came back into my joyless life with a typewriter and a grin. When next you write, Democratic One, dilate all you know on the cheapness of visiting Chez Hemmingstein. What about coming to England?

Yours always, Chink (registered cognomen!) alias Democratic General

Ratification of the Anglo-Irish Treaty in January 1922 and the installation of a Provisional Government in Dublin's Dáil Eireann under the chairmanship of Michael Collins, however, allowed the regiment to leave shortly afterwards.

First Battalion Notes by 'Chink'   Bordon Camp
Hampshire
England
February 1922

By the skin of our teeth, by the merest good luck, by the special intervention of the Regimental Patron, we have escaped. The final week was one of bustle and hard work. Pursuant to their plan of hindering the 'Provisional Government', the Irish railways and dockers intimated their intention of not playing at evacuation, choosing rather to go on strike. In consequence all the heavy baggage, escorted by the Machine Gun Platoon and transport moved to Dublin, which scattered the flock.

The garrison proper marched out at 08.00 on 7 February. Before departure steps were taken to render it impossible for the tricolour of the Republic to be displayed from the place formerly sacred to the Flag of the Union, and during the night the flagstaff was snapped off. In the grey dawn a representative of the District officer of the Royal Engineers was found surveying it and murmuring, 'Somebody would have to pay.'

The Detachment was played to the station by the Band and the early hour precluded any great demonstration of dislike or regret. A few waved tricolour flags, booed, hissed and cheered. The RIC mounted a guard of honour at the station, and friends braved an early breakfast and the chilly station to wave goodbye. The crossing was unpleasant, and no-one was sorry to arrive at Holyhead and be safely stowed into the trains which took us south.

That summer 1st Battalion, the Northumberland Fusiliers[13] was posted to Germany. They were part of the Rhine Army, in which the French were already less popular after provoking violence in their zones. Parades and marching competitions were a way of displaying garrison strength and proclaiming British military presence, as was promptly underlined when Lord Cavan in full regalia as Chief of the Imperial General Staff inspected the Rhine Army at the Dom Platz in Cologne. Snapped to attention there, Chink was aware of having an even higher standard of spit and polish to enforce. His style of reportage, meanwhile, was to lose none of its edge.

First Battalion Notes by 'Chink'   Marienburg Barracks
Cologne
Germany
24 July 1922

On 13 August 1914 the 1st Battalion arrived in France. On 13 June 1921 we sailed for Ireland. On 13 July 1922 we left Bordon *en route* for Dover and Cologne. The 13th would appear to have become our day of Fate. Then it was 'occupy by all means but don't hurt the inhabitants and keep off the Flower Beds.' Now – well, we did lick the Huns,[14] didn't we? So we're all trying to look fierce and martial and domineering, and it doesn't wash a bit. The Hun knows all about domineering, and it's practically certain that we'd lose heavily in an open contest. It's hardly possible after being booed out on home ground to come here and try to exult in the success of the fixture before last. This is, of course, pure hypothesis.

As the British Nation is always engaged in a detailed study of the lessons of the last war, as opposed to the more theoretic task of forecasting methods of waging future wars, it's our custom to regard the latest defeated enemy as our victim. Up to 1914 we had our hands full with Brother Boer. He was always attacking convoys or occupying kopjes[15] while we plodded about in surprisingly successful pursuit. By the time we had defeated Bruder Boche we were far too fond of him lightly to leave him. We still gave him credit for past glories and struggled with his visionary resistance in trench to trench attack. But then we found that there were those who had waged war in such unheard of spots as Afghanistan, Waziristan,[16] Persia, Russia and Ireland. Each had opinions of the war of the future based on his most recent experience, so our combined schemes have become a trifle confused. Fortunately the polyglot enemy never has tanks or low-flying aeroplanes, so invariably we succeed. The query as to what does an attacking Company do if it is itself attacked by a tank remains unanswered. The tactical scheme of today puts the art of Jules Verne and Rider Haggard in the shade.

The posting to Europe had the bonus that Chink was now much nearer Hemingway, who was currently based in Paris, although his own responsibilities in Cologne were intensifying to the point that even letter-writing took more attention than he was often able to give. That Christmas he sent Hemingway a regimental card featuring three officers – including 'Adjutant E.E. Dorman-Smith MC' – and scrawled 'Chink' beneath. Contact made, in brief.

He was spending his spring leave in Paris with his parents when, without warning, Hemingway appeared in his hotel room one morning to announce that they were off to Switzerland straightaway. His old friend had married in the meantime, but Chink's apprehension about the new wife, Hadley, proved groundless, and the instant rapport of all three would set the pattern for Chink's subsequent Jekyll and Hyde army leaves. That first holiday culminated in a final nightmarish climb over the St Bernard Pass in deep snow, saluted in *A Moveable Feast*. 'He had always been my best friend and then our best friend for a long time. He takes care of us.'

To Whom it May Concern   R.T.O.,[17] Cologne Main
                                                                                        8.4.23

The Bearer Mr E. Hemingway has been staying with me.

He is going back to Paris. I would be obliged if you would do anything in your power to assist him on his journey.

E. Dorman-Smith, Capt. 5th Fusiliers.

First Battalion Notes by 'Chink'   Cologne
                                                                    24 October 1923

Out of Germany always something new. At present peripatetic bodies of
the Rhineland Republican Army (it has a familiar ring) are drifting about
the occupied territory intent on proclaiming a patriotic, pro ally, reparation-
paying Separatist[18] government in the teeth of assistance from benevolent
foreign powers. The situation is obscure, but nothing very Separatist seems to
have got into Cologne. Anyone shot by the Cologne police is shot for non-
political motives and must not fear imputations of Separatism being added to
his injuries.

In other towns it must be galling for an honest, straightforward, hungry but
patriotic looter to have his corpse branded as that of a Separatist. Headquarters
of the political spasm are at Duren,[19] and the reason why the revolt takes so
long to spread is attributed to the fact that the 'Rhineland Government' has
based the march of its columns on resuscitated operation orders used by the
British Army in their recent manoeuvres in that area. The good people of
Coblenz have intimated their allegiance to the Reich by hounding the heroic
Sturmtruppen, the *Rheinwehr* [a parliamentary unit], out of town.

Hadley's pregnancy had drawn the couple back home across the Atlantic to
await the birth, supported by Hemingway's freelance work on the *Toronto Star*.
Letters, however, filled the gap.

To Ernest Hemingway   Cologne
1599 Bathurst Street   9 December 1923
Toronto, Canada
Forwarded *c/o American Express Co., 11 rue Scribe, Paris*
Forwarded *to 44 rue Cardinal Lemoine*
Forwarded *to 113 rue Notre Dame des Champs*

It's about 8am, Pop, and I, the world's most active man of business, am getting
in my one hour's real work a day before the Brazen tang of the military diem
opens out on my devoted *tête*. You recognise the style I trust. It's catching.

I got your letter yesterday morning just as I was going out shooting. Nearly
let off my fusil in the hall of the mess with delight. Couldn't shoot a thing all
day – but I fired several salvos to the health, luck, happiness and success of John
Hadley Nicanor. I feel rather like the old Lady in the New Testament (with the
suitable masculine variations) in that I foretold this event pretty completely.
Ask Hadley if I didn't …. If he was a triple deformity I'd stand godfather to
him for all time. I think it's as well though to have him properly christened.

I haven't conversed since last we met. Not got anything to talk about or anyone to talk to if I had. My brain is steadily becoming atrophied and its place is being taken by a mixture of Rhineland fog and short drinks. This is about all I can do this *morgen*, Pop. The brazen tang is about my ears. I must go and mount a guard.

I anticipate your book with suspicion and delight. All my love to Hadley and her offspring.

Ave and *Vale*, Chink

To Ernest Hemingway   Cologne
                                                        10.15pm. 15 December 1923
I've just been reading your three stories[20] again, old friend. It's warm, steam-heated and fuggy in the ante room which you know and the winds howling like the devil outside.

I must have been a fool when last I read two of them. 'My Old Man' of course I read at Chamby. When I read 'Up in Michigan' and 'Out of Season' before I couldn't get at what you were driving at. I see it now. I'm sorry, Pop, for if ever there was any value to you in my criticism my previous clammy reception of the booklet must have been a trifle depressing. However – forgive me – it's funny how one can quite suddenly get at a thing.

Just reading your stuff made me want to talk to you. I wish you two were sitting in Gang's[21] little room with a decent fire going and the room thick with yarns and cigarette smoke. What a lot I'd give to get that back! I shan't be quite happy till the three of us are anchored in one room again with all of a long evening and nothing to do. Fools go to the theatre. If anything I can write or say would serve in any way to accelerate your movements, consider it said.

I know it's beastly to stop a letter when the paper is ended but I've so much to say to you that it's no good writing.

Love to you both (all three, I mean),

Yours, Chink

He stood godfather to Hemingway and Hadley's son John, soon nicknamed Bumby by them all, during his next Paris leave on 10 March 1924. Their new flat was over a noisy sawmill at 113 rue Notre Dame des Champs, as bohemian to fastidious Chink as their previous smaller one had been in rue Cardinal Lemoine with its open pissoirs on the landing. Hemingway's first book, *In Our Time*, had come out that winter, dedicated to Chink and containing the Mons Cameo based on his first experience under fire at Mariette Bridge in 1914. 'We christened the baby with Chink's aid Sunday', Hemingway notified the absent Ezra Pound. 'I've got nothing further to worry about him.'

First Battalion Notes by 'Chink'   Cologne
                                                                January 1924

We are greatly in need of optimism. Here Leave is stopped by the railway strike, and a change of masters at home. According to the papers there's a blockade of the British Zone by the authorities. We don't believe in the wilfullness [*sic*] of the latter and prefer inefficient management: its employment as a punitive measure was a delicious after-thought.

Oh, it has been cold. Brrrr – One appreciates, in the full, the necessity for winter quarters. It's extraordinarily difficult to keep fit. Flu has added to the delights of the season by taking its toll of the Rhine Army. Most members of the Officers' Mess have been affected to a greater or lesser degree, but that is probably attributable to the steam-heated ante-room.

The cost of living continues to soar. One hears of Hun magnates who find it *cheaper* to go and live in Paris than remain in Cologne. How is that for a Paradox? The Mark has been stable for some time, and catches us just where it hurts most, in the pocket. It is quite impossible to dine in a German restaurant. Berlin, we hear from a colleague who was there on duty, is so expensive that he was lucky to have the return half of his duty warrant. This is an impossible country.

In 1924, increasingly serious about the need for Army reform, Chink successfully applied for a Sandhurst instructorship, and reported back there in plain clothes that autumn. Here he became an admirer of 'Dick' O'Connor of the Cameronians, and friendships included 'Boy' Browning, the Sandhurst Adjutant, and fellow-instructor Alec Gatehouse, in charge of the new Tank Corps uniform. Consulted about a hat which had to protect the hair from oil but not take up valuable space in the cramped interior, Chink tossed across his Basque beret, kept since a bullfighting visit to Pamplona earlier with Hemingway. The design was adopted, giving him wry food for thought in future when worn by Montgomery. At Sandhurst Chink was encountering Montgomery again, which did not improve their low opinion of one another, and he made a new enemy there by taunting Browning's predecessor, Arthur Smith of the Coldstream Guards.

To Ernest Hemingway   Royal Military College, Sandhurst
113 Rue Notre Dame des Champs   3 June 1926

It's like this, Maestro – Sandhurst plays St. Cyr at the Rough Rugby game on April 15 at St Cyr. I am going to support the shopkeepers. Wanted: a suitable name!!

Reasons   1. I get over cheap.
             2. It should be amusing.

So I'll arrive in Paris on the ~~15th~~ 14th. I don't quite know when. The night of the 15th the French officers here are preparing an Anglo Frankish orgy which will probably beggar description – I hope so. A colleague will probably not join me till the 16th. That will hardly be fair 'cause he'll start fresh while I'll have been once round the course already. Be that as it may!

I am rejoiced to hear that you're feeling so sure of your literary position as to be able to be respectable again. But we'll have lapses I trust. Certainly there will be lapses!! Get us rooms at the Université if you can like a good *A. de. C.*, decent rooms with a bath adjacent. What will they cost? I may have to appear at this function in uniform. Will the quarter mind? You may if you wish avoid me at that juncture. Stories – I've got heaps of stories. We'll get busy on them when you please.

My love to Hadley. Ever, Chink

If they ask how long we stay for, say a week!

But a sense of loss, not rumbustious gain, was in store. Hemingway duly reported to watch the match, intrigued as ever by the combination of tough physical sport and a behind-the-scenes view of Army life at senior rank. But Chink soon sensed that their old relationship was no longer the same. Rumours abounded that Hemingway was involved in a serious affair, soon confirmed by a weeping Hadley who had all Chink's sympathy as she confided that their marriage seemed to be over. Rather than being a companionable *A de C*, Hemingway was preoccupied or absent, and Chink was ill at ease with emotional tension. It came as a relief to return to his masculine world.

His career was beginning to gather speed, taking him to the Staff College, Camberley on the strength of being awarded 1000 out of 1000 in the entrance exam. There his growing lack of patience and arrogance made an unforgettable impression, particularly on colleagues in his syndicate of four, two of whom were Ronald Penney of the Royal Signals and the gregarious Oliver Leese. But it was the Senior Lecturer, Bernard Montgomery, who drew most of his sarcasm, and the disdain was mutual. 'Dorman-Smith', Montgomery announced in class one day, 'allows cleverness to precede thoroughness.' Passing out Grade A in the top four, Chink shrugged and publicly burned his lecture notes.

On leaving Camberley he married Estelle, Lady Berney, innocent party in a recent high society divorce case, and contentment with her coincided with Hemingway's divorce and re-marriage, drawing a line beneath their previous companionship. Estelle, in her twenties, was widely described as the ideal army wife, and her popularity made up for his own short fuse. Aware that she could have no more children after the difficult birth during her previous marriage of her young daughter, Elaine, he hid his regret and took her child under his wing. Bellamont became their mutual refuge on leave, where she encouraged him to be himself.

In 1929 Chink was the first infantryman to be appointed instructor of tactics at Chatham, the Staff College of the Royal Engineers. Innovative experience as Brigade Major to 6th (Experimental) Infantry Brigade at Blackdown in Hampshire in 1931 followed under Brigadier 'Archie' Wavell, whom he came to unreservedly respect, and leapfrogged him ahead. 'To be brevet Lieutenant Colonel at the age of 39 is a great achievement' commented *St George's Gazette* in 1934. 'But those of us who have experience of his enthusiasm, knowledge and industry will know that it is no more than he deserves.'

His next posting was to the War Office in London that year. Once installed, regular scanning in his spare time of *The Times* for its daily graph of rising German nationalism confirmed the criticisms of its crusading military journalist, Basil Liddell Hart.

To Captain B.H. Liddell Hart   The War Office
London SW1
18 December 1934

Dear Liddell-Hart [*sic*],
Like most of my trade, I have only a slight acquaintance with the jargon of psychology and am consequently no judge of the 'ologies'. But I would very much like to see the reactions of some generals to your references to the 'corporate psychology of the warrior'. I cannot see GOC-in-C. Aldershot regarding 'these psychological phenomena' as of scientific interest. Still, in the interest of Truth and Mirth, go forward.

Curiously enough the professional mind in all professions is an amusing study. It is really the second rate mind defending itself – the soft-shelled crab going into any borrowed shell. The main characteristics seem to be reliance on traditionalism and precedent as a defence against having to face facts. It usually means that the professional has dropped behind in his trade. Newcomers in an emergency frequently develop 'professionalism' of a particularly dangerous kind for the same reason. As you know well, the real thinkers in every profession are too busy and too realistic to advertise themselves.

Will you be free for luncheon at the Naval and Military some day early in January? Yours sincerely, E. Dorman-Smith

Agreement with Liddell Hart about the urgent need for mechanisation instead of reliance on vulnerable horses in battle proved a tonic, and complicity extinguished caution. Chink came to enjoy needling – and permanently antagonising – those he considered too traditionally-minded. The incoming Inspector of Regimental Artillery, Alan Brooke of the Royal Artillery, became a favourite target of his whenever War Office proximity brought opportunity.

To Captain B.H. Liddell Hart   The War Office
                                                                 26 January 1935

Dear Liddell Hart,

Re your enquiry about the higher training programme in 1935, I fear that 'the answer is a lemon'. Nobody, not even the DSD[22] will let me say what is going to happen. The MT[23] people are particularly anxious not to anticipate the Estimates speech. Apparently they get into trouble over their training grant if they do. So my lips are closed. I'm sorry, but I know that you will understand.

By the way my master was delighted at your note of appreciation of his pamphlet.

Yrs, E Dorman-Smith.

In 1936 Chink's reputation for soundness resulted in his being posted as an Instructor at the Staff College. To his dismay, though, his subject was Staff Duties, and the lectures he was expected to give were from his own days as a student. The Commandant was traditionalist Lord Gort VC, who, although legendary to the public, was to Chink the personification of military backwardness.

To Captain B.H. Liddell Hart   The Staff College
                                                                 Camberley, Surrey
                                                                 1 February 1936

Dear Liddell Hart

I don't think I like my new job very much at present. Having been ground to small powder to produce this reorganisation for what it is worth, I am being even finer ground to revise all the Staff College schemes which the new organisation has rendered entirely obsolete. That is something to be thankful for, and we're beginning to discover a number of amusing and interesting reactions to the fighting habits of that Dodo, the infantry Division, as the result of our gland grafting. It would be amusing if the old gent began to breathe victories again....

The habit of developing our MG Coys in the Rifle Bns regardless of the situation is bad. By giving them something which is not so easily 'issued blind' people may learn to put MGs where they're needed, and to do without them when they're urgently required elsewhere. Myself, privately, I incline to one Coy of 24 guns per Inf Bde, organised in three pls [*sic*] of 8 guns. That of course eases the carrier question still more, yet keeps within the figures dictated by Great War experience.

What is our hope of getting any money?

Yrs, E. Dorman-Smith

To Captain B.H. Liddell Hart

18, Bungalow
Staff College
Camberley, Surrey
15 February 1936

Dear Liddell Hart,

I have been following your battle which is all our battles with interest, amusement and considerable satisfaction. As a member of a body which has been stung so often, I was delighted to see you sting a First Lord on his most vital spot.[24] But have you also joined forces with the Big Bombers? And do you subscribe to the theory, so well and clearly propounded by your sister journals the *Daily Mail* and the *Daily Express*, of counter bombardment? I can see aerial trench warfare eventuating pretty quickly, combined with an evacuation of the stickier areas. In which case I may still be able to offer you the chilly comforts of a spare bedroom in a Staff College bungalow. But I see very little hope of decision by bombardment.

The present theory that all will be well if we build sufficient bombers is no more tenable than the heresy of over-gunning a ground force at the expense of its assault components. I do hope that because we haven't been able to achieve the sort of army we want, we will not be led into believing that we do not want an army at all. I'm probably old-fashioned, but I believe that granted the premise of a new war in which France, GB and the Low Countries were allied against Germany, the best defence against the full force of the German air offensive would be produced by an invasion of Germany. But of course there will be no invasion of Germany if we don't contribute our help on the ground.

This raises the next point. We have now – owing to air power and the submarine – all the disadvantages of being an island with far fewer of the advantages than before. The Channel which used to keep the French from us now keeps us from the French. Any offensive into the Rhineland must take place early in the war, and at present we can't get there in time. If we're to spend 200/300 millions, why not 30 millions on that remunerative project, the Channel Tunnel? A properly constructed tunnel would be an enormous asset, and it's a thoroughly peace-making project. We might have had one years ago – why not now? What about that for a fly on troubled waters!

I believe Martell [*sic*],[25] from a different angle, has come to the conclusion that offensive ground action in conjunction with the French should be the basis of our initial strategy against Germany. And curiously enough I believe that we're turning this corner, and getting away from the unnatural preponderance of Defence over Attack which has lasted since 1920 after ceasing to exist in 1918 and the 1919 that never happened. I'm making some surprising discoveries about our new mobility and offensive powers, and some day when you are not too busy we may talk about them.

Meanwhile, good hunting among the old cocks, but don't assume too soon that ground warfare is a back number. One or two of us have been surprised at certain new and heartening visitors! But the Admiralty ought to fork out £30 millions for safe communications with the continent. We now want the Channel Tunnel, without which our only contribution to Europe will be a daily load of bombs and, after a bit, people get used to bombardment.

Do you ever take the air during weekends?

Yours sincerely,

Eric Dorman-Smith

To Captain B.H. Liddell Hart   18, Bungalow
                                                          Staff College
                                                      Camberley, Surrey
                                                     18 February 1936

Dear Liddell Hart

Thank you for 'scientific doubt'. That is just the point at present! I've reached the position of doubting whether the confident assertions of the 'one arm, and that the air arm' school have any firm foundation. I'm not arguing about sea control. I agree that waters in which the ship has moved freely for thousands of years must now be labelled 'narrow waters, dangerous'. Beyond that I, being no seaman and not even knowing the first principles of sea control, cannot go.

But on land have we any real basis for your 'chaos of communications' theory, I wonder? Rail communication and inland water communications are indeed vulnerable. But road communications and track communications offer no dangerous model points except at Defiles, and road and cross-country traffic, as you've often said, is fluid and elastic. Western Europe is now well-roaded, and the Germans are improving their road network. I suggest to you, with proper deference, that the rail has had its noontide in war and the road is renewing its youth. I don't believe that air attack against Germany or a German army operating in Western Europe would cause such a 'chaos of communications' as to be decisive. It's not for me to quote your own war cry of organised dispersion. We're applying it here now and finding it perfectly practicable, thanks to modernisation and wireless. In fact modern armies operate in 'Artillery formation' from the base to the battlefield.

Now we come back to the combination of air and ground action. Can we nowadays advance in face of resistance? I'm convinced that we can, provided we increase our mobility and striking power – both being perfectly possible. We're discovering that we possess in the Corps reserves of tanks, artillery and troop-carrying MT [Military Transport] astonishing potentialities for staging quick, rapid and powerful attacks. In fact we're getting a bit worried about the powers of the previously unchallenged defence. I'd like to show this to you in chapter and verse on a map and to show you how the situation and our schemes here have altered since spring 1935.

We're beginning to rediscover offensive mobility – not with weapons and organisations of the future, but with things coming to hand which not a few of us understand. If we could apply that offensive mobility offensively, in conjunction with air power, I believe that we can hope for a greater protection of these islands from bombardment than by using one arm only. The Rhine isn't more than 250 miles from Calais – just 2 days march for a modern army – and the Rhineland won't be fortified for many years to come. If only we can get going quickly, we might carry the war into the enemy's country. *Encyclopaedia Britannica* estimates 30 million pounds and five years to build the Tunnel and possibly one could hasten this. The Tunnel Head itself shouldn't be too difficult to protect. Points, as distinct from areas, seldom are!

Lastly, 'superiority of attack over defence': 12 to 1 is nothing. Modern methods are massed artillery used well forward and early modern tank assault before the crust has formed. At any rate, think about that Channel Tunnel. It should appeal to you from many aspects, other than the throwing of a modern army into the Rhineland to take a province hostage. It's a nice conception. Shall we make Archie Wavell our Hannibal? But Hannibals are no good unless Hamilcars[26] pave the way. That is why I pass this ball, or plural of that ilk, to you.

Yrs sincerely, Eric Dorman-Smith

PS I am not a colonel and I haven't a DSO.

After a mere sixteen months, not the customary three years, Chink's obvious dedication to raising standards paid off. He was appointed to command 1st Battalion, the Royal Northumberland Fusiliers: every young soldier's classic dream of commanding his own battalion.

In 1937 the Battalion was stationed in Egypt with the Cairo Brigade, under the Anglo-Egyptian Treaty. Mussolini's invasion of Abyssinia in 1935 had opened official eyes in London to weak defence spots, and the resulting recognition of Egypt's independence meant that British and Egyptian armies were now allies.

Arriving with Estelle at his side and scenting impending war, Chink rated his 750 men as a flat-footed, lackadaisical, polo-playing unit, and he set out to modernise them – whether they liked it or not. Out went polo practice, the principal ambition to date. In came 5am exercises on the barrack square, intensive machine gun training, and compulsory motorbike and truck-driving lessons. From being 'on its feet literally speaking', he would write with satisfaction, 'by the end of the year it was a trained mechanised machine gun battalion.'

Three months later an unexpected War Office communication reached him, and he ripped the brown envelope open. The post of Director of Military Training for India was his if he wanted it. He did not hesitate. This was a plum job, given only to men expected to go right to the top.

To Basil Liddell Hart   Hilmi House
                                                           Abbassia, Cairo
                                                           20 April 1938

My dear Liddell Hart,

I fear that I've taken a long time to answer your flattering and attractive proposal regarding a book on the Infantry army in the future war. The trouble is that I've been unable to settle down to private work, for I still command. Combine that with consultant work on the Defence of Egypt and day-to-day running of a unit 750 strong, and there's little time left. Final orders arrived today to embark next Tuesday to take over in Simla on May 10th.

All this means that I'd greatly like to take the job on, [but] I couldn't get it done by June. There are a lot of chaps who know their stuff as well as I do and are less likely, perhaps, to fall foul of the General Staff censorship. If you can give me till September I could probably manage, but you may not be able to wait. Do not worry about terms. If the book is likely to be of any use to the suffering infantries of the globe, I'll do it for nothing.

I'm sorry to be leaving Egypt so soon. I would have liked another year in the command, and Estelle and I were plotting to ask you and your lady here next year. There is much to interest you professionally, and I daresay the change would have done you good. I don't suppose wild horses would drag you to Simla or Delhi, but if you ever do feel like bearding the wild Cassells[27] in his lair, I hope you'll give me a chance to view the encounter.

I rather dread this business. For ten years now I've been connected with modernisation and advanced soldiering and I'll find a return to the middle ages extremely difficult. It will be a sort of military Berkeley Square without any romance and half the G2s in the Directorate are Staff College contemporaries.

However, I can't afford to retire just yet and must do what I'm told – I'm already getting personal letters from contemporaries in India which make disquieting reading.

I bet you've had a lot of fun during the *va et vient*[28] at the War Office and the aftermath. I think that things have worked out not too badly, though I much regret the de-mechanisation of eight infantry regiments and the disappearance of four of our few fired-brigades. I suppose that circumstances beyond their control forced the Army Council to do it, but it's a step backwards for which the old gentlemen would have been heavily blamed had it happened in their time. As it is, nobody seems to have commented upon it, or on the fact that we seem to have returned to a mixed infantry battalion again. Out here, of course, any unit which isn't fully mechanised is completely out of the picture. Walking infantry are just stupid anachronisms.

Yours ever, Chink

To Basil Liddell Hart   Hotel Imperial
Queensway, New Delhi
6 November 1938

My dear Liddell Hart,

I was most interested in your latest salvage from the well in which the truth lies hidden. It only goes to show that moral courage is a rarer military attribute than physical courage. All services in all armies largely function on fear – fear of the next higher in the chain of command. In the senior ranks that's a more potent force than the enemy, and it's a deadly dangerous one for chaps in the front line.

The alternative is the 'Band of Brothers' idea – teamwork on the moral, as well as the physical, plane. But you don't get 'Bands of Brothers' out of our military system unless you have higher commanders of unusual moral courage and efficiency. Archie Wavell[29] got something of the sort going when he was a brigadier at Blackdown[30]. All his four Cos were promoted Colonel – a most unusual thing – and quite fairly so, because by removing the fear complex from their minds he doubled their efficiency permanently. It was a most interesting study in psychology.

In a small way, and under difficulties, I'm trying to change the atmosphere here. Cutting out statistical returns and demanding judgement from results is one way to help people feel that they're being trusted, but the army here is still very much concerned with seniority, and where that plus length of service is the criterion – instead of the qualities we all know to be essential – then it's inevitable that the less efficient elders will bolster their weaknesses by using fear. I really think things are better in the British service now because we're all in on a new game, and the seniors are forced out of their dugouts of precedent.

You have done us all a great service by your relentless hunt after truth. Keep it going, however much it makes us all uncomfortable. Another target for arrows is our habit of secrecy about subjects which should be widely discussed in official and unofficial circles. It's very bad out here at present, and having all sorts of undesirable repercussions as you may well guess.

I have begun to see this army from close up, and had a very useful tour of the NW Frontier. How Lewis Carroll would have revelled in that Land of Paradox where all the political officials are standing on their heads and many of the soldiers too. The only sensible people are the tribesmen who have discovered how to make money out of our obtuseness. I wish you could find time to have a look round here, especially on the Frontier.

You ought to have been on the Chatfield Committee.[31] I've now motored and flown over a good part of northern India and never seen better country for modern forces. This army is ripe for modernisation on its own lines to suit its particular problems and I hope to Heavens we get some Chatfield results, otherwise we'll just become more and more obsolete unless the Viceroy[32] and Cassels do something dramatic to enable us to make a modern army out of our present meagre 30 million a year. It isn't an easy position.

I'm much afraid that I'm not going to be able to make a job of that book on Infantry. It's not the sort of thing one can write during daily work, with all its official and social distractions. I had hoped to go to Cashmere [sic] for a month, rest my brain and then kill the back of it there, but I've not had a day off duty in the last 18 months and it's extremely unlikely that I'll get any leave at all next year either. So all I've got is a series of disconnected jottings, some written in the train (that interminable train business) and none that I feel are of much value. I'm afraid I took the business on without knowing what I was walking into here. Had I realised, I wouldn't have hesitated to say there was no hope of doing any good work.

Incidentally, this lack of leave and over-driving-the-willing-horse business is becoming something of a menace in the army. We're rapidly reaching a stage when it will consist of workers and drones. Ask any of the key people at home when they last had any decent holiday, and the answers will surprise you.

What makes it all the more difficult is that the next war is going on all the time. Ethiopia – China – Spain – the occupation of Austria and Sudetenland – the Russo Japanese encounter in Siberia – Palestine – recent events on the NW Frontier – are all part and parcel of this 'next war' business. We're all living in the 'next war'. Witness our recent and far-reaching defeat in the moral, financial and international fields. The day before yesterday I visited the Residency at Lucknow. Something seems to have happened to our race – or perhaps we are suffering from leadership by Cawnpore wheelers and have yet to find our Lucknow Lawrence.[33]

I'm afraid this is a long and rambling letter, but I have so much to say which can't be said in 'official circles' (and there are no other circles) in Delhi. I wish you would come out here next cold weather. We would be delighted to put you up. Wouldn't *The Times* play if you suggested it? You could look-in on Egypt on the way back and be home again in time for the military summer in England. That is to say, if Hitler hasn't pushed our old gentlemen beyond even their dirt-swallowing capacities. Think it over.

Our love to you and yours and again my apologies for failing to shape events here so that I could get some freedom in which to keep my promise to you.

Yours ever, Eric Dorman-Smith

Liddell Hart did not come out to Delhi, and Chink was to have ample time to become more discontented – but not yet. Estelle, Elaine and his terrier 'Mac' were with him in India, a soothing background on which he relied, and at work a bond of fellow-feeling was developing. The Deputy Chief, General Staff, Claude Auchinleck, held similar views on the urgent need for modernisation, and mutual indignation led to a useful routine of uninterrupted dawn walks.

Chink saw no point, unlike 'the Auk', in keeping his opinions to himself. He raged about 'out of touch' chiefs of staff, Indian Army 'know-alls', 'amateur' officers (usually Cavalry), the danger from India being the Empire's poor relation in military expenditure, and pounced on any evidence of complacency or bigotry. 'It is quite impossible', he fumed in his lectures, 'to get over the obvious fact that any war will inevitably expand to include India'.

As soon as Britain declared war on Germany on 3 September 1939, within two days of the invasion of Poland, his frustrations mounted. 'Nobody,' he declared publicly, 'can conduct a real war from the stagnant pool of Delhi/ Simla.' Soon Auchinleck was given a Corps command in England, and as the war expanded Chink continued to fret at being left on the sidelines. 'A pathetic spectacle,' he judged the campaign in France, and loss of India's unique breathing space made him more provocative than ever. 'Coma in Simla,' he liked to say when pointing out that they were still wearing plain clothes on duty in August 1940.

One day a file reached him as DMT [Director of Military Training] for a successor of brigadier's rank to become Commandant of the new Staff College at Haifa in Palestine. The pay was less, but the prospects far greater because the request came from the man whom he respected above all, Wavell. His erstwhile boss at 6th (Experimental) Infantry Brigade was now Commander-in-Chief Middle East, and impulsively Chink applied. No one at AHQ India protested. He was selected without delay, and his formal request that Estelle should join him was overruled.

War Diary
1940–1944

To highlight Chink's experiences at the time, as well as to avoid potential confusion between past and present for readers, *the Editor's words are in italics from now on* in all five chapters of the War Diary.

Explanations of the early fighting in North Africa and an assessment of the historical importance of the First Battle of El Alamein are by military historian John Lee[1]. For the same two reasons, *his contributions are also in italics.*

The Middle East:
1940–1941

As Commandant of the Middle East Staff School, Chink arrived in Haifa – third largest city in Palestine after Tel Aviv and Jerusalem – in the autumn of 1940. He saw himself as a failure put out to pasture, and his mood was bleak. The state of the war in the Middle East looked decidedly bad. The Allies, under Wavell as Commander in Chief, faced an oncoming enemy ten times their number. If the Italians broke through and captured the oil stocks in the Persian Gulf, Iraq and Mosul, all Allied tanks, trucks, aircraft and ships would grind to a halt.

Libya was the most vulnerable point, and on 13 September the enormous Italian Tenth Army had moved into Egypt, heading towards the Nile Delta and the Persian Gulf. Air reconnaissance showed that the four divisions were consolidating into a semi-circle of defensive camps. Eighty miles away at Mersa Matruh, ready to take them on, waited Dick O'Connor, a like-minded fellow instructor at Sandhurst in 1924 with whom Chink had kept in touch: O'Connor now commanded the small Western Desert Force, and flying to take up his schoolmaster's role in Haifa, Chink ached to be at his side.

On arrival at the whitewashed, lemon-scented port on the Mediterranean the contrast was heightened further. His room at the Windsor Hotel was cramped; Telsch House, the staff college on Mount Carmel, had clearly been a holiday-resort hotel before the war; the standard of teaching was as poor as expected. No allowance was being made for the desert fighting his students would have to face. Never mind. He knew the team he wanted and what he wanted taught, and he appeared as arrogant as ever.

Two bright spots stood out. The Senior Lecturer he was inheriting, Freddie de Guingand, had been his best pupil at Camberley, and a flirtatious young blonde was billeted in the same hotel. He made enquiries. Her name was Eve Nott and her husband was at the front. Introduced, he played the aloof Commandant until the impact of a summons from Wavell for an urgent mission in the Western Desert broke the ice and they became lovers. It proved a sexual revelation to Chink, and when all wives, including Eve, were deported shortly afterwards for safety to South Africa he stayed erotically and emotionally in thrall.

The summons, however, meant action at last, and impatiently he set off behind the wheel to liaise with his old friend 'Dick' O'Connor, briefed only that the Western Desert Force was held up from making a crucial attack on the massive Italian Tenth Army that defended the strategically important seaside town of Sidi Barrani.

Greeting O'Connor and his team, Chink learned that an entrance gap had been seized in the Italian' semicircle of defensive camps, but further progress was ruled out by an extensive minefield. His order was to advise, so deriding the orthodox plans on the table, he called for aerial photographs and began to concentrate. Soon he turned to O'Connor, and showed him. Enemy tank wheel-tracks were clearly visible, so by following them a direct 'pounce' before daylight could work. The lethal minefield would be avoided, they would use the Italians themselves as cover, and a simultaneous barrage of fire would hide the noise of their own tanks with lorried infantry close behind.

'A sound plan,' O'Connor would state. 'A bit complicated perhaps, but I had every reason to rely on my commanders, their staff, and the troops.' Wavell's approval was necessary first, so Chink drove straight to GHQ Cairo with the draft paper 'Assault on a Desert Camp' to get that in person, envious that others would be carrying out his plan. Returning to Haifa and the frustrations of fulltime theory, interesting news soon came of the appointment of his previous colleague Claude Auchinleck[1] as C in C India. The move boded well.

30 November 1940 Middle East Staff School, Haifa, Palestine 'The young turks have won a big battle there and he'll fairly purge India before he finishes. Am gradually getting a few of my ideas into circulation in these parts which pleases me if only because of my powerlessness at Haifa. Toned and de-sanded myself sitting in a bath yellow with Western Desert. Last night sat up late gossiping about the over-centralisation of government at home in the hands of Winston and Beaverbrook.

8 December 1940 Middle East Staff School, Haifa, Palestine This morning went into the Persian Garden.[1] A lovely day and I was all shiny in newly cleaned belt, polished chin strap, gaberdine suit, red tabs. A party of locals – two men and two women (one not bad-looking) with a cine camera asked if I'd be photographed in their group and it would have been churlish to refuse. Passed on, saluting and bowing, and the gardener persisted in giving me sprigs of jasmine for my room.

His plan, had gone like clockwork on 7 December, despite its unorthodoxy. Within three days, O'Connor's men had not only taken Sidi Barrani, but 38,300 prisoners,

237 guns, 73 tanks and over 100 vehicles, at the cost of 634 casualties. It was the first British military victory of the war.

12 December 1940 Middle East Staff School, Haifa, Palestine *6pm* The excitement of victory as news comes in of more and more prisoners, 20,000 now, and enormous captures of weapons and gear! The battle of Sidi Barrani may well be the turning point of the war and I curse my stars that I was not permitted to take an actual part in it. But that is the personal side which doesn't really count. What matters is the immense effect the victory is bound to have in the Middle East area, politically and militarily. All the little quislings who've been fence-sitting between Cairo and Calcutta will be busy climbing down onto our side. Our prestige will be enormously strengthened everywhere.

The Italians must be in a bad way in Greece and Libya. Anything might happen in Italy, including a revolution and separate peace. More power to little Dick O'Connor (I hope they knight him) and his merry men. They've already captured as many enemy as we put troops into the battle. The whole face of the war in these parts has been changed by one successful blow which wrecks Mussolini's dream of a New Roman Empire. The chaps here are frightfully pleased and almost too excited to work – considerable contrast from the first term when reverses in France created an atmosphere in which steady work became almost impossible. Busy myself about the affairs of this institution, but every instinct revolts against being here. Such an anticlimax to my life for the last ten years.

13 December 1940 Middle East Staff School, Haifa, Palestine News continues better and better. Don't think the Italians will hear the last of this licking for a long time. We ought to march the 20,000 prisoners through Cairo so all the Levantine world can see what happened. Rumour that the wireless reported the capture of three more Italian generals and large numbers of soldiers–wildly beyond our hopes. Excitement so great I've postponed the evening work till 9.45pm so chaps can listen to the wireless.

To General Dick O'Connor   Telsch House,
                                              Middle East Staff College,
                                                    Haifa, Palestine
                                                   15 December 1940

My dear Dick,

The heartiest and warmest congratulations on your magnificent and unprecedented victory. What your force has done in the last week will change the whole face of this war, and already even in Haifa there are marked signs of a release from tension – even among civilians. I only hope that you and your leaders and staff are not too exhausted, and I also hope that the Force has not suffered too many casualties. If things have gone as I hoped they would, that should not be so.

It will be a matter for regret to my dying day that I had no part in that battle, which is in my opinion the first truly modern battle in that all the troops on our side were modern-type formations and handled in a modern manner. As such, a full analytical study of the battle in all its aspects – planning, tactics, movement, maintenance – should be made at once and passed around the outer Army, for on the results of this we modernists can confirm all our hypotheses.

As soon as your headquarters are less busy, I would very much like to visit you again – walk the course of the battle, vet the orders – and get on with the first rough historical analysis. Somebody's got to do it. MidEast, BTE [sic] and your own staff will all be too busy. I am grossly underworked. In any case, I want the material for the School here.

When you have time to consider anything, will you consider that? This sort of record should be done by people who know, when the material is still molten, and not left till after the war and to Liddell Hart.

Again, my very best congratulations.

Yours ever, Chink

20 December 1940 Middle East Staff School, Haifa The Italians are suffering pretty badly from the malice of events. We seem to have the better part of two divisions penned up behind fortifications in Bardia[2], and I bet the poor devils are getting hell from bombardment by land, air and sea. Don't think Graziani[3] has much hope of relieving them either. Only reason why we might draw off would be shortage of water and difficulties of supply. Hope we don't have to, for obviously Bardia with its little harbour would be useful once we've put it to rights again. It must be in a fair mess now. Isolated Germany has got to hold down Norway, occupied France, Rumania, Poland and the Czeks [sic]. She's got to watch the Russians gradually increasing their strength, and put us out of the war before we strangle her. Japan is a broken reed. Whatever she undertakes against America and ourselves can't be decisive. Germany is still strong and capable of much damage and mischief, but already she has ultimately lost this war.

26 December 1940 Middle East Staff School, Haifa I see 1940 go without regret. Professionally, it's been a disappointing year. I've stood still, or rather gone backwards. Archie[4] writes personally to say he 'does not intend to keep me here too long if any opportunity offers for employment in close connection with operations'. Hope in '41 to get a chance of being more use than just giving, as he says, 'encouragement and advice'. Anthony Eden – Foreign Minister again! Once I got the Viceroy's[5] opinion of Churchill, Eden and Duff Cooper and they were highly uncomplimentary. That won't stop Eden embracing their posteriors in due course. In a 1940 broadcast I said how we'd lick the Italians in Libya.

3 January 1941 Middle East Staff School, Haifa A tale from Sidi Barrani. During the first days fighting, a gunner officer careering about the battlefield in a car ran into an Italian post on the edge of the S.B. defences. Taken to an officers' mess, he listened to a vivid discussion as to whether they ought to surrender. A minority demurred, believing we would operate on them in a manner highly unpopular to an Italian. When he assured them that we weren't those sort of chaps, they threw their arms around him and said he was their friend for life. Next day we took S.B. and he returned to his battery HQ followed by 30 Wop officers crying 'Remember what you promised us....'

4 January 1941 Middle East Staff School, Haifa Dick is doing gloriously. There won't be much left of Italian Bardia by midnight tomorrow. 2000 prisoners already – say another 2000 casualties on top. A pretty good dollop of the garrison already gone. Big stuff. Am glad he felt strong enough to assault the place, it will make the Italians elsewhere feel less secure behind barbed wire and mines and concrete.

5 January 1941 Middle East Staff School, Haifa I ought to see more people, of course. Suppose a decent Brigadier would call on the Commissioner and entertain, but I'm only a bird of passage and not a very agreeable one at that, so by denying myself to Haifa I deprive it of very little. Daresay my reputation is that of a conceited, casual, offhand, standoffish blighter who thinks himself too good for local society, which isn't true. But of course one does get rusty with people. I'll probably show up badly at Govt House. They've asked me again to stay, so am going. Don't think I can run out twice.

10 January 1941 Middle East Staff School, Haifa Dined indifferently as usual while others talked golf across the room. Tried to read philosophy and gave up, so sat quietly and ate and wished those dullards could know, as I do now, that the contents under my hat gave Dick the key of the doors to a place initialled S.B. That last is a fact. The paper drafted by me is in the official report – but anonymous and, as I'd no official status, unacknowledged. Nevertheless those sort of things give one a kick. If I'd been one more day delayed on the road down the battle might have had a very different result.

13 January 1941 Middle East Staff School, Haifa Drove away from Govt House with my usual bitter reaction after being among strange Englishmen and women. Feel I've said and done all the wrong things. Am not decently drilled to conventional country house society and unhappy in it. There was none of the pomposity of India, but everybody knew everybody else and I nobody. The McMichaels[6] were free from fuss and he's intelligent, but how much of his thought is borrowed from others? The atmosphere is reasonably cultured, but why do English people merely talk about other English people of the same set? Still, I survived. What they made of me I know not. 'Dull' I think they'd say! Dropped with relief into my hermit life again.

22 January 1941 Middle East Staff School, Haifa An amusing letter from Dick tonight in reply to mine chiding him for not taking me as his BGS[7] when he was getting a new one. Apparently he wasn't allowed to have me. What the devil are they all playing at? Couldn't ask for a better job.

24 January 1941 Middle East Staff School, Haifa Tobruk has gone the way of Bardia. How I'd like to see our people take a real Bisque and 'ride' straight for Benghazi 20 miles across the desert. Mustn't overreach myself, but this astonishing series of victories has made me greedy for more. If all goes well I'll be with Dick on Tuesday next. What fun!

Two days later, acting on Wavell's fresh instructions to observe and report, Chink flew to Cairo and drove for 48 hours through dust storms to reach O'Connor and his team.

1 February 1941 Continental Hotel, Cairo Dick out of a sudden urgency took me away from Haifa and sent me in to see Archie here. The day before yesterday I was in Derna[8] with the advance troops as they entered, and yesterday I lunched in Cairo. Go back to Dick by air tomorrow. Have seen lots of interesting places – Bardia, Tobruk etc – and travelled many miles of desert since last wrote. Flew from Palestine to Cairo and motored from there to the front through pretty foul sandstorms. Directly one pokes one's nose closely into a corner of this war, one ceases to see any other part; even a week in Libya has clouded my broader view.

On 11 February, glorying in the pursuit, he reported to Wavell in Cairo to get permission for O'Connor's attack to continue as far as Tripoli. But it was not forthcoming. On Churchill's orders the majority of troops had to be switched to Greece in the unlikely hope of uniting the Balkans. It was a fait accompli, and Chink's imagination was with O'Connor waiting at the head of his Western Desert Force. Meanwhile a new German commander, Erwin Rommel, was about to disembark at Tripoli, having been sent to rally the Italians with his crack troops, the Afrika Korps.

13 February 1941 10pm Middle East Staff School, Haifa Flew here this morning from Cairo, having motored the day before from Tobruk – 570 miles in 18 hours. On Wednesday, Thursday and Friday was with Dick in the final round up which finished off the Italian army in Cyrenaica. Stirring times. We were with the troops which crossed the desert and cut the Italians' line of retreat into Tripoli. All with complete safety, few casualties and only dust storms, rain, cold, hunger and sleeplessness as active enemies.

Have lost ½ a stone and now weigh 11½ in blue serge uniform. Also pretty weather-beaten; only cleared myself of sand on return to Cairo. Now to write an official report on the lessons of the campaign: not an easy job. So much to say but nothing I didn't know already, and the final move – across the desert, most exciting! – he and I had already discussed. Cyrenaica's rather like Palestine, though it was raining hard as I passed through. Palestine is a delight – fresh and clean and spring-like. Could hardly wait to unpack before a run to breathe clean air again.

15 February 1941 Middle East Staff School, Haifa *7am* Dawns over the Libyan desert on those last hectic days of pursuit and victory were incredibly lovely – lovelier because of the storms they carried with them. On the day the Italians surrendered, Dick and I motored onto the battlefield through an unbelievable sunrise and bright clear daylight with the air full of sea smells across a 'velt' just flowering into asphodel and covered with herds of gazelle and flocks of great slow-flying bustards. So then we came to the dreadful road on which the escaping Italian army had been stopped and wrecked – the Fascist autostrada of North Africa. As the car threaded its way through man-made wreckage one knew this was ephemeral and that the same dawn and quiet antelope-owned plains had seen the passage and end of other armies and races, yet remained untouched by the passing convulsion of events. Like Helen's loveliness indestructible, beyond man's handling.

This morning as I shook under a cold shower it came to me that what I'd seen wrecked wasn't just the Italian army, but any army which lets itself get rigid, obsolete and supine. There but for the grace of God went the King's army in India, and I've no regrets that I made myself disagreeable to the knaves who would have prevented modernisation out of their ignorance and vanity. This course ends next Saturday. Most of the chaps have got good jobs which carry promotion, and we're changing quickly. But so is the pace of the war, which may swing north and east for we've yet to meet our real enemies, the Germans, or our new enemies, the Japanese. Now the Germans have come to the Mediterranean things will warm up, but it's all to the good. The more we stretch them the better.

17 February 1941 Haifa It's 9pm and a long day is ending. Lectured this evening for two hours on the Cyrenaican campaign. Ten days ago was the crux of the last operation. I was cold, hungry, tired, yet I was with Dick. He'd succeeded incredibly, and I was completely happy. No woman counted then. It was a stark world where men were happy being themselves – mono-sexual. Kipling has it in 'The passionless passion for [sic] slaughter'. What it is to be a poet and know how humans feel by intuition.

22 February 1941 Haifa *10pm* Slept heavily and dreamed my standard dream – a deserted and dilapidated Bellamont.[9] How that place lives with me always. At 8.30 am supposed to say goodbye to the students. There's one of the many matters in which I fail. Can't bear Pep Talks, comes of being a sceptic and a derider. Still, it's expected of me, and am willing to try a new drink once.

28 February 1941 Haifa Fly to Cairo on Sunday with my report. Thank heavens I've finished it, though it did mean writing all day from Sunday last till yesterday. But that is done, and I'm well on my way to becoming a first grade Cyrenaican bore. The epilogue is 'We gambled magnificently and pulled it off'.

Chink went on to South Africa to give lectures on the recent campaign and to research an idea for a Higher Command Staff School for those of generals' rank, run jointly with the existing South Africa Staff School in Pretoria. He spent time with Eve, who was billeted there, and left Durban on 22 March 1941, sure about the need for advanced training and of a future with her. The German threat was clearly building up under Rommel's leadership in the Western Desert, but for once he did not have to stay on 'the sidelines' for long.

6 April 1941 Haifa *7.30pm* As I left for dinner a cable: "Brig. D.S. report forthwith MidEast – Temp duty BGS.' Presumably meaning a spell in Cairo answering for Jock Whiteley[10] or Arthur Smith.[11] Means leaving my deputy in the lurch at the beginning of a new course with a school double the size. Hell.

As Brigadier General Staff, Chink's Cairo office was in GHQ Mid East, and he was allocated temporary accommodation in the 400-room Continental Hotel which boasted a top-floor midnight cabaret. Chink's small room was directly below, and soon he cursed the loud dance-band. At GHQ ME his immediate boss was Lt-General Sir Arthur Smith, judged 'Amateur' by him at Sandhurst in 1924, and now Wavell's choice as Chief of Staff. Rapidly Chink concluded that Smith was not only running the office inefficiently, but giving Wavell poor advice.

8 April 1941 GHQ ME, Cairo May not go back to Haifa for two months, if ever. Meanwhile am working under A S, so office hours may be anything from midnight to midnight. Struggling to get into the work, helped by the able Freddy de G[12] who's brilliant. With the Balkans aflame[13] and the Western Desert shrouded in dust and the smoke of conflict it's not easy to see all possible repercussions.

9 April 1941 Cairo Roused at 6am for a constant impact of new problems – but no data or background. No matter, I'll get square in due course. It's a question of knowing the form here, too. I tread like Agag[14], delicately.

1.45pm Hectic morning and more hectic afternoon waiting. Blast! Fly off somewhere tomorrow morning, hoping to be back next day. Grim news of the capture of Dick O'Connor, with Neame[15] a night or so ago.

O'Connor's capture behind enemy lines on 6 April, due to faulty map-reading by his inexperienced successor (and consequently fellow-prisoner) Neame, raised the already tense atmosphere in GHQ ME. The day after Chink's arrival Mechili fell, Salonika was captured by the Germans'[16] and Churchill commanded that Tobruk be held to the death. Chink was ordered to take up instructions for its defence first thing the next morning.

10 April 1941 *In flight 10am* The Air Ministry omitted to put a desk in the back seats of Lysander aircraft, decorating the rear with a useless machine gun instead, so am writing on my brown box on my knee. My poor fusilier Appleby arrived from Haifa on Tuesday, saw me for one minute and got no orders save to expect me back when he sees me. Am in highly inappropriate uniform for where I'm going, so likely to be commented on with ribaldry by hardbitten toughs, and piling up a pretty incredible air mileage, but with a chief who lives in the air, what is a BGS to do? How silly to aspire to the fire when in the frying pan! But I serve a Great Man in APW[17] and am honoured with his trust and confidence, neither of which he gives easily. Wish I were more worthy. Am prepared to back my military judgement against most chaps, but wouldn't back my powers of decision or physical courage so strongly – a shaming thought when all over the world our young Davids are moving into the dawns to face Goliath in full armour.

Tawny desert on my right – a sea the colour of cats' eyes on my left and it's not a clear day, portending dust storms. The Boche is getting his first taste of them. The desert's impartial in distributing its charms. Last time I motored down the road below me (that non-stop run of 580 miles), thought I was saying goodbye to this part of the theatre of war. How little one knows. How pathetically futile are human forecasts – hopes, fears, joys and sorrows – yet they give depth to life. We try to deceive ourselves by saying 'Yes, this may be so for my generation, but I'm building for a future people'. Odd how we Westerners worship the future, while the Chinese preserve continuity by veneration of ancestors. Which is right? Or does it matter whether you look backward or forward so long as you look out *from* yourself. Or, a cynic might add, *for* yourself.

3.15pm Beating homewards, out of range of fighter interference. Everyone at HQs seemed in good heart. Glad I got the chance of going up to see them.

Rommel opened his assault on Tobruk the following day. The fortress would succeed in holding out for 240 days, supplied by a navy lifeline.

Friday 11 April 1941 Continental Hotel, Cairo *7.15am* Left the office at 9.50pm last night, so 5am – 9.50pm with 6 ½ hours flying. Bed midnight, 6 hours sleep and now to GHQ. The remaining props of western civilization – weak props being Balkan states – are tottering.

12 April 1941 Continental Hotel, Cairo It's hell having lost our poor little Dick.[18] He really <u>is</u> a loss because though ordinary generals can be replaced, his particular qualities are difficult to reproduce. Well, thank goodness he's alive and we may get him back some day. No good glooming. What we have to do now is to get on with the job.

13 April 1941 Continental Hotel, Cairo Still living like a troglodyte from 8am to 9pm daily and hating the life like hell. Would give a lot for something more like war, though I sit at the centre of the whirlwind. The immediate 'flap' is over. Don't mind as a staff officer being a gentleman's gentleman, but to be a gentleman's gentleman's gentleman makes me feel like a plum stone sticking in a giraffe's neck: just an obstruction. As soon as Whiteley returns will make my way back.

But Wavell's trust in Chink was reinvigorating. On 15 April he was instructed to work on the WPC, and the Worst Possible Case for Wavell meant if the Allies were forced to quit Egypt. Analysis confirmed his suspicions about their lack of military resources, and his empathy grew as he comprehended Churchill's unceasing pressure from London to produce a master-stroke. Four days later he was ordered to join the planning for the small Operation Brevity designed to help Tobruk, and his spirits rose.

21 April 1941 Continental Hotel, Cairo The war is at an intensely interesting stage, particularly where the enemy is making a vast effort to dislodge us from the Middle East. Might the effort and energy expended be too costly in the long run?

23 April 1941 Continental Cairo We'll soon meet the full force of the German war machine, and it will be a stern experience. Luckily, thanks to Dick's victories, we've put the enemy in a far worse position than if he could deploy his armoured forces in peace behind Graziani's armies at Sidi Barrani. Now if he comes on it's from 'bases' in a devastated country. The Navy and RAF did a good job at Tripoli yesterday. We've caused the enemy much loss about Tobruk, and will cause him more. Dick will be well looked after – we hold too many of them hostage. But what ghastly bad luck and what a loss.

25 April 1941 Continental Hotel, Cairo Beginning to see daylight through some of our most pressing problems – just a little time to breathe is what we need. This is rather a sticky baby to have handed to one – all ends wrong at the same moment.

28 April 1941 Continental Hotel, Cairo *7am* Am afraid my poor chief will look on April 1941 as his bad month. When it began we were still well south of Benghazi and there was a firmish front in the Balkans. Now the Balkans are sewn up and we still have the Western Desert. If ever I write the history of this war its title will have to be 'Mother Hubbard's War'. It would be so good to find a nice juicy bone for the poor dogs of war, in place of a few dry splinters carefully collected from the last 'dogs dinner'. Roll on 1942 when, we're assured, the garden will blossom with tanks and planes. Meanwhile we'll do all we can, but my heart is sore for my master with his overwhelming responsibility.

2 May 1941 Continental Hotel, Cairo *7am* I can see the minarets of the Citadel and the green freshness of the sporting club. Through the morning mist, over my right shoulder, gleam the sun-kissed flanks of a pyramid. Cairo is slowly coming to life. Stress here fairly acute, but at least we've got 80% of the chaps back from Greece[19] though we've lost much useful gear. The same old tale: Allies with obsolete armies utterly unprepared for modern war, and the enemy all out with every modern inconvenience. I believe the Boche soldier isn't the chap he was in 1914–18, but he's well led by ruthless men with one-track deadly minds which makes him dangerous. With luck we'll get around this rather awkward corner.

10.30pm Spent the day trying to prop a crumbling house with brown paper and string and a little glue: most inadequate. Australians are dying in the Western Desert because Treasury officials and Cabinets wouldn't spend money when they could. Guns or butter? Spend your money *before* a war, not during it; in any case, there'll be nothing to spend afterwards. Am in a managerial office dealing with an immense variety of problems, all of which could be solved on elementary first principles with 100% hope of success had we even 20% of the means disposed of by our enemies. No wonder I'm angry. My work is complicated, endless, and a profitless employ.

Was standing in the outer office for visitors to CGS (Arthur Smith) or BGS (me), in my shorts and shirtsleeves. Enter a small fussy Colonel of Marines: 'I want to see the Brigadier at once.' Myself: 'Yes, on what business?' Colonel: 'Never you mind. I want to see the Brigadier.' Marine staff officer in a horrified whisper: 'Sir, this *is* the Brigadier!' So haven't yet attained dignity – it will come.

4 May 1941 Continental Hotel, Cairo *10.30pm* In the office I fiddle about on a very high level and daresay men live or die on my decisions. But although as second footman I arrange a number of significant matters, above me there are layers of decision into which I hardly penetrate. The Chief remains a pretty lone wolf. Don't think he has anything else for me – not having commanded a brigade during the last two years I can't look for a command. No matter. I'll remain an onlooker at the centre of the typhoon – an interesting, if not dramatic, existence. Tonight go onto night duty again, meaning that I'm bound to be rung up by anxious duty officers, but when 'swift ones' come rolling in it's hard to blame them. More at home with this work, but even now not sure of myself. Personalities come into it all so much, and processes I'd regard as scientifically or technically clear-cut get bedevilled because somebody relies on A and doesn't like or trust B, which makes staff work not easy under our system.

3am Telephone just woken me. Waiting for the duty officer to arrive with a 'fast ball' he hasn't been able to field.

Decryption at Bletchley Park in England of Germany's secret Ultra (Enigma) signals had revealed a new menace. The island of Crete had been chosen as the temporary refuge for the huge numbers of British soldiers fleeing Greece, and Ultra now showed that a mass airborne assault on Crete by German paratroopers was planned within six days. Urgently, Chink was despatched to alert Crete's commander, General Bernard Freyberg VC[20] of the New Zealand Division, and to 'indoctrinate' him into Ultra. It would be ominous news because no defensive measures could be taken in advance to prevent a hint of Britain's ability to intercept signals reaching Germany; once suspected, the current decypherable code would end. Staunchly, Freyberg replied that he would do his best.

12 May 1941 Continental Hotel, Cairo *10pm* Last night I was in Crete in moonlight so clear that colour came through it. The blue of the middle sea. The green of olives. Add an all-pervading scent of aromatic broom and thyme. For delicacy and surprise, Crete with its mountains still snow patched, its deep ridges and valleys, its amazing promontories and its lovely sea coast is unique. For half an hour in the evening I stood on a high unfrequented place and breathed the scented coolness, feeling the whole spirit of Greece and Minoa inspire my vision.

But oh, so deaf! Blenheims are noisy, if swift, transporters. This is a magic war for me. Here, there and everywhere in any old kind of aircraft – Lysander, Proctor, Blenheim – and am learning their manifold discomforts. Strange how fatigue changes one's handwriting. Another busy day coming but no complaints, remembering chaps in the sandstorms of Libya or stifling heat of Basra.

14 May 1941 GHQ Mid-East, Cairo Arthur S. is away, leaving me in his shoes, and today is likely to be hectic. Wars don't stand still, and we're up against an energetic desperate enemy who must get results and has such a long start on us. England has become very remote. Even my lovely Bellamont has faded – not that I can't see every stick or stone of it in memory.

15 May 1941 Continental Hotel, Cairo Things are on the march and I toddle after them like a small boy being towed along overfast by a forceful nursemaid. War always feels like that when the initiative's with the enemy who has the means and will to pursue it. Archie is quite wonderful. They unload all sorts of unnecessary burdens onto him – worrying things which need a delicate exercise of judgment – and he deals quietly and effectively with them all. But he's perhaps too accessible and too ready to handle burdens. He's a great man, with a cruel job to do.

That day Ultra revealed that Goering had sanctioned a forty-eight hour delay, and on 17 May Chink flew to Crete again. Bidding Freyberg goodbye, the massive assault was due at 8am the following morning, but a further twenty-four hour postponement was intercepted.

19 May 1941 Continental Hotel, Cairo Highlight of yesterday: APW meeting me at 8.30pm in the passage. 'Well, Eric, if there isn't anything else I can do to get a move on this war I think I'll go home.'

20 May 1941 Continental Hotel, Cairo Yesterday found myself at a moment's notice hurtling into the air in the gloaming and landing into blackest night. Seem to live in that hostile element; it's more natural to fly than walk or drive a car. A demented shuttlecock, I've moved around from Alex – Cairo – Alex by air (Tuesday), Wednesday: Alex – Cairo, Cairo – Alex by air – Cairo by road. War's a grim business, and in some ways grimmer at the silent, deadly heart than for those at the violent outer edge of the hurricane. Have had a bit of both and like neither, but this least. One sees and knows too much.

4pm Haven't left the office all day. I don't underrate those parachutists. We've killed a lot of them but there are lots more, and though doped, at the prospect of action they're brave and dangerous. But it was the bombing that finished it.

22 May 1941 Continental Hotel, Cairo Crete is an epic struggle[21] with both sides flat out. The worst is not yet over but Freyberg is a fighter and an ambitious fighter at that.

23 May 1941 Continental Hotel, Cairo A long day for Cairo but nothing to what's happening in Crete, culminating in the Germans having mastery of the air. We pay that price and many others to the crowd who ruled England before this war and who will, given a chance, rule England again. To me the whole wicked business is not 'past' at all; it's the future lying in wait for us if we don't face facts.

24 May 1941 Continental Hotel, Cairo *7am* What a night. Went to bed at 11.30 and it seemed that the telephone rang from then on and a procession of duty staff officers filed through my room. Then there was an air-raid warning and an outsize mosquito, so am in grand condition for twelve hours of concentration ahead.

11.30pm An endless day. Slept for 30 minutes this afternoon which enabled me to put in another six hours in a sweltering office. The sequence of telephone calls and orderly officers begins again. Things still critical. That lovely town I looked at fourteen days ago is now largely in ruins. What a wicked thing to do to beauty, unchanged since the beginning, and now fouled by the Hun.

25 May 1941 Continental Hotel, Cairo Hardly been out of the office today, but the longest hours don't compensate for the feeling that one is buying a good dinner and soft bed at the expense of the magnificent men at Tobruk and Sollum and Canea – my beautiful Canea is smashed and desolate – who are being crucified for the stupid cupidity of the older generation.

29 May 1941 Continental Hotel, Cairo A day of grim decisions. Poor Archie.

30 May 1941 Continental Hotel, Cairo *7am* Afraid our enemies are crowing this morning, blast them. Only hope I'm there to see them take it when we in our turn get a superior air force and learn how to use it. There's going to be a considerable outcry over this unhappy business. We're doing our best to bring chaps back, but that isn't easy.

11.20pm A dull grim day. The shadows creep a little nearer, that's all. Haven't had time to give Roosevelt's speech[22] my full attention. It seems he hopes for a newer world. World-making seems all the vogue, even among those most concerned with keeping the old world where it is – or was. But the speech is a turning point. The Yanks are getting scared.

31 May 1941 Saturday Continental Hotel, Cairo Last day of a bad month for our army, brightened a bit by concluding operations in Abyssinia[23] and the sinking of the *Bismark*; otherwise all the brightness shines temporarily on our enemy. How wonderful to be able to make war with his resources and a single mind. Whether we could achieve his criminal ruthlessness I doubt, but we may yet throw up a hard genius like Wellington. Our generals have to make war with nothing. One can do it on land, but against a greatly superior air force the odds are seriously adverse.

1 June 1941 Continental Hotel, Cairo The *via Dolorosa* of Crete is consummated. We've paid a very high price for middle-aged and medieval parochialism and the enemy have paid almost as high a price for under-rating us as fighting animals even when overwhelmed. Blame myself not a little. Should have been quicker on the uptake, less entangled in the muddle hereabouts. Am only just realising how much there is to be done before a mechanism is established in which clear and forward thinking has any chance of development. Am cross and bitter because the enemy knows this game and we don't. The only reason for his extra-knowledge is his more open and receptive mind – and the fact that his air force and army don't compete; they cooperate. There are other reasons like the enormously long start in production, but the fact remains we've had a grievous loss.

Later I don't place 100% trust in life. Actually, that isn't altogether the case. The distrust is also in myself, and is going to limit my usefulness. It comes, of course, largely from fighting a series of losing battles with seniors. In this war the only enemies I've fought have been British!

3 June 1941 Continental Hotel, Cairo Arthur S saw me this morning – mumbled gratitude for work done during the succession of crises, and as soon as I've reinstated Jock W into the saddle I go back to Haifa. Gather the General Staff is to be reorganised and put on a proper basis, which is something achieved here, though it will never be admitted. Grousing again!

5 June 1941 Continental Hotel, Cairo Released from responsibility for the conduct of the war out here, have undergone the usual reaction. The strain of working in an atmosphere of muddle and disorganisation when attempting to meet crisis on crisis was more considerable than I thought while it was going on. Am not pugnacious enough to command troops in the field, not cold-blooded and insensitive enough to be a senior staff officer. As a professional soldier I should look on war as an opportunity for advancement – though it's proved the reverse, so far. But can't bear to see men suffer sorrow and pain who've had nothing to do with soldiering till now, nor have – unlike the Germans – been trained in the tradition that the fighting man is the cream of society. War can only be fairly borne by men conditioned to it by pride, discipline and belief in a cause. Can these newcomers stay the course? They did in 1914–18, but there are stresses now with no parallel then. The air arm being one, armour[24] another.

Yesterday evening I bought a stack of Pelicans – *Philosophy and Living*, *The Personality of Animals*, *The Physiology of Sex* – and an Egyptian sitting at a table in the bookshop spoke the usual jargon of palmists. Had an idle half hour so let him read my hand. He was struck by the length of head-line. 'Only Ministers of State have lines like that.' He picked out my unusual index finger, longer than the third, as indicating purpose, power, leadership and idealism. I despised money, he said, and personal gain. 'You are a religious man and probably have a bible in your pocket.' A safe bet for 70% of soldiers but, oddly enough, true. The broad base of my index finger indicated great ambition to develop my powers, and he stressed that I'm purposeful and strong-willed, with a love of beauty. Confused patterns on the upper palm meant crises in my family life. 'It is all there in your hand. You received sudden promotions between 40 and 44.' Not bad that. 'But that is all over for me now.' 'Oh, no. There is much more to come. You are bound to be a Minister.' (How Egyptian that last.)

'You are never content till you have thought a subject out to the end and found the truth. But you are too lenient and kind to the wrong people. The area X' – he pointed with his pencil – 'is higher than the area Y. General Wavell attacks – I am sure his area Y is big. You are a defender. I would give you a city to defend.' If a person harms me, he added, I never forget – very true. 'You have two children.' No, I told him. 'How long have you been married?' '14 years.' 'Has there been no conception?' A little embarrassing that, but 'No'. 'Well, you will have two children. Take my card and you will remember what I say and write to me one day to say that I told you the truth. Not one in a thousand has lines like you. Take two regiments and look at all their hands. You will not find a man whose first finger is longer than his third.' Well, how quickly he found the strength and weakness in my nature, and how neatly they cancel out.

6 June 1941 Continental Hotel, Cairo Am a bit resentful at leaving the hub of things, yet glad to be gone, mainly because I feel I'm not being employed to my full use, yet quite see why. I'm basically constructive, but in an uncongenial group out of accord with my views I'm an awkward and destructive component. Well, that I can't help. Can't accept the repose of letting things go wrong for fear of upsetting people or vested interests, so until I run into a favourable environment am likely to feel militarily frustrated.

Said goodbye to Archie today, looking very tired he was. Wish he'd take some leave. This unending responsibility is so utterly wearing and he gets pretty well badgered from Home, not to mention the complicated troubles from the heterogeneous body of troops out here.

Operation Battleaxe, which Chink opposed, began on 15 June and was intended to relieve Tobruk and capture eastern Cyrenaica. It was conducted by 7th Armoured Division and a composite infantry force. The infantry were to attack in the area of Bardia, Sollum, Halfaya and Capuzzo, with tanks guarding the southern flank. The Tobruk garrison was to stand by, but not to sortie until the Allies drew close. But the Halfaya Pass attack failed, and by dark on 15 June only 48 British tanks remained operational. The next day a German counter-attack forced the Allies back on the western flank but was repulsed in the centre. The Allies were now reduced to 38 tanks. On 17 June they evaded encirclement by two Panzer regiments, and ended the operation.

25 June 1941 Middle East Staff School, Haifa, Palestine How I wish we could have a real good crack at the enemy somewhere where it really hurts him. Wish Dick were here. Believe that he and I in combined cogitation could shift the Boche from Cyrenaica, conceited though that sounds. Could work well with him and there's always a way of doing these things, given a reasonable amount of means, imagination and guts. I'll produce the imagination, at least.

1 July 1941 Middle East Staff School, Haifa Russian news not encouraging. Think our new 'friends' are making the mistakes made by the French and Poles in pitting a horse-drawn immobile infantry army against a modern air-land armoured mobile striking force.

What a ghastly price a nation pays for inefficient War Offices, higher commanders and staffs. 'Brains' aren't getting much of a run in our own show. There's a tendency to justify bull in a china shop behaviour by the apology that after all toughness counts in war and its tough to run violently on your enemy from in front. Yes, but only tough on the brain pan and one's own troops. Have never liked 'soldiers' battles', letting the soldier retrieve the mistakes of the higher command may be magnificent but it isn't war. Wish I was sufficiently in the picture of our scientific research and development to know of any innovations which might be put to new and successful use. All the more reason for thinking out new ways of using the old things, but thinking's sissy. Toughness is all!

'Battleaxe' had been costly. The Allies had had 969 casualties, 91 tanks were knocked out or broken down, and the RAF had lost 36 aircraft. German losses were 678 men (Italian losses unknown), twelve tanks and ten aircraft. Wavell was among those sacked in the aftermath, to be replaced by Auchinleck[25] and on 22 June 1941 they formally exchanged commands: Wavell became CinC India and Auchinleck CinC ME. Listening to the news, Chink's feelings were mixed. He approved of 'The Auk's' military thinking, but the swap raised the likelihood of a personal dilemma. Wavell might wish to retain him, and a return to India would reunite him with Estelle, jeopardizing a shared future with Eve.

2 July 1941 Middle East Staff School, Haifa 7am This change in command here and in India is a good decision. Archie was getting tired, while the Auk is rested and fresh. Have written to both because they're as much my friends as any man, but that isn't saying much says the 'cat that walks by himself'. What's interesting to small fry is what happens when big waves like this occur. APW will arrive in India, the general idea then being, I suppose, that I should go back too. Auk may well think I should be taken out of my soft job into something harder. It looks as if am going to have to make a case for (A) remaining in the ME and (B) staying at Haifa, but as a middle-aged gent who's surrendered his future into Eve's gentle hands, I can be 'choosy' – if it's only a question of me, not the good of the cause.

But what to do, when the *denouement* to India comes, with all those high commanders who know me and my family intimately and will be severe judges? Well, 'they' just have to judge. Anyhow, am not going to India yet, or until I've thrashed out the next moves in this complicated long-range minuet to a music of bombs, cannon and aeroplanes. No Victorian spinet stuff for this change of partners! Why can't things stay put? How I hate the 'Dynamic of Events'. I'd rather have Eve than every crown, star, baton and sword they can cram onto my shoulder straps. The loss of those pointless insignia I can face calmly. Couldn't face the loss of her, because it would be the loss of everything worth having.

Estelle cares for me a great deal, both becoming middle-aged and lacklustre. Am sure that's exactly what she hopes to happen: just quiet and comfort in Bellamont, the occasional trip to Dublin, visiting friends, shooting. Well, perhaps I could have done it if I hadn't met Eve. But how is it possible now, with every sort of adventure calling?

12 noon A rare pre-luncheon pink gin. Presume the Auk has already reached MidEast and I can't help laughing. Simla will only just have acclimatised itself to Jessie Auchinleck,[26] a very gay and cheerful lady and no respecter of persons, and now my dear old shut-eye Queenie Wavell,[27] snob of snobs. She'll love it! Also will Jessie A come to Cairo? Every sort of intriguing possibility opens up. Less than three years ago I called her 'Lady Auchinleck' at a party. 'Why do you say that, Chink?' 'Just a hunch and coming quickly.' Good pre-war prophecy, but she loved it.

5.15pm However long this war business goes on – provided Eve and I survive it – the end will come. So what we've got to think of is what's going to happen when it ends. When that time comes, we can do one of these things: (a) Get together as quickly as possible and stay together. (b) Get together and having done that, for a little go to our 'spice'[28] and break the news. (c) Break the news first and then get together. Can't help thinking that (a) which is so natural will win. It isn't as if there was so awful much of living life together left – Eve's lookout for choosing a middle-aged man and my sorrow. Our big responsibility now seems to be to each other. My difficulty is that I might be ordered to India straight into the arms of Estelle with an applauding multitude of friends and enemies gathered to welcome our reunion.

Damn. In my emotion I've kicked over the waste-paper basket.

3 July 1941 Middle East Staff School, Haifa *7.30am* Yesterday evening, through Archie's *adc*, a 'by hand' letter from Estelle. Must have been brought by Auk – a very charming, sad little letter. Hell!! Oh well, what will be, will be.

12 July Middle East Staff School, Haifa Wrote to the Auk today saying that, being short-handed, I couldn't get away before the school closed down, but that I'd be in Cairo on the 25th and will get in touch with his *adc*. Don't want to appear too keen to sponge on an old acquaintance, particularly when he's so high and I'm so particularly 'unhigh'. I'll take no man or woman's pity – nor will I ask for anything except 'justice'. Won't even *ask* for that. But he's friendless out here and may well need me for a bit, as little Dick needed me. If so, it's childish to be standoffish just because he's become a great man. That last meeting was amusing. I was still Director of Military Training and 'somebody'. He commanded a District as a Major General and had just heard he was to go to England to command a Corps. Very diffident he was about it all, and he fixed a walk with me in New Delhi all round that lovely ruin the Pmana Killa [*sic*]²⁹ to voice his fear of the unknown. That was only 18 months ago and seems a life-time. He hopes his Jessie – whom he adores – will come to him soon, so do I for his sake, and because she'll warm up Cairo.

17 July 1941 GHQ Mid-East, Cairo Saw Auk last night – quite unchanged, rather lonely for Hindustan. Lunching with him today at Mena³⁰ because he wants to talk and I guess have a quick peck or two at the old brain pan.

24 July 1941 Middle East Staff School, Haifa *10pm* A letter from Liddell Hart³¹ in which he's read the script of a lecture I gave in India and decides that, 'It is evident that your military vision has not contracted but rather grown during your sojourn in India – a place so restrictive to most soldiers' thoughts. I have heard at intervals of the good impression you have been able to make on the state of training there, but I have been hoping that you might be sent to a bigger post in the Middle East where your strategic and tactical ideas could have more direct effect.' What a hope! Also a note from the late chief written from India which ends 'I hope you will fulfil your ambition to become historian of the Middle East. You have full leave to criticise my shortcomings. Till then, yrs ever, APW.' Rather sad, I feel.

25 July 1941 Middle East Staff School, Haifa When I reply to Liddell Hart am going to sound him out as to a history of this Middle East. Already think it worth laying ground bait. The official history will be a long and slow affair, not to be written for years. There's much to be said for getting down early to an unofficial tale. There would, even then, be two years work in it.

27 July 1941 Middle East Staff School, Haifa This morning a visit from the Deputy Chief of the General Staff – a major-general much my junior, so the strange experience of facing somebody who's beaten me in the race for honours and promotion. Up to now it's been my lot to be faced by men I've passed. Today I didn't care a damn – all that seems another world. We talked sensibly about this and that and, as men do, made a difficult interview as comfortable for each other as possible. Surely my life as a soldier is finished, because any life from which ambition goes is finished. I dispassionately recommended reforms to this place – joint land and air institution – and then he went away. From a hint or two he dropped, seems that a talk with a certain great one at a lunch may bear fruit.

At the end of July Chink visited South Africa on a month's lecture tour, making useful contacts to bring forward his proposal for the Higher Command Staff School. In between, he was able to stay with Eve. Might some role together, semi-political but philanthropic, be the answer to his disillusionment with the Army? 'When going down to Ladysmith,' he would remind her in 1945, 'you asked me what I was going to do about the future and how I was going to discharge my responsibilities [and] you began something inside me. What you said implied that neither of us could ever go inside our ivory tower for long, while outside there was distress and suffering.'

To Basil Liddell Hart   Middle East Staff School
                                                            Haifa, Palestine
                                                            29 August 1941

My dear Liddell Hart

I'm no longer in India. Last September I was sent here to run a small staff school, the powers that ran India at that time being concerned to move me on. Cassells, de Burgh and Roger Wilson have since left the army too, so they didn't stop the clock. But far from attaining any position where I could use any brain I possess to affect Middle East affairs directly, I'm engaged in a task which in England falls to a retired officer and inevitably only occupies part of one's day.

But I read (not about war) and I'm very hard and fit through walking and swimming, and have gradually become resigned to taking no direct part in a war in which one might have hoped for a chance to do useful work. On the other hand, indirectly I daresay, I <u>have</u> influenced events. There's a good deal of unacknowledged 'Chinkism' knocking about, so I observe with amusement and can claim the brainwave of Sidi Barrani!! But now that poor Dick is a prisoner I'm out of touch with the fighting people.

I've travelled a lot and seen much of Africa and the Levant, and my air mileage in the last 12 months must be nearly 40,000. Being debarred from active part in the war and, I think, out of the scheme of promotion (I've had as many 'passovers' as a Jewish village) I'm trying to think of the postwar world, for (whichever side wins) the end of the shooting will begin the new work for mankind. I don't think we can allow the old school ties – whether with a swastika, hammer and sickle or Grenadier Guards tie-pin – a free hand in shaping postwar events on the basis of bigger and better wars. I'd like to talk to you, and will one day.

Meanwhile I also collect data for a history of the war in the Middle East – perhaps we'll write it together! I'm afraid I never got <u>Dynamic Defence</u>, though I read of it with interest, and I look forward to <u>The Current of War</u>.

I trust all is well with your family and yourself. After four years' absence I've nearly lost touch with England, let alone the military world there.

Yours ever, Eric Dorman-Smith

31 August 1941 Middle East Staff School, Haifa Talked last night with an interesting chap on this course. He considers that the holders of the most advanced views in England today are among Regular Officers of my age and service, and that in certain circumstances the army might well take charge, particularly if reactionaries control affairs at home after the war. I'd hate to see a military intervention into politics myself. He held that more radical thinking is going on than anyone would believe – particularly among men trained at the Staff College. Curious. But, after all, if one trains people systematically to think, the product of their thoughts may well be unexpected.

3 September 1941 Middle East Staff School, Haifa *7.15am* Only about three weeks of campaigning weather are left to Germany in Russia before dark cold winter comes with snow and bad communications getting steadily worse. Not an attractive prospect for our enemies.

4 September 1941 Middle East Staff School, Haifa Last night a group of the senior officers course drew me into a discussion on the postwar world. The general idea was that a new 'deal' was possible but that, as in America, the money boys would fight it to the death. Suppose big money can still throw up Francos and there will be excellent material for private armies when we begin to disband the public ones. 'Black and Tans' will be cheap and black and tannery; as nasty as last time. No-one talks of the war against Fascism, a definite way of life with a system and a creed. We talk of the war against Hitlerism which doesn't mean anything to me for Hitler may die but Fascism would go on. We don't apparently fight Mussolinism, possibly because the word's too difficult and the man too unimportant. The danger is that by not focusing our battle on Fascism, it may 'win' this war by allegiance with Fascist minds in America and Britain. We've yet to understand the consequences of our allegiance with Communism – not, be it noted, Stalinism. Stalin is no less a bloodthirsty tyrant because he's now fighting against Germany. Nobody would raise a finger to prevent him from getting his throat well deservedly cut, but without Russia as a Communist power there's little hope of establishing a new deal in world reform.

6 September 1941 Middle East Staff School, Haifa The building is filling up with new arrivals for the long course. A horde of wide-eyed young men who don't look as if they'll ever settle down (and they do, very quickly) are pouring in from wild places ranging from Eastern Syria to Tobruk.

7 September 1941 Middle East Staff School, Haifa Last night, realising more fully that my professional life was over – that I could no longer hope to progress – I lay awake and mourned. Soldiering has been everything to me, and now the time has come when I'm no longer real enough for the men responsible to trust me with real responsibilities. Those are for safe, sleek-headed men who sleep at night. I've been weighed in some balance and found wanting. *Soit!*[32] I mourned the end of nearly 28 years of work and this morning feel swept and garnished.

8.15pm. Tomorrow am booked to talk to the chaps on 'Land Warfare' and be revolutionary and vigorous and all that. Have re-read the talk I gave on the same subject in Simla last summer, and think I was right. I went much further than anyone was prepared for then, but it's useless giving it again today. The trouble is, I can think of the deuce of a lot to say, but can't believe that any of it is worth saying. I'm in a bad state of disbelief – when I get like this I feel hollow – 'Empty, vacant, laid aside, like a dress upon a chair'. Well, I've been laid aside enough.

8 September 1941 Middle East Staff School, Haifa *7.15am* Always feel like Will Hay[33] when I address the lads. Wish I was a real one piece human and not a two piece person with each piece watching the other to see what a mess it's making of life.

12.15pm Well, that is that. Uttered an hour's worth of the banalities expected from a Commandant, and now the 5th War Course is well and truly opened. It was the kind of day in which one listens with detachment to oneself speaking and thinks, 'What a damn silly, obvious thing to say.' Strange that I set aside the world of friendship after the death of Lindsay Barrett.[34] Am reading John Buchan's autobiography *Memory Hold the Door* and interested in the importance of friendships in his life – other men. Don't think a soldier is put in the way of that. One doesn't make many friends at public school, Sandhurst is too short, and the regimental friends made on joining are dead! I moved from job to job too quickly to grow roots, and here I'm unsupported by men who think alike. But in some ways loneliness is strengthening.

9 September 1941 Middle East Staff School, Haifa Too many young minds are inelastic and unquestioning. Many men of thirty and less are only too glad to take life, particularly things military, for granted. If a lead from 'the Commandant' will encourage them to undertake loosening exercises mentally as well as physically, I've done my best to start them off in the way they should go. The really important work has to be done by my Directing Staff – my thirteen noble and true lieutenant colonels.

Finished Buchan's memoir. It's a picture of the sort of man one would have liked to be – integral, firmly-based in his beliefs, happily educated, a believer in the goodness of existing orders and an innate dislike of the questioner. He would have thought me unprincipled and unreliable. But 'high principled' conservatism, rooted in privilege and believing in an unchanging order, provokes the thug – at least, it tempts them to hold the old order to blackmail. Buchan was lucky in his classical and philosophical education: a great stronghold provided one sallies forth from it now and then.

12.55pm Have been round the syndicate rooms to shake hands with each student. Oh, I envy them their youth, and the hope they have because of it. Of course they'll probably not know what to do with either commodity, but how I wish I'd both to spend again.

11 September 1941 Middle East Staff School, Haifa *7.15pm* Listening to the BBC 4pm news, it certainly looks as if the Japanese are coming into this war. That means Germany is in greater difficulties in Russia than she's prepared to admit.

12 September 1941 Middle East Staff School, Haifa *4pm* Looks as if the Yanks are practically in the war; they're certainly in it at sea, an enormous assistance to the navy.

15 September 1941 Middle East Staff School, Haifa Monday 10th September 1940 was the last day of my pre-war life, for that afternoon I said goodbye to Estelle at the Simla house and Mac the terrier and I took a rickshaw to the station. Not a good journey that, through the lower bazaar – Pad-pad-pad-pad-Hut! – but Mac loves a rickshaw. On the platform were Elaine[35] and [her fiancé] Kenneth [Bols] to take him back. I was glad when the rail-motor left. It might almost be a century ago, so much has happened since. Dine this evening at the Peytons with Freya Stark[36] who has an eye for a pretty woman as well as a good-looking man. Will be disappointed if she turns out to be dull.

16 September 1941 Middle East Staff School, Haifa *7.15am* Couldn't guess Freya Stark's age – fortyish? She's surprisingly unlined for someone who's faced so much wind and sand. Not ugly but plain, with no physical attraction so safe for travelling, but a bright humorous face and there's lot of fun about her. She wore a red spotted garden party dress with an inadequate red wisp in her hair and looked awful. She talked amusingly about nothing much. In Baghdad during the rising she was interned with the others in the embassy, and all were allowed to send out shopping orders. The Arab policeman on duty was heard to say, 'How strange that these women who in a few days will be slaughtered should still be purchasing cosmetics!'

In this third year of the war it's right to compare it with the beginning of the third year of the last war against Germany, August 1916. We were still fighting on the Somme – we'd yet to undergo the worst of the submarine war, the agony of Passchendaele, the defeats of March '18 and the subsequent autumn victories. Even so, we'd turned the corner without knowing it. Germany's effort had not brought victory, and both we and the French were catching up the ground we lost at the beginning. Russia was to fall out, but the immense power of America was yet to be felt.

This time, though, France has fallen as did Russia, but Russia is still full of fight, and we've America in reserve, gathering strength. We – except for the sinking of our ships – have scarcely been touched, only by air where now we give as good as we get. Provided ourselves and the USA can keep Russia going, what can Germany hope for? She's gained immense victories at ruthless sacrifice, and we and the USA are only beginning to marshal our true strength. There must be Germans who realise how hopelessly they're trapped by their own victories.

17 September 1941 Middle East Staff School, Haifa This place is now a hive of industry in which I, as the Queen bee, sit back among the busy workers quite infertile – at least militarily so. The college has doubled in the last twelve months, and is now bigger than Camberley in peacetime and twice the size of Quetta.[37] The new lecture hall's a success, really a large hut with a stone floor, and it makes a world of difference to have two general lecture rooms instead of one. Plenty of scope for improvement still, but a different air. Twelve months ago I was flying from Karachi to Basra and taking my last look at India. I was feeling bruised and uprooted, realizing the authorities I'd irritated had ingeniously scored the last trick.[38] I was suspicious as to what I was coming to, and out here the war had only just begun.

Winter rushes on. Would hate to be the German administrators facing the problems of quartering and supplying 5 million men in the heart of Russia through a bitter winter, and I'd not like to be the people responsible for keeping the ordinary German quiet. Men and women there must begin to ask what the good of it all will be.

19 September 1941 Middle East Staff School, Haifa *7.15am* I used to set such store by solitude that if I wasn't alone for at least two hours a day I felt deprived – the need was as strong as hunger. Today I've let myself in for two talks to the chaps – 'Military Organization' and 'International Politics in the Middle East'. Neither's prepared, but at least the ideas are there. The problem is to find the best presentation so that an argument finds its way into their subconscious, gets absorbed and turned into principle, ready to emerge long after these talks as something they've originated themselves. Flags move daily on a large map in the No. I Lecture Room, and opposite the enormous map of Russia I've an even larger map of the ME – the Mediterranean and Muslim world – so big that I can build up the war graphically for them.

Hard to settle down to the future. Wonder what the Auk will say when I ask his leave to go! Will I be running out? I've men almost as old as me as students here, and they must have given up much to come back to the service. On the other hand, I've been working flat out during the lean prewar years, so perhaps have as much excuse to go as they to come back.

21 September 1941 Middle East Staff School, Haifa *7.15am* If the Russians can survive to the winter holding Leningrad – Moscow – Rostoff – Sebastapol and Odessa, the Germans will have failed this year.

23 September 1941 Middle East Staff School, Haifa Freddie de Guingand[39] proposes to come up here to shoot and take the air. He's my only contact with real GHQ scandal. Rumour has it that Arthur Smith has gone to England on 'leave'. Hope it really is a euphemism, because it's high time we had a change of incumbent.

Am being pretty downright with the instruction this time. We've a year of actual war, full of lessons, most of which I predicted before it began. Don't see why people who blunder should get off scot free. That report I wrote in March hasn't been published yet, let alone circulated. That's why mistakes are made again!

26 September 1941 Middle East Staff School, Haifa *11.30am* Today a letter from GHQ agreeing – exactly a year after I first proposed it – that this school should split into two wings: one for operations, the other for administration. So I've something concrete to do getting everything sorted out before this present course ends. The new dispensation begins in February. Amusing to win a battle, even after much delay. Had they agreed last October, I could have run two courses on the new model by now, and its effect would be showing. As it is, we mayn't get going properly before the end of the war.

7.30pm This evening worked on changes of organisation and drafted a letter to GHQ. Have recommended that the commandant should be a major general, but prepared a formal letter to the Deputy CGS explaining that I don't wish to go on being Commandant. They can't say I'm hunting promotion.

5 October 1941 Middle East Staff School, Haifa *3.10pm.* As do all the great, Pharaoh 2[40] [P2 was Auchinleck, since his predecesssor Wavell remained Pharoah 1, or P1 to Chink in shorthand] arrived 3 hours late. But not his fault: an RAF blunder. He's most affable and made me laugh. 'I've come here to see the school and you [but] I really must see these programmes first.' He hastily ran round and fell for all the traps I'd baited for him. He will come back – at any rate we're dining tonight at Pross's[41] and I'll then hear what he has to say. He's bursting to get something off his chest but exactly what I don't know.

6 October 1941 Middle East Staff School, Haifa *7.15am* With the greatest difficulty he persuaded his four military police outriders to go home, two stuck to us to the door and then we were free. Mild consternation of local bar supporters when they realised their neighbour's rank as he got outside a sherry. We had a very good dinner. Prawns, Chicken Maryland, mushrooms, coffee and a bottle of good Chablis. Afterwards he came back to the school and we sat on the balcony of my room in the moonlight and talked in a spacious, well-nurtured manner. But at the end of it all am no wiser. Moving on is in the wind, but don't think moving very far.[42] Men are great devils – we say little about ourselves. He didn't say, 'Chink, you are the predestined defender of Tobruk' or 'The Western Desert needs you' or 'Syria is your destiny, you and Catroux[43] will hold it for me.' Nothing of the sort! But it was a very good evening.

8 October 1941 Middle East Staff School, Haifa Have been thinking about my friendship with Lindsay Barrett in my youth. That was instantaneous and in the truest sense pure. Wonder if the subconscious is homosexual? Certainly there was no conscious tendency towards that – except our pleasure at being together.

13 October 1941 Middle East Staff School, Haifa *6pm* My trouble is of course that I must run a thing or it's no good to me. Can't submit to second best. Daresay I can hang on in the service and rise in ten years time to be a major general or Lt general. A knighthood? But *now* is the time to be used full out. I'm not, and nor is there any opening. Feel that years of study and application are wasted.

6pm This morning P2 rang, inviting himself to stay this weekend. He may wish to do a spot of brain picking and that makes me a little cross. Throughout my service people have taken ideas which have made them important, and entirely forgotten the poor bee (or B!) who made the honey. Am tired of it. I know people don't like me because I speak my mind and refuse to play the good clubman, and that because I don't play English games or pretend to the standard of toughness they understand, they think I'm not tough enough to command. If only I could get a foot in the jamb of the door. But I'm damned if I'm going to be a yes man or toady or pretend to like men whom I think are crooks just because they might give me promotion or go in on their mutual admiration groups. I am never – never – going to ask anyone for anything. *Grand merci NON*!!!

So how in the name of Heaven am I ever going to get an outlet for all the military knowledge that seethes under my thatch? Wish I wasn't so damn uncompromising and difficult to get on with. Wish I wasn't so often right. Don't care if they give me a Division – that's chicken feed. I must be nearer the big play. Wish I was in the German Army! That's the place for a professional soldier. Why did I chose this frustrating profession where all you need to be able to say is 'Yes Sir', 'Thank you Sir', 'Oh no, not if you say so Sir', 'Sorry Sir, it won't happen again'. with an air of optimistic hero-worship like a schoolgirl with a crush on the games mistress. Ugh!

7.30pm Oh Hell, wish I were connected with the management but on terms which would let me be myself and not something still in Eton collars. Am not cross or depressed, merely blowing off steam in ink! Brains – we just haven't damn well got any. We have personalities and prejudices and pomposities and politics, and I'm rapidly going down the drain! Just finished a life of

Kitchener[44] – there was an uncompromising blighter. Hope I'm becoming humble – certainly having it pointed out to me. There are lots of chaps who will never forgive me being the youngest DMT there's ever been in India.

17 October 1941 Middle East Staff School, Haifa P2 and his tribes not now coming this weekend. I'm confident there's an answer to our immediate problem, given a more intelligent handling than it's received since Dick left us.

20 October 1941 Middle East Staff School, Haifa At 12 noon was brought a cypher cable from Pharaoh I [Wavell] to meet him at Lydda at 12.30. Pulled out one of our new Ford vans and made it to the second. Pretty good going; forgotten I'd been a fast driver once. He looked well, said he'd recently seen Estelle, walked about as usual and didn't say anything except to answer queries from me about Russia. Then as he was taking off again he tried to say that he'd been talking to P2 about my future. I thanked him for his 'kind intentions' but said I thought I'd become unemployable through long unemployment, and anyway I rather fancied myself as an *embusqué* while Haifa was a considerable improvement on Limoges.[45] He laughed. Was afraid he'd pull something-in-India out of his pocket but glory be, no! Had an amusing companion on my drive back, one of my soldiers – an ex Covent Garden porter, well-spoken and well read. We discussed the world. Not sure he doesn't think I'm Stalin in disguise.

21 October 1941 Middle East Staff School, Haifa Wasn't fit a year ago and mentally very tired. I'd been giving all the best in me without a break for nearly ten years with practically no leave. Stupid of me perhaps, but the real work of a trained military brain should be done between wars. Once war begins there's very little scope.

Have only to think to see the lamplit library of Bellamont lit by an enormous fire. Myself slippered after shooting, a cold wet night out of doors. 6pm and the wireless on. Alcohol lurking an obedient demon in the background. Great curtains drawn across high windows and all the shabby comfort of a well-used evening room. No! Lovely as it all would be, it isn't for me. Estelle has been so kind and is so dear, but should we fade childlessly into a Darby and Joan inertia? My role would be to grow old in an atmosphere of kindness and care, a pottering purposeless background to E's fulfilment. Bless her. There is nothing 99% of men would like better, but to me the prospect appalls. It's so awfully unfair of me to say all this, more than unfair, unjust, to someone who's spent 14 years of her energy on my dull self. Bless her again. I'm very <u>very</u> fond of her and I hate the idea of making her unhappy – <u>hate it</u>. I feel most desperately selfish, but simply can't die in my socks while still walking about at my age. I understand why men die when they retire. The world would be flat.

26 October 1941 Middle East Staff School, Haifa *8.30pm* Note with amusement that the BBC keeps popping in a Western Desert threatening notice every now and then. There was one tonight. Bluff, double cross, treble cross, war of nerves – which?

31 October 1941 Middle East Staff School, Haifa Some I flicked on the raw when I came to ME are still raw, saying 'Chink is his own worst enemy'. It cleverly avoids saying why I make enemies. Lord knows, except for the two months when I was compelled to go to GHQ I've lain low enough, barring a few days with Dick. I ask for nothing, see nobody, defer to nobody and keep my own company. I say what I think and I dare to think. They've told P2 that I got a certain eminent one's back up, but they'll never say these things to my face or give chapter and verse. If the devils would say 'Chink is inefficient and we'll report badly on him and retire him', I'd understand. But what they're saying is 'Chink is dangerous to us. We can't report badly on him so we'll isolate him and keep him where his capacity for making us uncomfortable won't worry us.' Suppose it's a compliment. What a first-rate poison whisper, 'Chink is his own worst enemy'. Then if he isn't used or promoted nobody is to blame but Chink. Not us who pass the whisper on, or us who cleverly invented it. 'Oh yes, he rubbed somebody up the wrong way, Oh no, not me. I rather like old Chink.' Am rapidly becoming the man in the iron mask of Mid East. What a jest. And do I give a damn? No – except that I might get a chance to think, say or do something which would help to shorten this damn silly war.

5 November 1941 Middle East Staff School, Haifa *9.45pm* Freddie de Guingand arrived, delayed by engine trouble at Tiberias. Full of gossip but with the typical gallic flair for the personal. So we discussed that dull subject, myself, and apparently the outer world is regarding the subterranean struggle between Chink and a 'very holy' senior.[46] Could anything be more amusing, for I'm as impotent in exile as Trotsky, and equally in ignorance of the struggle.

10 November 1941 Middle East Staff School, Haifa We're all delighted at the beating up of the Wop[47] convoy between Taranto and (I assume) Benghazi. It really is a first-rate cop, and caps off to the navy combine with no heel taps.[48] The enemy are going to feel the loss considerably.

6.30pm There's a wintry howl in the wind, and though I know its menace is largely bluff, I automatically shudder, because as a child it really meant business. Bellamont – high, foursquare and Georgianly defiant above its lakes and woods – had a peculiarly penetrating banshee of a howl. Then it was fun to lie awake in an enormous bed and watch the firelight flicker on the ceiling, or to sit in the library – a haunted room – before a really magnificent blaze, by the light of several oil lamps digesting a super tea in the happy coma which follows a day bog-walking. Have felt the whole solid fabric vibrate.

12 November 1941 Middle East Staff School, Haifa P2's programme has come. Meet him at Lydda tomorrow and he's staying two nights. I'll give him my two rooms and office, moving into the guest room opposite. He'll want to dine out both nights.

Archie was frank about the defeat in Cyrenaica this summer when he spoke in Delhi, but he said no more and no less than I have. One is never defeated by the truth. Only by ignoring it is one in real danger of being caught out. They're just about to issue a bowdlerised edition of the report I wrote in *March*. Quick work by the General Staff! Most has survived, but some essential guts have been excised or perverted to justify the business now afoot. That's where I loathe the army, because its alleged men of honour are such scoundrelly liars when it pleases them. And there isn't a damn thing I can do about it, except pray that good men don't get killed as a result.

13 November Middle East Staff School, Haifa *10.50pm* What a day. Had P2 on my hands all afternoon so took him to the Megiddo.[49] Later we threw a drinks party for the distinguished visitors, Directing Staff and wives and certain senior students, then they supped in Mess and now I've put him to bed. There's my war contribution! When I got him alone said I'd determined to retire, to which he replied, 'We'll see about that. I'm pretty sure you won't.' Little he knows, bless his heart, so gossip hasn't reached very high for he was full of other goings on. He's not for wives appearing – Jessie is not going to be let come and she will be cross.

14 November 1941 Middle East Staff School, Haifa Think the Chief has enjoyed the lack of fuss. We had an interesting talk on shop this morning and he unburdened himself on a variety of problems. He's in great form and gives the impression of possessing huge reserves of power in his personality; extraordinarily natural, too. Tonight I'll take him to the Piccadilly.[50]

15 November 1941 Middle East Staff School, Haifa *7.10am* We didn't get to bed till after midnight and Lord only knows what indiscretions passed, including my opinions on men, mellowed by gin, chablis, brandy. He enjoyed the Piccadilly and its cabaret immensely. He has a great taste for life – these big men have, I think – and considerable vitality. They all go today but though two are my contemporaries and one a personal friend, the school gets a little swamped by a CinC, a major general and an air vice-marshal!

12.15pm Alone, thank goodness. The imposing party, flags flying, outriders, vanished at 8.30am and the institution shrank visibly. P2, without being too explicit, seems to dangle carrots in front of donkeys' noses. I believe they've asked the War Office to make this a major general's job, which if they intend to leave me here would be a consolation prize. But don't really mind. This place is thoroughly 'on the map' now as a real centre of thought, and if it develops as I think he intends there'll be plenty to keep me usefully used.

18 November 1941 Middle East Staff School, Haifa The chaps were working on a scheme which kept them up all night. They've another stiff day's work ahead and they'll learn that the life of the Staff isn't all sunny days. Times come when the mind must be summoned to work in a tired body, not alone but in a team with equally tired minds, and on the quality of that work may depend victory or defeat and many men's lives. Am working this course very hard on purpose, and if some crack it's better they crack here than in the field. We know the Germans are still 'inside their plans'. They've not yet reached the point of having to improvise everything and simultaneously conform to their enemy's initiative, which is where we've been since the war began. Don't believe they could successfully make war as we're forced to do – begin after war starts with nothing, and gradually coil the springs for response while fending off attacks. It's a costly business and to me incredible how <u>well</u> we manage it – it's like having to produce trained doctors after an epidemic begins.

7.30pm Had a visitor commanding an Infantry Brigade – large, red-faced, English, coarse-grained. The typical 'Leader Class' man of a very special world. Felt shy and small and shrinking before his powerful assured presence, yet in his line or my own can probably give him a lead. Have lost my assurance, if I ever had any. He takes about 90% of his world for granted, and I can't take 10%. If you doubt and question everything you become a ghost in the presence of solid people who are as utterly at home in their world as a Spanish bull when he enters the arena – only these chaps are nowadays as likely to be towed out by a team of mules. Whenever I look at his like, can perfectly well realise why I'm not employed on anything worthwhile. I don't exist for them; lived too long as an anarchist on a mental plane and am altogether too sensitive. Incidentally I'd win their B. battles for them if only they'd let me, but unless you look as if you were 100% at home on a London club bum-warmer there's no hope for you.

Just been earning the hatred of my Directing Staff by debunking some of their over-elaborate work so they'll have to begin again. Not always easy for them to see clearly. Fact is, I'm no longer tapping the best-trained brains and second best always takes teaching – frequently doesn't like it, either.

19 November 1941 Middle East Staff School, Haifa *9.20am* On the news heard with interest a good bag of 18 hostile aircraft in Cyrenaica, and with dismay a typically temporising speech by Leo Amery[51] about India. It's useless for London politicians to point out all the negatives – we know them all. We need positives. The bitter fact remains that no Indian is a free agent. They have no representatives who could constitutionally take effect, for India is run by the British Parliament – ridiculously anomalous. We've made modern India a unit even if we've failed to make 'Indians' unitary and now we've got to find a means of giving them responsible self-government, as nearly democratic as the composition of India will permit. It's NOT impossible, only we always begin from the dark face of the moon.

Major reshuffling among the great always stirs my professional instincts, which sit up and ask if their time has come. But I think not. 4 years ago Archie Nye,[52] the new DCIGS,[53] and I were running neck and neck as promising young men, I a little ahead but with fewer friends. Well, he too went to India but got himself back to England again, while I got myself nowhere. So at my age he's a Lt General. Life is founded on waste.

2.25pm The West's awake again and the Eighth Army is on the move. I drink a glass to Crusader's success and may that be far-reaching and conclusive. I went through the outline with P2 over lunch at Mena in July. I am, I believe, something of a talisman for him. Of course there's a gamble in it. Inevitably things won't go as intended; they never do in war, but think everything's been done to minimise the dangers of the unforeseen, and leave a margin for mishap. Wish I were with the chaps. Maybe the decisive fighting has already taken place. If we can lick the Germans we can expect a fairly easy time with the Wops. Our 2pm news will be interesting.

6pm Winston's statement on the Cyrenaican operations in the Commons was banal and misleading, particularly the comparison with sea operations. Much depends on the armoured battle. Clearly if we can down the Boche tanks without disproportionate losses we'll find ourselves once more in the relationship to the Wop army enjoyed by Dick O'Connor – armoured superiority.

9.15pm Waiting to hear about the armoured 'collision' which (according to Winston) ought to have taken place. Don't see why he should broadcast our intentions so plainly unless something decisive had already happened. Things should go well. The German and Italian forces were awkwardly placed for battle, having to form a front facing the desert with their backs to the sea on a long frontage. At least this is what *ought* to be happening from news reports.

Military Historian John Lee: Operation Crusader, now underway (18 November–30 December 1941) was the largest British attack on a German-led army to date in the war. Chink suggests in some places (see 17 July 1941, 19 July and especially 19 November 1941) that he had advised Auchinleck on how the offensive should be conducted. That advice would have insisted on the British armour being kept concentrated in order to deal with the main enemy armoured formations. Chink was in Haifa (1800 road miles/2900 kilometres from the battle zone) and could only follow events by the news broadcasts. It was a BBC announcement of the offensive that alerted Rommel to the real danger of the attack. Chink accepts the cheerful bulletins announcing British success but, after a few days, begins to wonder why the decisive victory does not seem to be forthcoming.

21 November 1941 Middle East Staff School, Haifa *11.15am* The news doesn't yet reveal the *dénouement* of the first phase, and I'm uncomfortable at the fuss going on at home. Can't believe troops need King's messages or blah from Winston. We know our job and if we've <u>enough</u> decent weapons we can deal with our enemies. No amount of telling us we're fine fellows will make up for a lack of planes and tanks. Nor will telling us we've enough if we haven't. It all depends on margins and the skill of our leaders. The show has been well launched – I know that – and I only hope the chaps in charge are big enough and quick-witted enough to make our margin decisive. Don't think the Germans east of Tobruk will be in a hurry to fight. I wouldn't. I'd turn their air force on to us for a day or two while collecting my forces west of Bardia, covered by the fortifications from Sollum to Bidi Omar. Meanwhile, I'd try to induce the Italians in the country west of Tobruk to move east and give battle to the British, whom I'd try to attack from both sides. That's Rommel's difficulty, having to rely on half-equipped and badly-trained Italians – like a pincers with one prong steel and one plasticine. However, if he does that we've got another jerk in store for him, I think, and not only one.

9.30pm News tonight not as definite as I'd like. Enemy high command in Libya is trying to take our advance from the south between those two prongs: a delicate and intriguing situation. Would like to hear that something decisive had happened because we went in on Tuesday and now it's Friday evening. Three days. Hope our communications and supplies are functioning without too much danger; that's where this fighting is *not* like naval warfare. Ships move self-contained for long periods. Tanks can't – they must be refuelled and re-ammunitioned frequently. The fuss over Tobruk is press moonshine, but can we dominate the enemy's air force over the area? Touch and go. Wish I had something to *do* with it all in place of sitting here in a vacuum. The whole school is restless, much like an asylum at full moon.

22 November 1941 Middle East Staff School, Haifa *12.15pm* The battle seems to be going as desired. Find it hard to believe the Boche would let themselves be caught, dispersed and trapped against their own defences, which seems to be what's happening and doubt the Italian navy will 'Dunkirk' them. Hope we haven't lost too many tank crews or aircraft. This time air/land cooperation's been 50% better. A doubtful story: not long ago the admiral commanding the Naval forces in Western Mediterranean was by error gazetted KCB and KBE on the same day. His C in C made him a signal: 'Twice a night is too much at your age'. Hope it's true!

6pm The bag of tanks mounts up – but we're 24 hours behind actual communiqués from MidEast here. Annoying. GHQ must know so much more than gets onto the air.

10pm From all to be gathered, still going well. Fortunate, according to the BBC, that the heavy rains where the enemy keep the bulk of their bombers have turned the untarmac'd landing grounds into bogs. Think we've overrun a proportion of their forward strips. Will go to sleep confident.

23 November 1941 Middle East Staff School, Haifa *7.30pm* Cyrenaican news goes well. We should by now have done in about 50% of the axis forces in Lybia [*sic*] which is a good start to the shooting season and the Boche will have difficulty explaining the fate of his armoured divisions. Wish the silly pressmen wouldn't talk about a Panzer Division – Panzer only means Armoured. We've amply revenged our defeat in the spring and may have more prisoners to show than Rommel had then. What sort of political muddle is Hitler going to be involved in with Vichy and Spain over North Africa? The last thing the Boche want is to increase their commitments here.

10pm. Won't listen to any more news tonight as its 36 hours stale.

25 November 1941 Middle East Staff School, Haifa Can't think why I ever became a soldier, except that it seemed to be expected of me and I didn't feel any particular urge to be anything else. Have made elaborate enquiries as to whether I could get out of the service, and am assured that it's impossible in wartime, which disappoints me. Was also assured that the 'South African job'[54] wasn't good enough for me.

26 November 1941 Middle East Staff School, Haifa *12.15pm* News still too vague to pronounce a verdict. This battle has been going on for 8 days now. We can claim to have held the Germans and Italians east of Tobruk in a vice, preventing the Italian forces west of Tobruk from intervening to help. But fighting seems to have been pretty lurid and the Germans show no signs yet of packing up. We haven't enough tanks to give the *coup de grâce*, as can be deduced from the battle lasting so long.

7.30pm News is non-committal. Afraid we've not succeeded in closing the gap between Sidi Rezegh and Tobruk through which the enemy can withdraw via El Adem. Looks as if the operation's been as costly to us as the BBC keeps insisting is the case.

Military Historian John Lee: '*Instead of fighting as a concentrated force, the three brigades of British Eighth Armoured Division fought separate, un-coordinated battles and were defeated in detail. The tank losses in the first few days were very high. Alan Cunningham, commander of Eighth Army, lost confidence and, when Rommel started a deep raid into the British rear (called 'the dash for the wire'), began to organize a retreat. Auchinleck flew up to join him and insisted that Eighth Army stood and fought the enemy where they found them. On 26th November he replaced Cunningham with Neil Ritchie in command of Eighth Army. Rommel, whose supplies of fuel were repeatedly cut by the Royal Navy, could not win the long, grinding battle of attrition and was forced to retreat out of Cyrenaica entirely.*'

27 November 1941 Middle East Staff School, Haifa *7.10 am* We must have more definite news today. The battle can't go on at this violence and sooner or later our enemies must crack.

4pm. The wind's gone SW which will mean sandstorms over Libya and help our enemy. The situation hasn't improved. He seems to have slipped out of our noose and to some extent regained the initiative at the expense of the loss of some worthless territory and a fair number of men and weapons. The crushing first move seems to have failed to reap the full fruits of the good preceding strategy, implying a pause to make new combinations. This is only conjecture from the BBC. They haven't the least idea of the real military situation.

9.40pm. Seems to have stabilised now with both sides exhausted. Can we get our administrative arrangements going sufficiently well to let us reorganise? We mustn't stop now.

28 November 1941 Middle East Staff School, Haifa Rommel has handled a difficult situation with great skill, and it remains to be seen how soon we can get a new manoeuvre started. Official press commentaries have been more than usually inept. Am looking forward to the official accounts – unexpurgated if possible.

29 November 1941 Middle East Staff School, Haifa *12 noon* Both sides are trying to sort themselves out and get something going afresh. Must be much confusion behind both lines – though 'lines' suggests a solid order, definite and visible, whereas the relatively few troops on either side are spread over some 40 miles of frontage and 80 miles of depth, intermingled at that. Behind the general line we've reached, running from Tobruk to some 50 miles south into the desert, there are not only the raiding force of German tanks and lorry-carried infantry, but also scattered bodies of German and Italian troops and administrative installations such as workshops, hospitals, supply dumps etc, including abandoned aerodromes. Many will have their personnel waiting to be made prisoners, thirsty and hungry since their own supplies are finished. It all depends on who can begin the next blow and how soon. Maddening only to know what's happening from the BBC. Wonder what the casualties may be, and if any senior people have been hurt, starting a chain of consequences that may affect me. Doubt it. There's nobody up there I can potentially replace. Chaps here have buckled down to work again, realizing that if they miss this battle they'll probably run into the next one.

2 December 1941 Middle East Staff School, Haifa *7.15am* Don't think they'll take me away from here, as the next course sees a complete change of curriculum, plus reorganisation of the staff, and possibly collaboration with the air force which would be as great a triumph as inducing a woodcock to perch on one's gun barrels. To tease the authorities, wrote privately to the Military Secretary people suggesting that since the recently appointed GOC Palestine[55] is junior to me as a colonel (they swore he wasn't, so I checked) I'd better retire, on the principle that officers passed over should retire. That will annoy them but the answer will be 'No'.

4.40pm Jove, it's raining – fairly pelting down. Wonder if they're getting this in Libya? Poor devils, they've had a pretty hard time of it for the last fortnight with no relief and little kit or food. Don't know how much longer either side can stick it out. Things must be very near breaking point for both.

6.30pm A note from P2 this evening expressing his confidence in the transactions, but it's going to be what the old Duke[56] would have called a 'damned close run thing' and casualties will be heavy. Hope people are using their brains. All sorts of 'novel' things will happen in this fight and the commander who can best adapt himself to the realities will win.

7.30pm The Boche are back at Sidi Rezegh and joined hands with their people from the west. It's by far the fiercest fighting in this war here – a mincing machine. Anyone who could produce 100 fresh tanks would scoop the pool. It's a bloody business. Tonight I dug out my corduroy battle-dress, curious instinct men have to want to get into these battles. With Dick at Beda Fomm I was cold, hungry, weary, but entirely content. It's only in this under-utilised life that one gets cantankerous.

3 December 1941 Middle East Staff School, Haifa *7.15pm* Convinced we need to reorganise our whole system of command, and consequently our methods of fighting. Felt it before the war and made some progress, but ran into vested interests. Am prepared to wager that many of the difficulties we're experiencing are due to our adherence to an obsolete system and our fear of 'swopping horses'. It's necessary to swop still more riders. We must eventually find the correct balance between intelligence and personality, forehead and chin. Our society as a whole isn't prepared mentally to be ruthless enough to attain efficiency. Physical courage, physical endurance we have in plenty, but quick reactions – vision – foresight, no! Am certain that many lessons of Dec '40–Feb '41 are being relearned at the cost of more lives now and haven't a hope of getting anyone to listen to me seriously because I'm too radical. We'll blunder along with our faulty organisations and tactical systems. Oh for a Cromwell or a Gustavus Adolphus[57] to build us a new model army.

4 December 1941 Middle East Staff School, Haifa *7.15am* Am preoccupied by the war. Much as I try to insulate myself, my thoughts are continuously with those undergoing the strain of nearly three weeks of fighting in a barren land, facing death not under a tree but by a lava rock, a stunted sage bush or a miserable camel thorn. All my willpower is with them saying 'endure!' The enemy are fighting desperately for time in the hope that reinforcements will arrive and turn the scale.

5 December 1941 Middle East Staff School, Haifa When casualty statistics are collected they go to England and are published eventually, but not locally. We never hear. Am prepared to bet that nobody knows what they are in this show yet. In the rapid, far-ranging ebb and flow many wounded must be left out unattended who in other wars might have been picked up. The ratio of killed to wounded will be very high. It is *Strengsten Verboten*[58] to mention names of casualties in letters until they've been officially reported.

6 December 1941 Middle East Staff School, Haifa Propose to lecture the school on the Cyrenaican campaign up to date. That's tantamount to finding a couple of rib bones and a foot and thigh of some obscure prehistoric animal and proceeding to build a plaster likeness. Of course, knowing the anatomy of war it's not difficult to make reasonable guesses – natural laws follow the chain of cause and effect. Won't be far wrong with the general appearance of my mammal when I've finished.

5.30pm Continuing my struggle to find out the truth about the battle in Cyrenaica, and my ideas begin to take shape.

On the other side of the world that day Japanese planes attacked the United States Naval Base at Pearl Harbor in Hawaii, killing over 2,300 American personnel. Twenty-one ships were sunk, beached or damaged, including two battleships, and 160 aircraft destroyed, leaving 150 others badly damaged.

8 December 1941 Middle East Staff School, Haifa *11.45am* The Banzai boys have fairly pulled the plug and the silly old Yanks were caught napping at Honolulu. It will make them as sore as the devil and much nastier customers to deal with. I'd love to be in the States to see the faces of the big money boys who thought they were going to be let off fighting in this world war. Well, this morning they know better. Hope our brains are adequate at Singapore because now they've begun, the Japs will press the pack good and hearty. Too early yet to begin to guess about the course of operations in the Pacific. We should hold Hong Kong. Malaya is OK. We'll probably have to get rid of Brooke-Popham.[59] He's too old and has forgotten warfare. It was an obvious racket when he was appointed to command and doesn't seem to have worked very well. The Indian Ocean won't be so healthy as we're bound to have business with Japanese submarines and surface raiders, but we're ready for that. Rommel's on the run, and when tired troops begin running they take a lot of stopping; besides, tanks can't run away. Glad we've had the guts to take our Boche prisoners out of the train at Cairo and march them to their cages. Good. They marched us through Athens, so we needn't bother to be genteel. Of course the Japan business has obscured everything else. Why oh why did the Yanks let the Japs catch them with their trousers down? I'd give the world to have a hand in it all – as it is I only talk about it.

9 December 1941 Middle East Staff School, Haifa *4.30pm* A GHQ visitor tells me what I'd already deduced, that the situation wants bold treatment now. Gather the 'lessons' being 'discovered' are all facts which I've been urging for years and told P2 about at length the other day. It's a nightmare to be watching mistakes, knowing the answers and powerless to intervene. Had I been Chief of Staff in ME this year I could have avoided our major disasters, but I'd not have been popular with my seniors, my equals, or the Embassy and civilians. The time is rapidly approaching when I'll have to take a stronger line. They must employ me where I can be of use. Hate pressing myself forward, but am prepared to stand or fall by my opinions. Can't bear to see things done wrongly by men I know to be brainless. All our crowd play for safety and take refuge in War Office rulings and textbooks.

10 December 1941 Middle East Staff School, Haifa The Japanese have hit us a sore blow in the loss of two battleships[60] from air attack off Malaya. We must now realise that the true world war is on – Germany and Italy have declared war on the States to prevent isolation developing in Japan, so our strategy must be conjoint, not just non-belligerently joined.

13 December 1941 Middle East Staff School, Haifa The whole tempo of the war has enormously increased. All preconceived ideas have got to be scrapped and one big fact stands out. Germany's in serious danger of an oil shortage.

A lovely major of the Greys[61] at breakfast, utterly satisfied that the whole work of creation from its beginning till now had been designed to create a world in which he could satisfy all his needs. True, he hadn't been given a brain capable of having any, except those foisted upon him by tradition and custom, but those suited him amply.

On 11 December Churchill had announced in the House of Commons the Eighth Army's change of command, revealing that Cunningham's replacement was Lt-General Neil Ritchie of the Black Watch. To Chink, that news came as a visceral punch in the stomach. Ritchie was younger by three years, and had previously been Wavell's DCGS, the job he himself had wanted. Now the chain of consequences he had speculated about on his own behalf, placed Ritchie at the head of the Eighth Army. A change at the very top of the army made less initial impression. On 1 December General Sir Alan Brooke had been appointed Chief of the Imperial General Staff (CIGS), making him Churchill's foremost military advisor, and mutual enmity between Brooke and Chink dated back to their War Office days. Ritchie's high command, though, eclipsed all else.

14 December 1941 Middle East Staff School, Haifa The time has come for me to put my cards on the table and have a show-down with higher authority about my treatment.

16 December 1941 Middle East Staff School, Haifa *9.10pm* In disgust at my permanent relegation, today sent in my grouse, pointing out that rather than serve indefinitely under a cloud it would be in the interest of the service to let me retire. They can't deny that I'm holding a less important job than when the war began, and since I came here numerous junior to me have been promoted. I can see no hope of change, and to be so treated is not conducive to efficiency. They must explain why I'm the only officer in MidEast singled out for professional punishment – largely thoughtless mismanagement but also anti-Chink fifth column work. It's good to get this feeling of being bottled up off my chest.

Official letters now contain 'lessons from recent fighting in the Western Desert' all of which we've been teaching here for a year. The Directing Staff are as blasphemous as I am about it. Can't bear that men should give the only thing they have, their lives, because other men dislike to use the greatest gift they have, their brains. Am in a rage today.

18 December 1941 Middle East Staff School, Haifa Just finished a long interview with an old acquaintance, Brigadier Angus Collier,[62] the new Military Secretary at MidEast. Everyone is 'aching' to promote me if they could find a job in which I won't disturb them! Nobody can find a job as remote as that because, tactfully, I am too 'brilliantly disturbing'. He hummed and hawed over my application to retire and said they'd never accept it. The pill is gilded by the fact that they had a wonderful job – 'first class' – waiting for me, but the War Office wouldn't authorise it. Have given my whole life to the army for 28 years and probably done the work of any two of my contemporaries, but I've scared the birds so much they daren't risk using me except in this side-tracked employment.

Listened to him talking about people in the UK whom I used to know was like a voice naming the dead. Never realised how remote that world has become, or what a gap there'll be between us out here who've seen the war and those in England. They're all together, not split up and dishevelled as we are.

19 December 1941 Middle East Staff School, Haifa *7.15am* A night of gusty anger at my wasted life. It felt like a nightmare in which people cry out for help and you're forced to remain motionless, pinioned by forces whose power you feel but who remain invisible. Damnation! I know more about warfare in theory and principle than 95% of British soldiers. I'm an infinitely hard worker, in good health and well disciplined. I never disobey an order, do whatever I'm told with a heart and a half, and could either organise or command. And because I stirred up a nest of torpid kraits[63] in India and showed up Arthur S. here, I'm forbidden to help in the prosecution of this war. At 46 I can only look forward to remaining mum at this tin pot establishment, watching things happen disastrously. The men who manage things have neither the brains nor the vision to see further than the newest corpse or the next battlefield. So I go to join Fuller[64] and Mackesy[65] and other intellectual misfits in a system which has no use for intellect. The joke is that they're right. Brain in an army is like a nail in one's shoe, out of place if it's too sharp and hasn't been properly turned down.

6.25pm. Have got all I wanted to say off my chest. Politely declined any suggestion that they should get England to offer me a major general's job there, saying I want no charity. Collier realises that I've a damn good right to feel aggrieved and insulted.

20 December 1941 Middle East Staff School, Haifa Today at 9am got an order saying that I was to go as liaison officer to Jericho, so I've cabled back to Mid-East and asked whether it's a practical joke.

4pm Rang up the MS people and asked what rank. 'No extra rank.' So let fly. They tried to persuade me it was magnificent employment, specially devised by P1 and P2 in consultation, so I said big men shouldn't go slumming and rang off, still expostulating. Though I hate the bloody place and all in it, must go to Cairo at once and stop this nonsense.

8.15pm Couldn't get a passage by air in time, so must leave at 6am and try to get to Cairo by dark – know I'm useless to my country or the Service in a routine sherry-swilling job. If I can be of value as myself to the cause I'll willingly give my life, but if somebody is merely giving me the sort of job which the ordinary man of my age would welcome, I can't do it.

22 December 1941 Continental Hotel, Cairo *8.30am* Roared into GHQ seeking whom I might devour and of course ran into everyone, all delighted to see me. Arthur S. produced a letter promising me the earth which I said I didn't want, and I was ushered into P2's sanctuary. We began a showdown. He was damn nice about it all and said I couldn't retire in wartime. Answered that I was technically a member of a neutral state and could probably do what I liked. Fast ball!! He then demanded to know why I hadn't gone to stay with him. Reply: "Failed DMTs and the great don't mix." We had a devil of a scrap and great fun it was. Of course he's a juggernauter. Rides rough-shod over one.

They want me to do a perfectly poisonous job, quite impractical – coordinating Armageddon between P1 and P2. I maintain that it can't be done. Even if it could, a person like me – discredited and passed over – isn't the type to inspire confidence, and I couldn't go to-and-fro from India in my present rank when everyone who was junior to me is now senior. We dined together and the struggle isn't settled. Have virtually left the Staff School. That's over.

GHQ Cairo:
December 1941–June 1942

*O*ver dinner that night Chink had simmered down, won over by Auchinleck's *patience. The new role was tactfully explained in more detail, his previous contributions praised, and by the final nightcap their good working relationship had been restored. Chink's optimism had surged back.*

23 December 1941 Continental Hotel, Cairo *7.15am* All goes well in Libya and all goes ill for Hitler in Europe. Will be proud that I had a hand in a man's work. My new part may be a considerable one behind the scenes. Am going to have a spate of work – related travelling – Cairo, Aleppo, Iran, Baghdad – for three months, but first back to Haifa to wind up that phase of my life.

24 December 1941 Continental Hotel, Cairo *7.15am* My headquarters are now GHQ MidEast, where I've an office and a telephone, but will often be away. The job is vague but interesting, though no firm ground to stand on. Am satisfied that everyone is out to make full use of my 'talents' but gingerly. Don't mind if public opinion sees me as another Henry Wilson[1] – sardonic, wise, erratic, Irish! What of it?

7.15pm Still in the office. Don't get any time to myself, naturally with things going on around all the time. Have arranged a room in the Continental Hotel. Everyone is being kind and considerate, particularly P2 who's using me haphazardly as part of his private brains trust. He's apt to collect a group of men about him who give his own mind what it wants, and he has little use for 'yes' men.

25 December 1941 Christmas Day, Cairo Tomorrow night I go with P2's party to Mena, will get 4 hours sleep and be like a boiled owl next day. Will be glad when I can live on my own again, even if it's interrupted by travelling. Worked all morning and walked twice around Gezirah[2] before 4pm, resisting gloom.

27 December 1941 GHQ, Cairo Can't see anything in this job, now I've had the chance to look at it. Can fuss about from A to B importantly, but already so many organisations exist for action and planning that outside the staff layout am in danger of being merely a messenger boy! Realise that P2 is holding me here to pick my brains and keep me on ice for promotion, as well as out of kindness. My protest coincided with selection for this job, and the general idea is that it will expand. It isn't the job that matters, it's the need to settle down to steady work. We haven't yet finished off Rommel. He's a skillful and determined fighter, and in spite of heavy losses has enough means to make things unpleasant for us. He's experienced what it is to reach the end of his administrative tether after a rapid prolonged pursuit, and probably thinks we won't have the means to attack or encircle him.

28 December 1941 Middle East Staff School, Haifa, for final ceremonies as Commandant Last night Auk talked to me again. He's so disarmingly friendly and genuinely sympathetic that I can't support my own point of view without feeling small-minded and selfish. He said he wanted the use of my brain and that I must be tolerant and 'learn to suffer fools gladly'. Why keep fools? They only get men killed! He won't say, and I'm too proud and stubborn to ask what's to become of me. Feel like a manikin held in a Gargantuan hand: helpless. But presumably the man thinks he's doing the proper thing.

31 December 1941 Lydda airport, en route back to Cairo Yesterday wound up the course with a final oration and we had a pretty good cocktail party. Afterwards dined with my G1 at Pross's. So abstemious these days that am out of training for serious drinking!

Chink's role began with another trip to South Africa, with the brief to research the practicalities of basing the Higher Command Staff School in Pretoria. His offer to be Commandant having been ruled out, he privately nicknamed it 'Tara'. General Smuts backed the idea, and escapism with Eve confirmed his commitment to her. On return the confines of GHQ ME rankled even more, as did the slow pace of the war.

1 January 1942 Continental Hotel, Cairo Tomorrow I must start again, and be careful in this town of hospitality and many contacts and drinks.

Thursday 5 February 1942 GHQ, Cairo If we feel strong enough for a riposte and can hold Mechili,[3] the fact that the Germans have come so far east may help us. On the other hand, if they can keep on advancing onwards to Gazala[4] we'll have to leave Mechili and lose our chance of threatening their L of C[5] south of Benghazi.[6] The whole problem hinges on our resources in armour,[7] Motor Transport and air. If we have to give up Mechili, then it's 'back to

Tobruk' – disappointing end to a costly struggle which will be difficult to renew, particularly as the Far East will absorb the air reinforcements essential to a new offensive. The enemy know how attenuated our resources are, so will maintain pressure over the whole Middle East zone.

7 February 1942 GHQ, Cairo Eighth Army is back at Tobruk and now it's all to be done again. We'll never control Cyrenaica until our line of operation runs from Jerabab across the neck to El Agheila.[8] When we've enough troops and transport to operate on that line Rommel won't dare go north or stay. To do permanent good in the Western Desert we must develop that southern line.

Sunday 8 February 1942 GHQ, Cairo P2 has gone off for a week and as I don't wish to create the impression in Cairo that I've become what the Chinese would call his 'Running Dog', I lie low. Have been drawing a Brigadier's pay since December with no more to justify it than a talk to Smuts, a lecture at Pretoria, and a report which took 2 hours to write. Add 14 hours with P2 and a certain amount of plain speaking. Brother Reg[9] has left his palace at Rangoon,[10] and he, P1 and that dull man Hutton[11] have been in conference on the defence of Burma. Java's finished. We must recast our world strategy on the basis that the Eastern shores of the Indian Ocean are largely Japanese and our position there is compromised, so we must maintain our position in the ME. It's not impossible that the Germans may stake all on an invasion of England, arguing that we've been so effete in Singapore and Hong Kong and so silly in the Western Desert that we may continue true to form. There's too much of the 'I accept the blame – pass the whitewash' spirit in these parts. I support P2 with qualifications, because he seems the only man capable of standing up to his task out here.

Funny how the WD has been an unlucky place for generals. <u>Captured:</u> Neame,[12] O'Connor. <u>Sacked:</u> Beresford Peirse,[13] Cunningham,[14] Godwin Austen.[15] There's a Hoodoo on it and I'd renumber 13th Corps. The man who goes on from strength to strength is Strafer Gott[16] and if he stays there long enough he'll suffer the same fate. He's been a tower of strength and when I write this war history he's one of the few who'll get away with it. Higher Selection board in Mid-East consists of the commanders in chief: Jumbo Wilson,[17] B.P., Ritchie.[18] Add Arthur Smith, GHQ. Each has to get something for their own subordinates so bargain one against another, and the Auk must accept recommendations. If I'm promoted I enhance nobody's reputation. They now say that I can't be used because 'I've had no experience of active service.' Whose fault is that?

10 February 1942 GHQ, Cairo *10am* Yesterday evening was shown into P2, who accused me of avoiding him. Same old stuff to the effect that I was a brilliant mind with twice the vision and speed of anyone else, far too quick a thinker for ordinary transactions and therefore difficult to place. My friends (gosh, the Judas kiss!) all agree that I'm 'too brilliant', that I 'don't suffer fools gladly' and 'make my seniors feel small'. So he couldn't give me a command appropriate to my status and hoped I understood. Pointed out that I'd heard it before and wished he'd get on with the war. Seems he feels he's in my debt (which is true). If I didn't fit into the British army at war, I said, I'd better leave. Oh no! Somewhere there must be the high appointment which only I can fill. Finale was a reference to my Bolshevik state of mind, and we then had a whiskey and soda and got down to his plans of campaign to which he sought my approval.

The whole body of my contemporaries and seniors have formed a tight ring like buffalo when a tiger's about, and are standing horns down to keep me out. Just like the old General Medical Council, when some unorthodox medico propounds a dangerously new idea which may damage their practices. To keep me quiet, I'm to prepare a report on the system of command and control in battle. This futile undertaking implies that I'll have *carte blanche* to go anywhere and ask any questions – so ensuring that I'll maintain this high standard of unpopularity. The first essential of a good system of command, of course, is good commanders with brains. Feel as if I'm the only sane man in a lunatic asylum, the prelude to insanity!

6.30pm. Lunched with P2 as if I was the bluest-eyed boy of the whole starry-eyed constellation, and returned to his office to discuss high strategy.

11 February 1942 GHQ, Cairo *3pm* Actually my make-believe task will lead to a far better understanding of our recent fighting. 'Never in the history of war have so many been chased so far by so few.' A hackneyed comment, but fair. The ferret goes into the burrow on Monday next and, though anything I write is probably destined for the waste-paper basket, the facts will help towards a Middle East history.

Am beginning to develop a 'give a dog a bad name' complex. Of course, I could retaliate among the war correspondents and politicians who team in this town, but it's *infra dig* to vent one's spleen. If what they say about me was only 50% true, then the powers that be are keeping me outside the management to preserve a safe standard of mediocrity – *vide* Singapore and Malaya.

Singapore news really distressing. Knew nothing of the military situation there, but did think something effective had been done to defend the island from the north. Those defeats are having profound repercussions on the future status of the white powers in Asia. Nehru's in a strong position in India today, and if he takes control the Viceroy can hardly put him back in prison. The great 'Post-Mutiny' game is up in India, and the Sahibs are too Brontosaurian to adapt. Sahib rule is finished, and it remains for us to help the Indian to become a real world citizen. Historians have been busy lately on the causes of the English loss of the American colonies, and the same state of mind in high places seems to be producing the same results now.

13 February 1942 GHQ, Cairo *12.45pm* Have bought a suit of battledress – largely to save my clothes which can't be replaced during the war, and also to prevent our people mistaking me for the enemy in my corduroy! Will be away for ten days. News not good, all boiling up for March to be our worst month. Singapore still fights, but for how long? Too much to hope that the enemy will shoot his bolt and withdraw. He's bound to keep up the present intense pressure, and we're in no position to counter him.

Had a brief interview with P2 and told him that he'll have to educate his senior officers in the 'simple little rules and few' which avoid disaster and sometimes even produce victory. 'Keep your forces concentrated for battle.' 'Don't lay yourself open to being defeated piecemeal.' Copy book maxims which none of our 'seniors' seem to know. He took it well. The Boche senior officer is, on the whole, far better taught than ours. Putting up generals' badges doesn't make one a general. We've got to train a body of younger men for staff and command who make mistakes as seldom as Rommel. It's vital stuff, and I've not wasted my life if I've collected enough data to make it clear to P2 that he must begin to collect a cadre of properly 'educated' senior officers before he can begin to be successful.

10.15pm. At dinner was handed a note to ring a certain number. The voice announced itself as Claude! P2 requires me to walk with him at 7am.

Saturday 14 February 1942 GHQ, Cairo *9am* Rasputin-like, took the opportunity and believe I've activated a line of thought already dormant in his brain. Feeling ruthless and unscrupulous now. He showed me a telegram to London, so perhaps we do some good on long-term policy, but the immediate task is to purge the cotton-wool producers. I'll be glad to get out of Cairo. Can't avoid eating and drinking too much here and generally deteriorating. Am going to the office via a bookshop, then a cocktail party given by the *News Chronicle*.

15 February 1942 GHQ, Cairo *9.30am* Didn't get to bed till 1am after amusing subversive dinner with sundry pressmen, talking over world futures and present problems. Don't remember much and nor I think will they, for with the cocktail party and a bottle of Chianti the conversation soon attained a Nirvanic quality. Am still prepared to trust P2 and will be in direct communication with him[19] by letter, so in certain respects his confidential agent on the battlefield. Am authorised to deal with him direct. Future entirely vague but can't believe things won't come right. The main point is the war. Don't mind about myself.

Chink was to spend that evening in the company of a man he would describe in his diary next morning as 'my friend', but who long after the war would be named as Rommel's invaluable 'Good Source' for British military detail. Chink, like many other senior soldiers in Cairo at the time, was unknowingly socialising with Rommel's 'eyes and ears', and in fairness to Colonel Bonnar Fellers, the pleasant American in question, he, too, had no idea of his use to Germany. His posting as US Military Attaché to Egypt in 1940 included unrestricted access to Allied plans, which on US State Department instructions were to be transmitted to Washington by impregnable 'Black' code. In September 1941, it is now known, the code was stolen, photographed, and replaced by Servizio Inforazioni Militari (Italian Military Intelligence). From then on, Fellers' messages were decoded and translated as soon as a German listening post near Nuremberg picked up his repetitive call sign. It is now believed that Rommel received a concise Appreciation of British operations in the desert – units, strength and morale – at breakfast every morning until 29 June 1942, when it was discovered that the 'Black' code had been compromised. Unaware, Chink faced an early start next day. It was time to investigate Eighth Army's system of command and control in battle.

Monday 16 February 1942 *In flight 8am* Am writing in a Lockheed speeding away from Cairo. Dined last night with my friend Col Bonnar Fellowes [*sic*], the American military attaché, and two American naval officers who'd just arrived. We discussed war and the English attitude towards enemies, which they found insufficiently ruthless. For the sake of argument I attempted to justify a policy of toleration, urging that it's always a mistake to bring an enemy to a state of mind in which he fights like a cornered rat. The greater subtlety is to give him the opportunity to retreat or surrender. They weren't prepared to discuss war aims, probably because their soreness at Japan's success at Pearl Harbor and their desire for revenge is at present sufficient. That won't last.

Later: My great snoop may reveal a leader who adds knowledge and intelligence to energy and courage, but I preserve an open mind and no great hopefulness. A month ago I'd only to expect extensive voyaging in Asia Minor, but here I am plugging westwards into this damn desert to which I always think I've finally said goodbye.

5.30pm As my car ripped out its inner tube on a nail, it failed to synchronise. Hung about a rear HQs waiting, and now it's too late to go on. Installed in P2's caravan-lorry for the night – spring bed and all. Tomorrow go on west to meet the army commander[20] and his staff. Little desert flowers are everywhere and beautiful, with long-stalked asphodel around the caravan. Just for a little the flowers remain, then everything is brown again.

18 February 1942 Western Desert A gusty day in this encampment near the coast, and cold. Useful talk with Gen Ritchie this morning, and even more useful with one or two contemporaries who see things much as I do, without collusion. Won't get down to the real meat until I see the smaller units and formations – the chaps who've felt the shoe pinch.

A letter from Estelle today, saying a pressman disguised as a soldier whom we knew in India and is now in passage between Delhi and Cairo told her I was engaged in an affair! She merely says, "Have a good time, it won't make any difference to me." Who that chap was, can't think.

7.10pm The snoop of snoops has temporarily ceased. Have made some surprising discoveries.

19 February 1942 Western Desert Heartbreaking to see so much effort expended on so little achievement. Am writing in a mess tent half underground. An HQ halts and digs itself in – puts up tents and creates an illusion of life in vacancy: tables, chairs, drinks, newspapers, electric light! Hard to believe that outside is nothingness. Staying next with Gen Gott [sic] who attracts me a lott! Then with the South Africans, Indians and others. Begin to see the form here, and as usual it fills me with keen interest because of its stupidity.

9.30pm Writing shiveringly in bed. Have undressed into pyjamas on the tacit undertaking that we'll neither be raided nor bombed while I wear them. Can hear a wump-wump from Tobruk, and this afternoon had to lie on my back trying to spot a highflying enemy who'd skittishly coughed up a couple of bombs. All peaceful tonight. Staying here tomorrow to jot down my ideas, interview Signals, Intelligence etc, and move on next day along the 'front'. Then off to the 'back' and my investigations end. Have met no man in any job up here whom I couldn't replace. Whatever they say about me, don't feel in the least modest. This war business is too easy.

20 February 1942 Western Desert A cold wet gusty day with more rain coming from the west, so Boche aerodromes will be more waterlogged than ever. Don't like this job much, snooping isn't really my line of country, and will be glad to quit. Makes me angry to think of all the waste of effort at Hong Kong and

Singapore, with 30 to 40 thousand good troops and immensely important gear gone up the spout. Poor Archie Wavell is attempting an impossible job in trying to control this business. Strafer Gott has the makings of a great man – I like his hook nose and broad view. He's much more to him than Ritchie.

4.30pm Motoring about this desolate land in the wake of the two big shots looking at their army is not an exciting undertaking. There's a quality in this landscape, possibly its vacancy, which makes for optical illusions or mind wanderings. If one doesn't concentrate on looking, the eye invents things which aren't here – turns camel thorn scrub into remote forests and garbs the desert with illusions. Then bump! You're awake again, looking at the dismal remains of a crashed plane or burnt out lorry. There's nothing to keep the eye active and mind awake in the unbroken ring of horizon.

21 February 1942 Western Desert Writing in a minute hole in the ground covered by a tent just big enough to take my camp bed and luggage. Am living with the Indians,[21] and outside is a bright, clear sky. Orion swings right side up, familiar, and the desert smells of flowers. Still hearing strange stories of the recent days, and more than ever convinced there must be radical and sweeping changes. Today saw some chaps left behind during the retreat who eventually seized a German lorry and its occupants and drove back to our lines. One German was a smart sergeant major, young, goodlooking, 100% military. His captors, with 3 weeks of beard, were a wild and woolly lot of bandits. They were feeding off bully beef, ration biscuit and jam and praising a 'civilized meal'. Bless them!

22 February 1942 Western Desert A foggy day with the sun trying to force its way through a ground mist. Ideal morning for delivering an attack. But the Boche is far away, and neither side keen on new adventures. As someone said yesterday, 'The annual outing in Cyrenaica seems to be over.'

23 February 1942 Western Desert *12.30pm* With the Afrikaners[22] since yesterday, where gazelles' water is sweet and the Mediterranean almost as blue as the spring flowers in the wadis. Have heard a great deal and been talked to very frankly. Now see why many things went wrong, and have my ideas about how they can be put right. But will anyone agree?

25 February 1942 Western Desert South next into the dry country, to be among the English again, then back to Cairo – not because the flesh pots are calling, but to see P2.

10.30pm Spent the morning in a *khamsin*[23] with the Afrikaners and in the afternoon drove 30 miles to stay with something ultra English[24] and only recently de-spurred!

26 February 1942 Western Desert *3pm* An interesting morning with chaps of the bluest blood. Better the devil one knows! Among the Afrikaners everyone is friendly, polite, hospitable. Among one's own, grudging hospitality, offhand manners, no friendliness. The Englishman moves everywhere as a villager: narrow, suspicious. They've no sense of humour, except the wizened humour of *Punch*. They can't laugh at life but only at each other – rather maliciously. Their world is a little one – the group they live in – and a stranger is an object for brick-throwing. Perhaps it's my trouble: I've grown unEnglish as I've matured.

200,000 men on both sides in these wastes must in some way be thinking about women and love. Some generally, just the need for a female, many in terms of their own family, and others like me in terms of the woman who one day will be the mother of their children, a real companion in life. But one way or another, though they wouldn't admit it, all here think and dream of women. It's a stupid unnatural life when men live womanless and women maleless.

27 February 1942 Western Desert *11pm* Tomorrow I set sail for Cairo – by road. The weather's bad so air travel's uncertain. Tonight heard that Jock Campbell,[25] who got a VC recently and was promoted a major-general, was killed this afternoon in a motor accident. He was on top of the world. Saw him only yesterday, and found myself wondering how long he had to live. Well, didn't have to wonder long.

28 February 1942 Western Desert *4pm* Left Army HQ this morning after a talk with Ritchie, and can't see him as the man of destiny. Another talk with the head airman,[26] who is competent. Now await Willoughby Norrie[27] who had much to do with earlier encounters and may feel like talking, but I know enough about the Eighth Army now. Cairo by tomorrow evening, and am determined that they decide for good how I'm to be employed. It's either a high appointment or nothing. Either suits me, but not this limbo.

2 March 8 1942 GHQ, Cairo *8.30am* P2 got me out onto the Nile bank and then said, 'Go on, talk!' At the moment I probably have more unofficial power behind the scenes than anyone in MidEast, but hate the position. Still, I'll use it flat out for everything but myself.

4.30pm This morning it was officially announced that Arthur Smith had handed over the dust of office to Tom Corbett[28] – little, red faced, Irish, intelligent without being intellectual. He was Brigadier adviser on Cavalry when I was DMT[29] India and we worked together. Guess eventually P2 and the new CGS[30] hope to put me in as DCGS,[31] but that won't be yet. Feel exhausted by the muddle of emotions involved in this 'Night of the Long Swords', because hate being mixed in intrigue and have been forced by circumstances to become an intriguer. Almost believe in my own evil eye, and don't like that.

Tomorrow can get down to writing my report on 'Command and Control in Modern Warfare', and am reluctant to write about something which, were I in full power, I'd merely implement. Two letters from Estelle – one angry, the other apologetic, both pitiful. Don't feel I can tell her how things stand yet, because our futures are still overshadowed by the war.

6.45pm. At GHQ ME[32] was ordered by P2 to parade for an early walk tomorrow. We discussed the higher war course and my suggestion that it should be located in the Union. He suddenly said, 'What have you got down there that you're so keen on South Africa?' My answer, which I hope disarmed him, was that my interests were 'purely scenic!'

4 March 1942 GHQ, Cairo *7.30am* Waiting for P2's car to fetch me. He's anxious to discuss the recent campaign – as anxious as I am to avoid that, because my opinions will probably hurt him since at times he was involved in decisions which I and others criticize. However, it will have to come out sometime.

12.45pm We discussed the future. We'll get to the past via the future, but evidently it's still raw. Don't propose to go on being his captive, flattering as it is. Can't forget that 3 years ago we were much of a muchness, and that my advice put him on his high road. Don't envy him that road – for most men a precipice yawns sooner or later. In a fit of impatience wrote off my report this morning. Brief, forceful and addressed to CinC, recommending direct action with those concerned, rather than a great deal of writing. Hope they'll swallow my revolutionary 'technique'.[33]

5 March 1942 GHQ, Cairo *7.30am* Last night saw me closeted with P2 and the new and old No. 1 while I expounded the technique for a new war. P2 is, after all, a lonely man and needs all the help he can find. He's certainly better than anything England could produce at present, though am pretty sure that in Whitehall there's a high power movement for his removal. Luckily the new Minister for War[34] is his friend.

9 March 1942 GHQ, Cairo *11am* With P2 away for a week there's unlikely to be much for me. Arthur S still floats ectoplasmically about the office but hands over on Saturday. (Peal of bells.) Went to Frank Theron's[35] office to tell him I was non-employed. He, being an orthodox ritualist, was shaken when told I felt no loyalties except to the world cause, for which Russians, Yanks, British, Indians, Chinese were fighting, and that I'd serve in any rank anywhere, at the front or 'down south'. Fear he's a diplomat rather than a revolutionary. The peacetime mind is paramount in British conduct of this war. Don't know how we can defeat our incredible flair for procrastination and indecision. Believe the Japs will get Rangoon at any moment. Reg [Chink's youngest brother] will leave his palace for a swamp – poor old Reg, he's been used too often and I too little.

6.30pm The papers tonight prepare us for the fall of Rangoon. Gen Alexander[36] has succeeded Tom Hutton in command of Burma and Reg is to broadcast. Where and when do the <u>British</u> come into this war, except with Dunkirks – Greece evacuations, Crete, Cyrenaica, Malaya, Hong Kong, Singapore, Burma! Against this dismal record we can place (a) The Battle of Britain, (b) Dick O'Connor's brilliant campaign, (c) The capture of Abyssinia, (d) The occupation of Baghdad, (e) The Battle of the Atlantic, which isn't yet decided, (f) The Defence of Malta. Not an inspiring set of victories compared to our defeats. Our higher policy seems hopelessly at fault and none of those cabinet reshuffles strike at the root of the problem. This is a WORLD WAR, not England's private war. The Axis[37] run their war from Berlin and Tokyo. We run ours from Washington, London, Canberra, Delhi, Quebec, [sic] Cape Town, Sydney, and a pretty good muddle emerges. The German-Japanese combine are 100% unilatarised and singleminded. Moscow works alone, and rightly repudiates all of us. We're adepts at compromise and futile at decision. We must hold what we can of the Middle East.

8.30am next morning. A note in my Egyptian mail says Sir R[38] is no longer in his Palace, which was a cross between a public lavatory and a Victorian railway station. A policy of scorched monstrosities is required.

11 March 1942 GHQ, Cairo Am furious about Hong Kong.[39] What the hell did they expect the Japs to do, kiss them and read them fairy tales?! Everybody knew about the massacre and raping in Chunking. The bland voice of Anthony Eden[40] – the man who is never wrong at the right time! Why wasn't the damn place evacuated, lock stock and barrel? Because the big money boys wouldn't have liked it. And why was it reinforced by Canadians, and has anyone been broken or shot for being so bloody silly? Daresay Singapore has been as bad. Hell! I'm so damned angry.

Between 12 March and 10 May Chink was again in South Africa. researching the potential Higher Command Staff School in depth. He stayed with Eve whenever possible, but researched and seized every chance to put forward the proposal in talks and articles. Smuts continued to support it, and he returned with a curriculum drawn up, potential premises specified and figures costed. It was only to face yet another blind alley.

11 May 1942 GHQ, Cairo *3.15pm* Saw P2 first thing. Absolutely no hope of the institution coming to the Union,[41] and even if there had been, I wouldn't be with it. So that is that. He's trying to get me to GHQ in place of Sandy Galloway,[42] the snag being that Wavell has already asked for me at AHQ Delhi. P2 has requested permission to cancel his gift of my body.

12 May 1942, Cairo *4.45pm* Wavell's reply came when I was breakfasting with P2. He'd 'hoped to have my original and unorthodox brain to assist with problems', but since I was on the spot here, P2 had prior claim. Good. The radio announces a big German attack in South Russia. Looks like the commencement of this season's fun and games, and the heat will be turned on with gathering intensity. The picture hasn't changed much locally since I left, though repercussions elsewhere alter the perspective.

In mid-May Chink was appointed to Ritchie's old job of Deputy Chief of the General Staff in GHQ ME, though the War Office in London had still to sanction his appointment. Promotion would be to the temporary rank of Major-General, which eased his irritation at the arrival in the office of cavalryman Richard McCreery, an armoured commander under Brooke's patronage. They had been uneasy contemporaries at the Staff College, where Chink had scathingly referred to him as Dreary McCreery.

14 May 1942 GHQ, Cairo *11.15am* Clambering into the saddle with that semi-sinking feeling with which one mounts at a point-to-point. Our parish extends from Tobruk to Teheran, so am likely to float about a good deal to justify my appointment to the dizzy heights of vital decisions. Without a doubt this is a more efficient show than last year. A far better staff – now I'm on it!

15 May 1942 GHQ, Cairo *7.30am* Yesterday snooped about the building looking at the layout and finding where chaps lived. Last night dined with Tubby Martin who commands the 5th Fusiliers and has just got a well-earned DSO for the fighting at Tobruk. On St George's Day the regiment dug out its colours and paraded through Cairo, red and white roses everywhere, to the cathedral for a memorial service. A good thing to have done. Phew it's hot. Am sticking to the paper as I write. Far too early to assess results, but it looks as if the Boche will seek a decision in Russia before going elsewhere seriously. He may try to keep us quiet down here, as a local show. Let him try.

20 May 1942 GHQ, Cairo *7.15am* Getting dressed for the office – shorts, bush-shirt – is easy. Girls have shed their brassieres for coolness and the streets are billowy with bosoms in every stage of deliquescence.

25 May 1942 Continental Hotel, Cairo *7.15am* Can't look upon this war business as anything but a ghastly game played with gusto by two groups of undeveloped children with dreadful toys. It all seems unreal. I too can play the game, but with the sense that I'm playing in a nursery as an adult. Perhaps there are all sort of mental worlds. The child's world is one, and there's this war-world where men behave like cruel uncontrolled children on holiday. They seem happy to escape the domination of their women, who from their babyhood governed life. Women (on whom they secretly depend for everything) are put in their place while the chaps go off for a happy day of smashing things up and generally raising Cain in the name of high-falutin' unrealities. Of course, now one lot of children is bent on smashing up the other I'm forced into the game willy-nilly. Wish there were some adult minds to stop them before too much damage is done. Am remembering my last fight with Victor.[43] He was a year younger but sturdier, and we must have been about thirteen. We had a terrific set to in the drawing room and pretty well wrecked it in the process. I remember Victor's nose bleeding and our attempt to put the furniture straight and tidy up before the grown-ups returned. Also remember that I enjoyed the encounter! The same sort of business is going on now, only I seem to have grown up and look at it with angry detachment.

27 May 1942 Continental Hotel, Cairo *7.15am* What a business war is, nothing stays put for one moment. Values are constantly altering. No factors except the desert, sea and air are constant and even they indulge in temperament. The most abiding quality is the mind and courage of man, and how to assess those? By guess and by God, as the greatest gambler may gain the greatest prize – or lose absolutely. The war is moving into a decisive stage. Don't know the truth of events in Russia, but sympathetic activity will flare up all over the Mediterranean. The Italians won't be let off by their taskmasters, and it's fair to expect a 'do' in the Western Desert.

7.25pm. GHQ. The party has begun with a German armoured movement round the southern flank of our positions, bypassing Bir Hacheim.[44] Too early to say if this is the main offensive or a feint to draw off our armour while the main attack comes against the Gazala[45] positions. In any case we've not been surprised and if we aren't too stupid we should give him a good licking. News comes in slowly and its no good fussing. Am confident that my own predictions aren't far wrong. Suppose the Boche intend the Italians to take their share of the 'fun', and think our dispositions are generally sound. To bed early in case the pace of things hots up.

28 May 1942 GHQ, Cairo *7.15am* Our loosely-knitted front runs south into the desert from the sea at Gazala inlet and the high cliffs above it. A strongly held area across flatter uplands, rolling and seamed with deep wadis and barren till it passes into near desert, flat and featureless to the real desert about Bir Hacheim. All that stretch, 50 miles nearly, is defended in the modern way – great 'mine marshes' etc. At Bir Hacheim the defences peter out, because one can't go on forever extending a front and have any strength behind, and there's an empty vast world patrolled for great distances by our mobile forces. In rear are other strong garrisoned places, including Tobruk, and free to move anywhere is our main mass of armour and mobile troops. All this the Germans know – it couldn't be secret. Further east lie our aerodromes and the railway and the long coast road to the Delta. The German problem is how to disintegrate that layout, and I think he's trying to do it by a combined feint round the south of Bir Hacheim and a break through south of Gazala. I will know in two hours time.

2.30pm Rommel has made a bold move, but by no means unforeseeable. Splitting his army into two, separated by our positions and mine-fields. He's chosen to fight in ground we know well, against all our armour. Luck plays so great a part in war that one can never predict a battle's course, but if luck's on our side for once we may be successful. Prince Henry[46] dines at Mena tonight with the chaps who were students with him at the Staff College, and I'm bidden to attend. Hope he'll not keep us late.

Friday 29 May Cairo *6am* Last night began pleasantly enough. The terrace at Mena, a fat full moon and seven or eight nice people, all of whom I like. The guest was the Duke of Gloucester. I laughed quietly at our fantastic system. He – who is about the world's stupidest male, created to exist in a vacuum by an intensive process of selective breeding – is a lieutenant general. One ex-student – a good, hardworking fellow – was a brigadier. Another, a full colonel and the rest lieutenant colonels in important jobs. Prince Henry was full of his experiences as liaison officer in France and England, and unwittingly gave a good picture of jockeying in the higher ranks. Nevertheless some good men, contemporaries of mine, are coming forward – orthodox chaps, naturally. The dinner went on interminably. HRH had been told pretty clearly that we were busy men who got up early, but Devil a bit! He didn't leave the table till 12.15, so bed at 1am. Was I angry! He was pleasant and natural, but how long can we go on with this childishness? The world must shed its last monarchs, just as it shed its dinosaurs. He looks not unlike a dinosaur – same uncomprehending blank peering look – little brain in big skeleton.

12 noon In flight Writing in the flippant tail of a Boston Bomber leaving Cairo, feet on the parachute exit and body wrapped in a not uncomfortable harness. Behind me sits the rear gunner and I hope he knows his job! P2 bunged me off at 5 minutes notice with a message of cheer and orders for Ritchie at Eighth Army HQ. Hope we get there tonight. Air and land situation naturally vague, but from all we've heard today most reassuring. The enemy have done a wide turning movement round our desert flank and come north towards Tobruk, where we were prepared to receive them. None have got near, and they're now going round like a bee in a bandbox in a restricted area bounded by our Gazala positions on the west, the sea to the north, and due south the Trigh Capuzzo.[47] Don't see how the enemy is to escape from a rather nasty jamb unless he hopes to burst a gap through the Gazala front from east to west. This machine *is* fast! We're already over Wadi Natrun.[48]

Rommel's expected attack had begun during the dinner. The static Gazala Line was fortified mainly by the isolated 'boxes' he hated, garrisoned by brigade groups which resulted in the fragmentation of artillery. He believed Rommel would opt for a combined feint around Bir Hacheim and a breakthrough south of Gazala, and early reports would show if he was right.

30 May 1942 Western Desert *8.15am* A fresh clear Mediterranean morning and am writing in an operations lorry not far from the sea. Back to Cairo as soon as I've seen the Army commander and got his dope on last night's arrangements and what he hopes for the future. Everyone up here in good fettle and confident, as well they may be. Victory comes to him who makes the fewest mistakes, and the enemy have so far made 80% to our 20%. Unusual for them, but something seems to have gone wrong with their planning and execution. They're under fire even now.

Sunday 31 May 1942 GHQ, Cairo *7.15am* Flew from Eighth Army HQ to Cairo in 2½ hours and at 7pm yesterday handed out the hot news of battle to a distinguished audience, including HRH, Casey,[49] P2. Later a small party of chaps in the know dined at the Mahomet Ali and we lifted our glasses to the first real resounding kick in the pants the Boche has received since the battle of Britain. He elected to take us on in the heart of our organised positions, and been soundly drubbed. His operation has been a costly failure and the whole axis fabric in Cyrenaica is weaker. Our losses are easier to replace than his, and his have been heavier than ours. But the morale loss will be the final test. His High Command elected to take this business on with Germans mainly, leaving the Italians the secondary task of holding us in front of the Gazala positions. Now the Germans have been licked – have failed to take Tobruk – and the effect on Axis unity will be serious. Our propaganda will make full use of it. I was full of admiration for the way things were going up there. The staff work, layout of command and administration, the whole atmosphere was efficient, confident and smooth. We seem to be getting towards the right technique at last.

2.15pm Battle news gets better and better. The enemy's shattered forces have begun to bolt though the gaps in our mine-fields, and though I don't yet know the total damage, am sure the Axis Land element has to a great extent been put *hors de combat*. Remains an amorphous and demoralized group of Italian Divisions, which may well be made to withstand a counter offensive. If we can cash in on this victory to the extent that the news seems to warrant, July may see the removal of that threat to Egypt. They'll hardly renew this highly unprofitable venture a third time. Victory here in June, combined with German failure to penetrate the Caucasus before October, would make the Middle East a peaceful zone – until we begin to dispose of Italy.

2 June 1942 GHQ, Cairo *7.15am* I've always felt the time would come when Rommel would be proved to be the tough bull-in-a-china-shop he really is. German operations have been markedly stupid, both in tactics and technique, especially in the way they left the Italians to attack us frontally, while the Hun came round the back to attack from the east. Did they really believe the Wops would attack? It's reassuring for the future. Seems to show that Axis superiority in materiel[50] is waning, in tandem with élan and military ability.

Military Historian John Lee: Rommel had conducted a brilliant move with his concentrated armoured formations around the open left flank of 8th Army and had defeated several isolated British brigades. But he was never a master of logistical detail and soon found his main force deep within the British defences and desperately short of fuel and ammunition and vulnerable to a well-organised counter-attack.

3 June 1942 GHQ, Cairo *7.30am* We worked late last night. The Chief laid stress on the fact that the fighting hasn't ended, and there's much more to come. The enemy is tenaciously trying to establish himself within our former lightly-held positions south of the main Gazala front, perhaps to resuscitate his original plan. The situation offers possibilities, and we've yet to see how they'll develop. The news and Auk's message to London give as clear a picture of events as could be gained at this distance, but they don't tell of the dust and drought and blistering sunlight, and they can't convey any idea of what actual fighting means to the man in the front area. I try hard to keep that picture in mind, so that in giving advice I don't forget the suffering humans behind the mechanism of war. Hope I don't, anyway.

2.30pm The dust was bad in Cyrenaica yesterday. Heavy fighting seems to continue.

As clear a picture of events as could be gained at this distance...' Ritchie's tone of confidence did not match the emerging picture from GHQ maps and Ultra decrypts revealing an inability to anticipate Rommel's moves. When Bir Hacheim was surrounded and Ritchie's counter-proposal was to make a frontal attack on the Cauldron, where Rommel was posted most strongly, Chink was incredulous. Eighth Army were handing the enemy gains their tactics did not merit, and Rommel's continued use of the Italians, in particular, was an obvious weak spot.

4 June 1942 Cairo *7am* The battle continues out west, a slogging match with both sides flat out and little likelihood of brilliance. The enemy are trying to eliminate Bir Hacheim which is a thorn in their backside – once away they may hope to swing their weight north towards Gazala and Tobruk, but so long as we hold firm their limited gains are little use. Meanwhile we sort ourselves out from last week's 'whirling spray' fighting and look for their most vulnerable part. The enemy is expending the greatest energy, and that may be his undoing as he'll be forced to call in the Italians to help at decisive points.

The War Office have at last agreed to my being DCGS with effect from 16 June and till then I continue as Brigadier Dorman-Smith, by now almost a hereditary rank! Perpetual Brigadier of an ephemeral army. So as long as I 'give satisfaction' and the Auk remains in power I'll stay wherever GHQ ME sets up its tents. My erring feet are back on the rungs of a ladder I've learned to despise.

5 June 1942 GHQ, Cairo *10.30pm* The battle is at a stage of arrested climax. The end is not yet but it will be favourable – that is, if Ritchie is even 50% competent. If he isn't, then anything may happen. I'd like to be up with the party to see for sure how he really handles a show and what the business is like, but I'd only be a nuisance so stay here.

Military Historian John Lee: Despite his highly confident messages, having obliged Rommel to go over to the defensive deep within a belt of British minefields, Ritchie threw away any chance of victory by allowing the British armoured formations to dash themselves to pieces, one brigade at a time, making fully frontal attacks on the enemy. German anti-tank guns destroyed the attacking armour, and then Rommel resumed his offensive, overrunning the isolated infantry brigades in their defensive 'boxes'. Ritchie's continuing flow of optimism to GHQ Cairo made it impossible to follow the true course of the battle of Gazala. On 13th June 1942 Chink would wonder why the enemy had been allowed to rest, reorganise and resume 'probing at our vitals. The fault lies in ourselves'. Soon 8th Army was in full retreat, and unravelling dangerously in the process.

Saturday 6 June 1942 GHQ, Cairo Battle in the Western Desert still rages inconclusively. The enemy is gradually exhausting himself against our obstinacy, though not against our brilliance, and he must be getting close to cracking point. He's tried to restore his initial mistake without appreciable profit and is now making a final desperate bid for victory.

Tuesday 9 June 1942 GHQ, Cairo *7.15am* There's no brilliance in our management. Courage, tenacity, obstinacy, but not the science which makes a battle work as a whole. There are so few men in our army who make war their profession, their hobby and the meaning of their life. Those there are, are rebels, and rebels aren't employed till orthodoxy is emptied and proved vain. I'm sorry for P2. He has stuck to and supported Ritchie and Ritchie hasn't the divine spark.

2pm Have got to clear my mind about certain organisations out here and can think better in my hotel room than in the office. This battle's been going on now for 14 days and is nowhere near ended yet. Think I could have finished it in a week, but then I always think I can do things better than other people.

11 June 1942 GHQ, Cairo *7.20am* Yesterday was a high record for Cairo GHQ, beginning at 5.30am when I was called, and ending at 1.30am today with scarcely a break. The situation continues in suspense, though I've no doubt of the issue. The enemy's feeling the strain more, and it's significant that the German press is now saying they only attacked to forestall our offensive – in which they claim to have succeeded. We have their order of the day and evidence to show their intentions went far beyond that. Of course the statement may be intended to fool us – but it may equally be intended to prepare the Axis people for failure. Only problem is how to make them pay dearly for their temerity and initial mistakes. It's an interesting position, and our press boys are being pretty lurid about that unspecified patch of flat desert which they've named – why the Lord knows – the 'Devil's Cauldron'.

12 June 1942 GHQ, Cairo *7.15am* Last night, to the whooping and crying of Cairo air raids, at last slept. We had to leave our southern strong place at Bir Hacheim because it became isolated and difficult to supply with water. The enemy couldn't take it in spite of all their aircraft, but they could and did sit round it and stop our convoys getting in. After 14 days Blitz turned to Sitz – the water supply of besieged places becomes the decisive factor. The Axis will make propaganda from it, mainly because of the hysteria of our own press about its defence, and one has to rely on Ritchie's judgement. We've made mistakes, and when you make mistakes you're liable to lose points. Points in war are men, guns, tanks and vehicles.

13 June 1942 GHQ, Cairo *7.20am* For two weeks now complete victory has been in our grasp yet we haven't achieved it, and the enemy, rested and reorganised, is afoot again, probing at our vitals. The fault lies in ourselves. P2 may pay a high price for his loyalty to a man who behind an imposing façade is a very ordinary occupational soldier, not a professional. You can't make war on professionals unless you, too, are a pro. Players can always defeat Gentlemen when they want, and this struggle on our side is Gentlemen v. Players – as in Napoleon's time. Then the French were the Professionals. The only real professional servicemen we have now are sailors. Hell take it! I could raise a team for the Western Desert which would peg back these one-track Boche with the greatest of ease. I regret my glowing vision of the possibilities arising out of the original Axis failure and resultant opportunity that temporarily obscured my mature judgement about the personalities concerned. If I can derive any consolation about being right, it's that one learns to back one's judgement in the big things.

14 June 1942 GHQ, Cairo *10.30pm* As pretty and hectic a flap as I've known since this time twelve months ago when I had to give sound advice – advice now listened to, too late. Poor, poor P2, this is a desperate disappointment. I'm sad for him.

15 June 1942 GHQ, Cairo *6.30am* Today we must begin new office hours and work straight through till 5.30pm with an hour for lunch as in London. And this is a prelude to our living in a tent in Mena![51] It's quite insane. Can't think who encouraged P2 to do it. Am about to walk with him now, before the day starts.

16 June 1942 GHQ, Cairo *7am* Yesterday was hell – 106 in the shade. Left office 8.30pm, dinner 9, then back to the office. In between tried to sort out 101 tangled skeins, the daily task of a senior staff officer in wartime. Also sweated! It's so easy to become unbalanced – tempers very frail these days. Heard Roosevelt's broadcast prayer for the UN, 'God of the Free'. It's the biggest thing since the Lord's Prayer, and likely to be as little understood. 'Our earth is a small star of this great universe, yet of it. We can make if we choose a planet unvexed by war, untroubled by hunger and fear, undivided by senseless distinctions of race, colour or theory. Grant us courage to begin this task today, that our children and our children's children may be proud.' We have it all there. The perfect spiritual approach.

Thread my way among the thin and crackling ice of personalities in these Headquarters – General X and General Y must have their say, be consulted, not offended, and so it goes. It's open to question whether we don't worry far too much about personal susceptibilities in our army. At present I've considerable influence but unfortunately can't influence past decisions, and these are having their effect on the present battle. It might have gone better. It also might have gone a lot worse, and the result is that Axis forces have left the strong ground in which we couldn't easily get at them and paid a high price for the breaking down of the Gazala position, without capturing the troops in it.

3.15pm Off to get my car. Don't mind the sun in the open, but the heart of Cairo becomes a furnace, fit to burn the life out of us.

17 June 1942 GHQ, Cairo At 6 o'clock yesterday the Embassy phoned to say the Governor of Burma was at Shepheards[52] and would I dine with the Lampsons[53] to meet him!? At 8.15 went round to find Reg. He looked tired after a hot air trip from India, en route to London to say a few words about Burma. We drank very good sherry till the Lampsons turned up, and afterwards all sat in the cool darkness of their garden by the Nile and talked. What a pleasant change from the evening session of the Gloom Club in P2's war room.

The Germans in Libya have gained a victory. They've forced us from the positions covering Tobruk, so to that degree have won the second round. But they've failed in their real purpose, which was to cut Eighth Army in two by isolating troops in the Gazala salient. Their initial defeat spoilt that manoeuvre, and they've gained their later success at the cost of the Afrika Corps' tremendous losses. Now the third round is to be fought, and conditions are still favourable to us – *if* we know how to use them. 8th Army is intact and fit to keep the desert, preventing any proper investment of Tobruk. The fighting has been terrific.

18 June 1942 *In Flight* *6pm* Am in the glossy nose of a Boston bomber halfway to Cairo from the WD.[54] P2 took me with him to see Ritchie and sort out a few things. A blue sea on my left, a tawny desert below and I look at the world as the bomb aimer sees it. P2 is in the rear MG seat way at the back so am alone in my little cabin. Nice to be alone, but it's a queer way for two High Officials to travel. In a properly organised world P2 ought to have his own fast machine and fighter escort. Am hungry, thirsty and sleepy, for didn't get to bed till 12.30am and was called at 5am. Everyone with the fighting men looks so fit, while we of GHQ look like high cheeses – pale and green-mouldy. We certainly feel it.

Have been ragging P2 about setting up a tourist agency in Libya after the war to show off the battlefields and he's fairly bitten by it. It would be a gold mine for 10 years. We'd need capital to build – somewhere by the sea, near Gazala perhaps as there's sweet water there. We'd start with one guest-house and several cars – a concession against competitors with the local government. Some old soldiers as guides. Arrangements with airways & BOA [British Overseas Airways]. What a draw P2 would be! He said, "Very well, we'll start in partnership. Of course the women will want to go to Cairo now and then." He thinks of Estelle and his Jessie – daren't disillusion him. Now we're flying over the Wadi Natrun with its hideous bloodstained lakes and monasteries. Plane a bit wobbly, making it hard to write. Reg went off early via Khartoum. The lad didn't look well – pale and fatigued, but he has great dignity.

11.30pm Arrived back Heliopolis 7.10pm. Straight to GHQ and work.

19 June 1942 **Cairo** *7.15am* My head is running wild with ideas for reshaping this army for victory. This time last year P2 had just been warned that he was to come here and I'd just gone back from GHQ to Haifa. Wonder whether I'd have achieved anything had they left Ritchie in England and made me DCGS last July? Probably yes, because I saw the answers then and they're the same now, only we've never dared to tackle them. Now the army is back practically where it was when P2 began. Tobruk is again beleaguered and all is to do again: wrong leadership, wrong organisation, wrong technique.

But how in Heaven can we have anything else under our social/political system? Our military leaders are recruited from the less capable members of a limited class. We soldier for an occupation – an occupation in a small way, and the results are patent. But it goes much further. It's hard to tell practical men to do things in a large way in their imagination – to think big and realistically about imaginary armies – when the army they're given is a scattered dust of minor units dispersed about a commercial institution called the British Empire. No wonder the system produces nonentities. The only thing to do is shoot or fish or hunt or play polo – make love and die like a gentleman when Nemesis approaches. The business now is how to rectify this in the hot heart of a war in which the enemy still has the initiative. I start again today with the theories I held this time last year and we'll see what can be done.

2.30pm My preoccupations now are how to sort out after this setback and get going again. Things must not rest as they are. We'll find the answers, and in certain important respects we're better placed than last year. Then we had very few troops; now we still have a large army. Then we had no prepared defences; now we have. There's no question of the enemy constituting a serious threat to Egypt; last year, yes. Tonight unless there's a flap will dine with Jumbo Wilson at the Mohamed [*sic*] Ali.[55] Wonder why the old elephant asked me?

Later The dinner was pleasant. A Major Whittaker of the Rifle Bde described his experiences west of Tobruk which thoroughly confirm my views. We've got to reshuffle this army and its weapons before we find the right tactical combination. If only people weren't so slow-witted and inert. The English don't seem to function beyond a certain pace, and all that elaborate consultative mechanism is designed to stop events from going too quickly. Guts, endurance and imperturbability are there – gear them up to a faster tempo. I honestly believe that half the army are delighted to be settled down in fortifications again. They understand that. But we've got to get away from fortifications – be able to dominate the desert for 500 miles.

20 June 1942 GQH, Cairo *10pm* P2 took me to lunch at the Gezira Club,[56] then straight back to work. The terrace was crowded as we ate club sandwiches and drank American tea (iced, never had it before). People were in bathing kit, and there were one or two streamlined women who gave the animal side of me an impersonal twinge of desire. P2 didn't want to be alone and wished to be consoled by something constructive for the future. Much the same conversation as 11 months ago. Then he was still full of illusions but today he's tasted success and failure so is a different person. Now there's nothing he can do except 'stoop and build it up with worn-out tools', one of which am I.

Greatly doubt he'll be allowed to survive long enough to stage a comeback. He'll be made a scapegoat for the failure of a system, and though he can't escape a measure of responsibility for the choice of subordinates, his removal would be disastrous. He has so many of the qualities of a really great man and the cause will suffer, for there's no one up to his calibre in England.

21 June 1942 GQH, Cairo *7.15am* All this hard fighting won't free the enemy from the immense impasse of geographical factors. We've only to keep going on to win in the end. We should have scored a victory in Libya, and we've failed. We must try again. Oh but it's hot. How can one think clearly in such a climate? Only one must, and believe my brain is working well.

At 4pm meet the press to face a barrage of criticism, to much of which I subscribe myself. Yesterday the enemy stormed Tobruk and we couldn't keep him out. I strongly advised against attempting to hold the place, fearing this might happen. Tobruk was never seriously attacked last year. Had it been, it would have gone. It's gone now, and our enemies will crow heartily. They've cleared Cyrenaica, but that doesn't mean they can take on Egypt just yet. Am sure of that. But it's a sad ending to high hopes, and I'd very high hopes indeed two weeks ago when we checked and turned back his first rush. There'll be a terrific outcry in South Africa about their losses. Poor Kloppie[57] – fancy being stuck with a situation like Tobruk just after being promoted major general. Smuts is in for a rough passage. Hasn't been a merry morning at work, and we've been too busy to feel for the good chaps who are suffering for all this.

10.15pm A pretty lurid 60 minutes with the press. They were kind, but they're out for blood. Hope I've put them on the right lines – Reuter's ticker had some of my slogans, such as 'Libyan see-saw'. We'd a frank discussion, during which I used the word 'bloody' before realising there was an American woman journalist present. Apologised to her, which evoked much comment as an amusing old world refinement – she'd lived with most of them one way or another (so gossip goes) and they all habitually said 'bloody'! Off at 7am tomorrow to see Eighth Army and hope (DV) to be back tomorrow night. This reverse doesn't surprise me as we suffer from an *'Embarrass de Ritchies'*. If P2 doesn't get onto this fact, emphatically he'll be out.

22 June 1942 Continental Hotel, Cairo *9.45pm* Got back from the WD tonight at 6.30pm. P2 in the rear gunner's seat and I the bomb aimer's in the glass-covered nose. The heat inside that tiny space – I know what frying in one's own juice feels like. Things were in hand. If there's no brilliance, neither is there panic. HQ were pleasantly placed near the sea and P2 and I snatched 15 delicious minutes for a bathe; hadn't swum for 6 months – glorious, deep aquamarine.

It's hot in my small hotel room. If I could get time off to sort out my kit I'd swop it for the CinC's camp at Mena, where nights are cool. When this desert pendulum comes to rest at a strategical equilibrium, which I judge to be when the Axis get back their old positions at Sollum and Capuzzo, I'll tidy up.

Free of the need to invest Tobruk, the enemy will keep stronger forces forward, and with its gradual development as port and advanced base will be able to wind up for an advance into Egypt. That will take time. This campaign wouldn't have been possible for him had it not been for the reduction of air to naval power, due to the intervention of Japan. The first time we had to withdraw from Cyrenaica was because most of the ME force were sent to Greece. Can't treat the Middle East as something in isolation. It's a draughty corridor down which blow hot and cold winds from all the 'fronts' of the world. Grilling in the glass nose of the Boston Bomber I wondered what the gallant befeathered major-generals on prancing chargers of old would say of their modern equivalent, sweating like a basted quail in a glass casserole.

On 23 June Ultra decrypts[58] showed that Rommel was poised to drive deep into Egypt. His troops had recaptured Tobruk two days earlier following Allied defeat at Gazala, and were sweeping towards Eighth Army HQ at Sidi Barrani. There was an acute danger of additional German troops on the Eastern front capturing the vital oilfields in the Caucasus by September. On 24 June Ritchie made his stand at Mersa Matruh, an area Chink knew well from his prewar command of 1st Battalion Royal Northumberland Fusiliers in Egypt. In those days he and O'Connor had agreed that only a token force should ever be positioned there, with strength concentrated on higher ground inland. Afrika Korps were approaching fast, sure that the battle ahead was as good as won.

'The historic moment,' exulted Mussolini in a telegram to Hitler, 'for the conquest of Egypt has come.' Egyptian stamps featuring them both in joint profile had been printed, and Rommel's appointment as Military Governor of Cairo negotiated in full. 'The fall of Tobruk and the collapse of the Eighth Army', as Rommel would subsequently admit, 'was the one moment in the African war when the road to Alexandria lay open and virtually undefended.'

24 June 1942 Continental Hotel, Cairo Slept till 4am when an orderly officer woke me with a message. Called at 5am. Events have proved me wrong. A higher command with any ability should have been able to reap a nice harvest, but little by little the enemy was too professional for us and restored the balance. The result has been defeat – but not decisive, and a tidy force which only requires sorting out stands ready to defend Egypt should he follow up. Tonight he'll be crowing at reoccupying his old defences on the frontier. But in fact he failed to trap us there, and Eighth Army remains afoot. Political repercussions difficult. Spent today trying to refute the major lies in the UN press. Tobruk, a ghastly dustbin, had become a symbol, and its loss has the effect of that – like a maidenhead in old novels. But it doesn't mean the loss of everything – we don't live in Victorian novels. I look on the bright side – one must! – and try to look at things objectively as a military scientist. I maintain that my optimism's justifiable.

11.30pm. Just been told guardedly on the 'open' phone that I must go to the WD tomorrow and be prepared to stay for several days. What that portends, don't know. Worse situations have been retrieved by bold and courageous action. We'll see what can be done, but it won't be very orthodox! Depends how quickly the enemy comes on. Must pack, pay bills etc in one hot helter-skelter. What a life. I've no idea as to my mission – it's bound to be exciting.

The First Battle of El Alamein:
June–July 1942

Early the next morning, bringing Chink as his principal staff officer, Auchinleck sacked Ritchie and took personal command of the Eighth Army. Far away, in London's House of Commons, pessimism was deepening. 'We are at this moment', Churchill warned, 'in the presence of a recession unequalled since the fall of France.' At Eighth Army HQ, conversely, Chink's confidence was growing. 'From the moment we suppressed Ritchie," he would tell Liddell Hart within the year, 'I had the strangest feeling of certainty about all I did or advised. The certainty that I could see what the enemy was about to do and how one could damage him. It seemed as if all I'd read or thought about war came to my aid. I didn't feel I could go wrong.' The last ounce of optimism, as Wavell had taught at Blackdown in the 1930s, was more useful than many men. He identified four main reasons for the previous run of disasters: Ritchie's poor judgement, the dispersal of artillery, the 'separatism' of the cavalry, and the tradition of pinning troops down in 'boxes', which Rommel's army could bypass. He gave no thought to the fact that his presence at Auchinleck's side was breaking the traditional chain of command because Jock Whiteley[1] was still in situ as the Eighth Army's principal staff officer. Using Chink for GHQ ME staff work and Whiteley for Eighth Army business, it began to be whispered behind his back, was wrong. It crossed the wires.

26 June 1942 Western Desert *4pm* Writing in a sleeping lorry-caravan close by a delicious cove where I swam with Dick O'C. Then the battle had yet to happen. Now that's history. We've only just avoided the fate we dealt Graziani[2] and stand again on the old ground, only not to meet Italians as in December 1940, but efficient and success-inflated Germans. Our job now is to remember that as late as March 1918 we faced a similar disaster in France on a larger scale, and weathered the storm. By August we were able to break the Hun into small pieces, and that act we'll repeat now. We just need time to sort things out. Our enemy knows the value of time as well as we do. He'll give us damn little. We must make the most of split seconds.

His immediate advice to Auchinleck was succinct. Control of the artillery had to be centralised. Isolated boxes were a handicap. A co-ordinated retreat to El Alamein was necessary to buy time and the chance to redeploy. Armour was weak, so the 60 Grant tanks should be regrouped together, weeding out lighter vehicles for a mobile Light Armoured Division. The Eighth Army had to be pruned down so it was flexible, with composite battle groups formed on the German model. Rommel's weak spot was his use of Italian divisions, so they must attack them hard because the Afrika Korps would have to be moved to support them, and that would dilute the enemy's striking power.

28 June 1942 El Alamein *12 noon* No sleep on Friday night. Saturday and Sunday were busy, but in spite of loud noises off last night slept like a log so fresh again. Can't write about the battle but there's no doubt about the ultimate result. We've got a reasonable grip on the situation, and though we may have to give ground for a little, it's the power to fight the last round which counts. We'll soon have the power to fight and <u>win</u>. P2 took hold manfully and we've straightened out our plans. I remain an optimist. Have a tiny 'living' caravan which was nearly wrecked yesterday when the Boche did a low-flying moonlight shoot-up of the road it was moving on. It's something to have a place to sleep and keep kit without constantly packing.

Midnight A long and busy day. We haven't begun with the Wop Hun yet – we will. If we weren't taking hold in a near disaster I'd be enjoying myself enormously. As it is, it's my business to help P2 through a difficult time.

29 June 1942 El Alamein *5.30pm* A moment of quiet. Still sorting ourselves out prior to giving battle again under more advantageous conditions, which will cramp the freedom of his armour and give our better artillery a chance. He's got Matruh,[3] but that means nothing provided he doesn't get Eighth Army inside it, and he hasn't. Eighth Army is free to fight again. By and large we've given him two of the nastiest days he's had since his offensive began a month ago and we're cooking up another rod in pickle.[4]

Am already looking like a scoundrel – sun-blistered and dusty. A bit sudden to be dropped into this imitation of a debacle, but think I'm competing with the intricate task of being a personal adviser to P2 and a principal staff officer at his HQ, which is also a functioning field HQ. We'll make out somehow, and already the party seems to be sorting out fairly neatly. The enemy is flat out for victory, and if he doesn't get a decisive one he's going to experience a tremendous reaction, as in 1918 when his magnificent effort in March led on to the disasters of August.

Am told that in Cairo life goes on normally enough – so much the better. The 'wind-up' would impede us considerably. Gather there's a terrific political row going on in England about Tobruk, with a demand for scalps. If that goes on, the PM may have to sacrifice P2 and that will be the end of my soldiering too. As for the effort to patch up a deteriorating campaign, I'd sooner be here than anywhere else in the world.

30 June 1942 El Alamein *5.30pm* In a lorry on a most unpleasant patch of desert in a dust storm. Bangs some miles away indicate that our enemy, the tireless and ambitious Field Marshal, is at it again. Blast him. He's determined to get to Alex and driving his people as troops have never been driven before. Hope they break under his impulsion, but they seem hard efficient blighters. Been watching our own people at war and don't believe many of us have the brains or body for the business. We're too slow and idle. Cromwell found the same thing and had to remake an army before he got a proper instrument for war. I've always known the sort of army I want, and if we get away with this party I'll damn well see I get it. Still think, given luck, we could have won the battle of Gazala and avoided the fall of Tobruk. Cannon are still at it towards the sea. Expect we're in for a busy two days.

Midnight Waiting for the next set of dramatic events to develop. The enemy are faced with having to go on. They've come the devil of a long way in a very short time, and must either finish off what they've begun or perish. We stand between them and the Delta, and it's a question now as to who can stick it out in the battle which may begin within a few hours. I've seen all my prophesies fulfilled in the last few days and don't like them coming true when they need not. Aircraft are about. Ours, I think, touching up the Hun – so much the better.

A nice night, not too cold. P2 has been sleeping soundly for three hours now, and he'll need his rest, poor man. This is a damn critical business and the highest commander's responsibilities are endless. Will be glad to get him away from direct command of this party. It's been announced that Ritchie's been superseded. Remains to find a successor, but not too easy filling gaps in high command. Having seen at close hand all the grief the great have to bear, particularly under our system, I've no envy of greatness. It's an empty business, except when called to be disinterestedly and unambitiously great in the course of duty. Pretty platitudinous but also pretty true. If we're lucky, it mayn't yet be too late, only how I wish for the Commonwealth's sake that my ideas were given a chance earlier. The morning will be critical.

From the start Chink set out to relieve the lonely strain for Auchinleck, careless of how it might appear to other military eyes. During the day they were rarely apart, and at night they slept beside one another on the sand in a lean-to tent attached to one side of Auchinleck's caravan. He understood why Auchinleck reached for his hand in the dark and sometimes fell asleep without letting go.

1 July 1942 El Alamein *7.30am* The battle seems to be on and the enemy is again attacking away to the NW. But at least we had yesterday to sort ourselves out after Matruh, and we're in fair fettle – certainly as well off as the enemy. What a thoroughly muddled affair a modern battle is, unless you've the key in your head in terms of space and time – space is everything. The desert's so vast it's possible for large forces to weave in and out of each other, rather like battleships at sea, and that's been happening. Control is immensely difficult, for wireless is still untrustworthy and entirely unsecret. Speak in clear, and you give the enemy your plans. Speak in code, and you slow everything down. Use cipher, and it takes hours encoding and decoding. You pay your penny and choose your inconvenience. Distances are immense. Our relatively tiny army is fighting in an area 50 by 50 miles, 2500 square miles of country, and in that vast space small mechanised forces churn round each other bewilderingly – unless one holds that key to their movement.

Hell but I'm tired. Didn't sleep last night being too busy, and the night before was up at 3.30am. Sometime today must get some sleep – half asleep now. Have two days' beard as missed my chance to wash yesterday and doubt I'll wash today. The Auk is in great fettle. This is a decisive fight for him and he revels in it. He's far happier in the field than in an office; so are we all, I fancy. There's a big formation of Bombers overhead – can't see through the roof so hope they're ours. Gunning is getting quieter. Sounds as if the attack's slowed up.

9.45am The day seems to have been going on forever already, and it's not 10 am. Am doing a Brigade Major to P2 with a great deal less staff than a BM would have. He takes a big chance when he tries to handle a decisive show through one staff officer – me. Odd that this should have happened twice – I was the only officer with little Dick in 1940 at Beda Fomm.[5] Yesterday when I went to the mess lorry for a glass of water the waiter filled a tumbler from a can with a flourish. I gulped it down with a great swig – it was neat paraffin! Burned my mouth and throat and cleaned me out all night, so full of inner cleanliness today. Not good in the middle of a battle, though, and belched paraffin all afternoon. Gunning still goes on in the distance and we're doing our best to meet and defeat any contingency. Am nodding – must snatch a nap sitting up.

7pm Phase 1 of the El Alamin [*sic*] battle has been on since early this morning in a particularly unpleasant dust storm. These desert battles are rather like the old Waterloo battles fought at 30 times the range. You can never tell when a tank will turn up in your HQs, yet we're over 20 miles from the front. But that's just an hour, nothing – say 4 miles or even 3. Like Wellington, we seem to be holding them well and if we lick them here they're lost. It's going to be a close run thing. Am in good company. One of the Corps BGS is damn good – sensible, tireless, with bags of initiative. Freddie has just gone after a useful visit. Would like to see the troops and how they fight but can't get away. Am pretty horrid to look at – not shaved or slept for 48 hours. Just pathetic! Everyone else finds time to spruce up, and to have no servant is the devil. Will be up most of the night. We hear they're still playing polo in Cairo but the tide of war is seeping close so there must be much anxiety below the surface. If everyone does his duty here during this next 48 hours we'll lick this blighter properly. P2 is proving to be a fine commander in the field. Full of decision and grip and determination. Wish I were as good a staff officer for him.

Military Historian John Lee: Rommel's first attack at El Alamein, known to 8th Army from Ultra decrypts, ran into perfectly co-ordinated artillery defences that pummelled his troops to a standstill. It was reported by Germany's 90th Light Division that a real panic set in, causing a flight to the rear - a first for the Afrika Korps. Over the next two days well-positioned Allied anti-tank guns of much greater quality than heretofore were able to inflict crushing losses on enemy armour, leaving them vulnerable to sharp counter-attacks by a newly-confident 8th Army. By 3rd July 1942 Rommel would admit to himself that he had lost this battle and the drive for the Nile was over.

Thursday 2 July 1942 El Alamein *7am* Got some sleep last night, dozing in a chair in spite of interruptions, and have shaved and rubbed off sand so feel a lot livelier. This is going to be a heavy fate-filled day, and by tonight we'll know the answer to a great many questions about the future of this local war. A victory for us today would solve so many problems. Am equally prepared for every sort of ending, for this is a sort of Crécy.[6]

As it happened, Rommel was changing his plans that night, aware that his offensive would be able to continue for one day, at most. 'General Auchinleck," he would write later, 'was handling his forces with very considerable skill, and tactically better than Ritchie had done. He seemed to view the situation with decided coolness, for he was not allowing himself to be rushed into accepting a second-best solution by any moves we made.'

4 July 1942 El Alamein *7am* Such a beautiful dawn this morning – quiet, with a low whispering desert wind, and the eastern sky one great streak of gold and crimson. Above in the clear darkness was the morning star and its little follower. Just the desert horizon, that incredible band of colours – the sky deepening from light blue to sepia at the zenith, and two glorious day stars. The simplicity of design was breathless, like one clear high note of music.

This is the fourth day of the battle of El Alamein, and we've at least gained the indispensable time to put Egypt and the Delta into a state of defence. More than that, we've administered a severe check to the enemy at the moment when I feel certain he thought us down and out. We're very far from being down and out – in fact we're full of fight and may yet stage a decisive comeback. We're fighting to that end. Bet the 'flap' in Cairo is unparalleled, certainly among the foreign element. The 'gippy'[7] is apparently calm and the government doing well. If I'm right, the enemy is rather fed up and far from home. The first flush of victory gives place to a despondent feeling that everything will probably have to be begun again in disadvantageous circumstances. He's fighting a different kind of Eighth Army now – one with restored morale, thinned of non-essentials, alert and coordinated. A little sleep now and then, not a lot – just isn't time.

Too busy to write yesterday. What a reversal of fortune to find myself turned from being somebody about to leave the army to being Chief of Staff to P2 in this battle. The tide is surely turning. A visitor from India was at our HQ yesterday when a less agreeable visitor from Germany touched us up, so some pretty lurid stuff will go back to Delhi about the dangerous life P2 and I lead. It's a relief that Winston got his majority in the vote of censure,[8] for it means P2 will be left to conduct this campaign himself and that's what we need. A new man at this stage would be fatal. He's doing it admirably, too. Don't know what happened to Ritchie. Gone home, I suppose – reversal of fate for us all. This desert air is wonderful – not hot, strong sun, dry wind. Am brown and sunburnt, every wrinkle etched by dust – grizzled but fit.

2pm Some forward troops have been hard at it, but today has so far been peaceful for me. Will try to get a little sleep this afternoon.

3.30pm Sleep be damned! Just possibly we're beginning a dramatic moment, for the report is in that Germans are beginning to surrender on our main front. The rot may spread, and if it does we have the opportunity we've been fighting for.

8.30pm We've got a number of tired and demoralised Boche, but though we can claim a second good day the battle of El Alamin [*sic*] isn't over yet. A certain amount of excitement from the air this afternoon, quickly over. Not too close, but enough to put us into our slit trenches. Have a whole stack of crowns and stars in a chaste arrangement now, but find great difficulty in speaking of myself as General Dorman-Smith. Still live in the clothes I wore here 5 days ago and haven't dared to take off since, except to wash. All my kit is lying about my room at the Continental, as I just locked the door and came away – hope it hasn't been looted. Sleepy, but can't go to bed yet as prepared to bet P2 is hatching something which I'll have to translate into an order about midnight. Should get some sleep though, unless the enemy attacks or runs away because in either case we'll be busy.

5 July 1942 El Alamein *4am* Writing in a lorry office, having for once done enough sleeping on hard sand. Just seen reports for the night and find them extremely hopeful. NZ troops have done a magnificent night attack and killed many Italians. Poor Wops. I bet they're hating their visit to Egypt like the devil. Up to now the Boche has protected them, much as a spear head protects the haft. They haven't really been used, except to contain and extend us. They've never joined battle, and usually been put beyond our minefields. Now it's different. This is an open battle on a front of some 20 miles, and the Germans are fully deployed and fully *employed*. They have to look to the Italians to protect their flank to rear, and so the Wops are left to fight us in the open. We've already dealt with one lot roughly, taking 24 guns, and last night got into another crowd. If we go on like this they'll crack, and leave the Germans in a pretty hole. The battle isn't over yet, but the day before yesterday the Germans broadcast that they'd be sleeping with the Ladies of Alexandria that night, and they're still here! We're a good deal stronger than we were four days ago, when we decided to give battle at El Alamein. More guns, more tanks, more troops have arrived. The enemy will have to make a supreme effort to reverse the situation and regain the initiative. It's questionable whether his tired army has the means or willpower. Things look much better than even I imagined when we started out on this adventure eleven days ago. It looked then as if we might be bundled out of Egypt altogether, unless some miracle stopped the rot. In war miracles come from the hearts and brains of men, through their will to victory.

I think the army was stunned at its defeats – sore and puzzled. Couldn't make out what was wrong or why it was being licked. We've changed that atmosphere, and I go so far as to say that had the battle in front of Tobruk been fought on the same lines, Rommel would have been signally defeated after his first rush. Unfortunately Ritchie had no ideas as to how a defensive battle should be fought, and never really concentrated all the powerful means at his disposal. Result: piecemeal defeats, local but cumulative, leading to an enforced withdrawal from Gazala and the isolation of Tobruk with that large body of troops inside. It's a sad story and an unnecessary one. What we can do now in the teeth of disaster could far more easily and effectively have been done then. Mine was the optimism of a professional who saw no medical reason why the patient shouldn't survive the operation, provided the surgeon had a minimum of skill. Really did think the surgeon and his chief assistant had sufficient between them to deal with a situation which initially had so many advantages. Unfortunately both proved to be slow-minded and stereotyped. Nice chaps, but …. These last four days have shown that the technique used here could equally have been employed in the Tobruk fighting.

Today should be most interesting. Rommel will try desperately to sort out and get another attack going. He dare not retreat, for if he once begins there's nowhere to stop. Doubt he could get his Italians back to the Frontier. So unless he can attack again and drive us off this ground, he's foiled. Even if he does, we've had time to reorganise Egypt and the Delta, and that's a big problem for an army as tired as his to tackle, after fighting its way across 400 miles of waterless desert. We're giving him hell from the air. Hour by hour great fleets of Bostons – great for the Middle East, 20 in the air at once – sail across our heads and unload on his. He's been exposed for four days to a heavy concentration of bomb power and shell power, with nothing like the same power of reply. Doubt his troops can stand the strain much longer. Ours too must be tired, but they've got water, rations and petrol, and they've stopped going back. All heartening. It's the sum of things that count, not items in isolation. The balance sheet of war is never closed until fighting stops. At any moment an item may appear on the credit or debit side which alters the whole account. Tobruk was a disaster I strove hard to avert and thought I had. In fact, not clear yet how Ritchie let himself get his troops bottled, or whether P2 agreed before the event. But that's past history. The problem now is how to retrieve the mistakes we've made, for which I don't feel guilty having warned both against the acts and actors. Expect a busy day – probably some bombing. He must know we've a headquarters here. He had a go at us yesterday evening; quite close one bomb came.

7 July 1942 El Alamein *4pm* The bitter aimless desert wind blows time along as aimlessly as itself. Skipped a day as things were too hectic yesterday for anything except eating and sleeping – very little sleeping, at that. Work just beginning again.

On 7 July the arrival of additional Eighth Army staff took over some of Chink's more time-consuming chores, enabling him to concentrate on a design, based on Worst

Possible Case thinking, which he was planning in conjunction with the Eighth Army Chief Engineer, Brigadier Kisch.[9] The defence plan stretched a tactically deft safety net in front of the cities of Cairo and Alexandria, so that if Rommel bypassed one trap he would be angled into the other. Three sectors five miles apart converged as bait, and minefields radiated like a spider's web from each and stretched to British fortifications on the Alam Halfa ridge.

'Some genius,' wrote the BBC war correspondent Denis Johnstone when he came across it, 'had planned our principal minefields lengthways along the slope of the Ridge, rather like a ship with its bows facing the incoming waves. This left the Barrel Track to Cairo wide open. If Rommel had driven straight ahead he would have been in the clear but without communications, for the Eighth Army would still be dominating the bottleneck between the Ridge and the Quattara depression. A nice weighing of risks and possibilities. Here for the first time we had the spectacle of an ace German general not only being outfought, but outfoxed.'

8 July 1942 *10 pm* We're still barring the road to the Delta at El Alamein, both sides feverishly preparing for the next move. On the face of it all goes well, certainly a lot better than we'd any right to expect. Now it's up to us to profit by our respite and reap some reward for our hard fighting. Am having a busman's holiday tonight as the bulk of regular performers on the staff of this HQ have taken hold, so am released from attending to detail for the larger purpose of thinking ahead and helping P2 on his bigger work. So much the better. After being a prisoner in the range of a little group of vehicles for two weeks, can get out and about the battlefield at last. We're abreast of events as regards plans, and nothing the enemy does can come as much of a surprise or find us unprepared with a riposte. These are solid gains.

P2 is delightful. He really seems to enjoy his return to active campaigning, and is one of those rare birds, a real general officer with a mind and will of his own which he's determined to impose on the enemy. It's meant having to be pretty ruthless with our own people – changes at the top – but this is no time for gentleness. We'll all be the better for it. Wonder how long he'll be able to free himself from the calls of higher policy in Cairo. Suppose I'll stay with him until then, or until I get sacked. At any rate, this taste of campaigning is what I needed. No one can say now that I've not had war experience. I'd like more of it, but feel I won't be let stay here if P2 goes back.

9 July 1942 **El Alamein** *6.45am* Just done a most luxurious thing – sponged all over in a tumblerful of water. Being relieved of direct responsibility for detail doesn't mean idleness, but am less tied to a telephone and free of having to write orders, messages etc. Last night was quiet from our point of view. Not sure the enemy found it so quiet. We're raiding him nightly and he doesn't like it. His big idea is to get a good night's sleep and he hates our deep searching patrols which on this enormous front may turn up anywhere and are difficult to keep away. Besides, we bomb him heavily all night. I think the 'new management' puzzles him.

3.30pm P2 keeps me as a sort of 3rd gear brain into which he changes when the situation's too uphill for his own gears. He's a commander and a fighter, but definitely *not* a deep thinker, although he has the power to grasp a new idea and the energy to put it into practice. He's a very good soldier and a good manager of men, though less good as a judge of men. He's ruthless now and full of offensive spirit and energy. He expresses himself well in writing and orally, and is not afraid of anybody. In some ways I'm more forthright and uncompromising, and I'm a quicker thinker, with perhaps a deeper grasp of basic principles. He relies considerably on my judgement, which so far has not, I think, been false. We go over all possible developments of the situation together, and then I interpret his views into policy memoranda on which the staff proper prepare orders. In that way we manage to keep well abreast of events or, indeed, ahead of them. Naturally, we both watch our enemy like hawks, and we study his mind and his idea of our minds. We lead pretty laborious days because P2 won't let up on things – quite properly. That habit of accepting the enemy's acts as the acts of God has done us much harm, so we intend to the best of our capacity to mould his will to ours.

Gradually my mind is being engulfed by the sand I sleep so soundly on. The days for work, the nights for sleep, and deaths for everybody, anyhow – only one doesn't worry about that side of it until it happens, and – except as a problem to be defeated – one doesn't worry too much about the enemy. He's there to be defeated and fought at till he *is* defeated. The check here is significant. If the momentum stops it's not so easily started again, and time is short for victory. We've got to attack to win, and go on attacking everywhere. We aren't faced with the dreadful alternatives of the Germans. They've no reserves, or the great reserve of United Nations, and the USA is coming in quick and strong. Glad I haven't got to advise my nation as the German General Staff must soon. They're up against the Four S's: Sea, Sand, Steppes and Snow, and there's no way around any of them. We'll suffer yet, necessarily and unnecessarily, but I'm certain of the outcome. Am going to write a revolutionary paper on ultra-modern tactics.

9.30pm We're making this HQ keep sun hours – sup al fresco at 7.30pm and go to bed at 9.30pm or 10pm. Much better for everyone – rise at 6am, breakfast 7.30am and the rest of the day for work. That's unless there's terrific pressure, or things are going very wrong. Wasn't as easy when we first arrived. Tomorrow should be exciting. We've had a day of playing tig round the southern flank and I hope the enemy thinks he's done wonders – they're vain chaps, these Boches. *We* are unconcerned. The old brain is beginning to work clearly and lucidly. It's the cooler climate – last night needed a second blanket. If only we could polish off this blighter here and now, only it's not that easy and one looks around for the troops lost at Tobruk. The papers at home continue to fuss about that. They've been so misled about its defensive strength, but it was never a strong place and just not attacked last year. Ritchie was wrong to leave a single man in it, quite wrong. They're belly-aching about Matruh, too. Thank Heavens we never intended that stupidity.

10.30pm Was just off to bed when the war intervened. Now trying to sort out a project concerning men's lives – not a little tricky. So it goes on. Sitting at this phone tying up RAF and Corps and all manner of loose strings. That's something else to be got right. Well, that's that. Rather untidy but we know where we are.

Military Historian John Lee: Having defeated Rommel's offensive, it was still Auchinleck's intention to destroy the enemy army and this required 8th Army to pass over to the offensive. He was under constant pressure from Winston Churchill in London to get on with this task. Chink, using Ultra decrypts, singled out the Italian divisions holding the line opposite and subjected them to a series of heavy attacks that pushed them back, taking thousands of prisoners, and forced Rommel to deploy his German units to shore up the position. This went on until late July with somewhat diminishing returns for 8th Army. Ultra decrypts sometimes gave only a transient opportunity to attack with advantage and the exhausted staff officers and troops of 8th Army found themselves checked by a still-formidable enemy fighting from increasingly strong defensive positions. By 30 July, Auchinleck would stand the battle down and began the long preparation to resume the offensive at a later date.

10 July 1942 El Alamein Another day gained, and we're worrying our enemy to stop him getting over complacent. Morning news pleasing, but it's only the morning, and it's a long day.

11 July 1942 El Alamein *10.30pm* Still on the patch of sand we came to nearly a week ago, and on the whole things much improved, both personally because I've gradually got together the irreducible minimum of gear for desert life, and in Eighth Army, because we've dealt our adversary a number of shrewd and telling blows where they're apt to be painful. In fact he's having none too good a time, and not gathering strength as quickly as he needs if he's to go on East soon. Situation much rosier than I expected seven days ago. Continue being a sort of deputy commander plus staff officer-in-chief to the CinC, and for want of a better title call myself DCGS-in-the-Field. Reminds me of an old-world London church! My duties I devise for myself in altercation with P2, for discussion is frank and forthright and if I have *carte blanche* to speak my mind, he doesn't spare my feelings if he disagrees. But between us we roll out the plans and policies, and on the whole they've worked. There've been anxious moments – one forgets them quickly – and wonder if I'll ever acquire a true perspective of events of the last shattering fortnight. It's 17 days since we rode our winged horse to the battlefield. To bed – it's late and P2's already asleep in the lean-to tent we share beside our office lorry. Must try not to wake him as I go in.

12 July 1942 El Alamein *11.30am* A good night except for a bit of work entailing a decision by P2 about 2am – bad hour for decisions. Back to sleep to the noise of bombing by our planes a long way off on the German lines. Quite satisfactory results from the little attack yesterday. Enemy in not too good a position and were it not for his very good tanks he'd definitely be badly placed. We'll find a way to deal with them.

1.15pm Lunch in the open air, a fresh day and the sun strong enough to make the table red-hot. Everything hot – water etc. A healthy life, if only there was more exercise. That isn't possible until we finish off the enemy in front of us.

4pm Drowsy from the afternoon heat and a disturbed night. Guns going hard. Almost like the old days in France, and since ours do most of the roaring it must be a pretty depressing noise for the enemy.

10pm All quiet tonight and situation favourable. Day by day we strengthen ourselves after that dangerous and difficult retreat, and now we're a different army. Can't believe our enemy's position is improving at the same rate. We'll see in due course. Got a run in the country with P2 this evening in one of those silly-looking little American four-wheel-drive 'Jeeps'. They're marvellous cars, go anywhere – soft sand, wadis, rocks. Nothing stops them. They've extra low gears and could push their way through a bog. They look Victorian but are utter simplicity and as strong as the earth. We fly the Union Jack from the diminutive bonnet, and go to see the soldiers and the ground on which we may have to fight.

13 July 1942 El Alamein *7.45am* We're wrapped in a soft sea mist which covers the battlefield. So all is silent for nobody can see to fight on the ground or in the air. The sun will suck it up eventually. Meanwhile everything drips moisture and it might be Sussex. The situation shows no change. We're not complacent. It's now our business to make changes and we'll do so according to our plans, not his. He doesn't seem inclined to do more than rest, reorganise and reinforce, and lies on a very wide front in close contact with us. Tempting – very. We plot and plan, and hope for even 50% results. In war wise people don't expect more, which doesn't mean that at crucial moments one shouldn't bid up to 100% as I believe Ritchie could have done in front of Tobruk. Am more and more convinced of his stupidity as I analyse the course of his campaign, but P2 still has a pathetic loyalty to that imposing façade of a man.

14 July 1942 El Alamein *6.30pm* We're causing our enemy much distress and have definitely stopped triumphant procession No.1. We hope to reverse it, just aren't ready yet. Rapped him soundly on the knuckles, and yesterday he returned the compliment – and got soundly rapped again. So he's rather *piano* today and not so certain as to when he'll get to Alex. Estelle has been staying with the Wavells. Says Delhi is much the same as before the war and laments that I'm not DCGS there with a major general's house, but adds that I'm probably happier where I am and everyone has faith in the combination of P2 and me. My lorry office is stifling hot – blacked out and airless. Tonight we try again at the enemy. Must keep on and not let him settle down. He's cracking at us again now – I hope without success. If so we've scored heavily.

15 July 1942 El Alamein Slept undisturbed, though elsewhere our plotting meant little sleep for a great many tired people on both sides of this unnatural no-man's-land. Woke, P2 and I, to dubious news of results which gradually expanded with the sunrise into something more warming. Not out of the darkness yet. Situation still obscure, but at least something on the credit side. Keep at him, give him no rest and wear him down till we're strong enough to act decisively. Progress is being made, that's all. Can't tell in war with any sureness where the breaking point is and we've got to be penurious – small profits, quick returns. Today is going to be one of crises demanding a cool appreciation of the value of events both to us and to our enemy.

After lunch This is when it gets really hot and flies become active. P2 commands more power than anyone else in the ME and from here we dictate all ME policy, directed towards immediate victory. We've taken a good showing of prisoners, gained important ground, beaten off strong attacks and the Boche is a lot worse off this afternoon than yesterday. We surprised him again, not the first time this week. Must expect a violent reaction this evening, but we're ready for it.

6pm Writing under difficulties in the little map lorry. P2 is in and out and staff officers constantly intrude with reports on the battle, which is active. A distant but distracting stuka attack, and the ever present Egyptian fly. There are moments when one dares to hope. The clouds part. There's a little blue. Then they close again, but each time the blue patch is larger and tonight there's a good showing in the west. A good day for our arms, better than expected this morning. The enemy's evening effort to put things right hasn't begun – we must wait till 7pm or later. The tally of prisoners taken mounts, but P2 refuses to believe any story until it's been verified by counting. Any bag's an agreeable change from our own losses which were heavy before P2 took over, as has been officially announced by the PM.

Hear there's unkind talk in SA about General Klopper who had to surrender in Tobruk. Kloppie I met and liked during my visits south. He was – is – young (38) and inexperienced. If he wasn't in a Dominion army he'd not be a divisional commander, but we have to accept the shortcomings of the military system of our allies. However he was absolutely loyal and I'm certain did his best in an infinitely difficult situation, too much for any but the very best of generals. I did query his suitability purely on experience, but too late. The Germans realised the one thing they must not do was to let Tobruk get organised for a siege, so they struck at once with the utmost violence. Had Ritchie been able to stand at El Adem[10] as we've stood here covering El Alamein things would have been different.

6.30pm Return effort on its way. If we defeat this, more blue will show, but the next hour will be anxious. It always is when the enemy gets going and we have to resist. Still, the point is that he dances to our tune, not we to his, at present anyway. Our chaps have fought magnificently today. Pray they're fit to go on. They must be tiring, but so's our enemy. Tired and disillusioned at the cup of victory being withheld when two weeks ago they announced with fanfares Rommel's entry into Alexandria.

9.30pm No definite news of what happened during the enemy's riposte, but taking everything into account a good day, for which am grateful. Early bed as tomorrow will be important too.

16 July 1942 El Alamein *11pm* It's late and it's been a long anxious day, but all the Boche attacks have been beaten back with loss to him. Last night his dusk attack did well, but today we've equalised if not improved things, and we're well up on the balance of three days fighting. Prisoners adding up nicely. Happier tonight.

17 July 1942 El Alamein *4pm.* Future planning and seeing people all day. Gets hot in these lorry offices and we began at 5.30am. Enemy came on as we expected very skilfully, and had some success but not all he wanted. Since then he's been butting his head against a stone wall in his effort to regain lost ground at a cost he can't afford to continue. We've had at him elsewhere, taking more prisoners than he can afford to lose with equanimity, so gradually we're wearing him down and wresting back the initiative. A prudent man might decide to clear out. Not Rommel. He'll stick it to the last hoping for something to turn up, and I trust get further into the mire. Sat up last night making a [rough] Appreciation[11] as to what our future policy ought to be.

5pm P2 has just gone off in a jeep to see troops working on a backward position, so am free for a little. Have as much liberty as a paid female companion and collect about as much abuse, for whenever I suggest a new thing his first instinct is to force me to justify it by oaths. He usually says No, and then goes off and does it. But I understand the strain of high command, and sympathise. I know he looks upon me as a companion with no personal axe to grind, who won't tell the world who had the bright ideas. He knows I don't give a damn what the world thinks. If we succeed here it's possible that he'll feel like getting rid of me as a reminder of an unpleasant phase of his life. Perhaps that's unworthy. He may equally glue me to his war chariot as bringing him luck. I do bring him luck when he takes my advice, and when he doesn't he frequently bogs things. What will happen after this show is over – provided it goes well – can't see. If we lick Rommel here and the Russkis keep the Boche north of the Caucasus, the ME would get a rest. I might be allowed to go away and put the story of El Alamein on paper, but first I'll have to reorganise and modernise this army. That won't take long to set in train, and then he might let me go.

If we lick Rommel here, the effect will be much bigger than the size of contending armies warrants. It will go round the world as the beginning of the end of German victories, giving fresh heart to all sorts of people. We strive for total victory here now. Our only aim is to attack and destroy because this is one of the world's decisive battles. To some extent it's already won, since we've gained the time we needed to gather new strength, organise positions in rear, and make new plans. But that isn't enough. Total victory here might well crash Mussolini. If Musso crashed, a great many other plaster statues would fall too. It's a big vista, and we're well aware of the importance of what we're doing. The only thing I grouse about is my inability to get round the front and see people. Am pinned by the leg to my maps and plans, and have hardly walked a yard since I left Cairo. Wonder if the enemy will attack again this evening? He's had three tries today, bless him, and been kicked hard in the pants each time, so perhaps he'll chuck it for a little. But Rommel is a persistent beast. The question now is, which of us tires of the game first? I think he hopes we will. He's <u>wrong</u>!

18 July 1942 *3.45pm* Still at this so-bracing seaside resort, El Alamein. Our enemy is further from Cairo than ever and I think we can say the impetus has gone. His offensive, when it begins again (*if* we let him begin again) will be a new attack, not part of the rush we stemmed here. That's over. He must stage a new campaign for the autumn – that is, if we let him. Am waiting for a brigadier who was at my first prep school to come and tell me his damn tanks won't run, or some other friction which makes the waging of war such a difficult business. 'The path of true war never does run smoothly'. Blast it. However, it isn't going any more smoothly for the enemy.

10.15pm Another long day of the futile 'this and that' ness of war. Glad we're far from the hunted days when we thought more of preserving our army than of using it against the foe. I plot our enemy's destruction – maybe wishful thinking but sincerely anxious to see the contaminators of the sacred soil of Egypt destroyed utterly as a warning to all who might follow. If I knew of a proper incantation to the old gods – Osiris, Set – would get going to marshal their assistance.

19 July 1942 El Alamein *11.30am* Early, and it seems as if the day has been going on interminably. Not been out of this little area round our camp for a fortnight. Still, it's worth it. More time to think about what's going to happen next.

2.30pm P2 is curious. Find him watching me as if for my approval, even borrowing my words. He knows I no longer give a damn and he finds that uncomfortable, but he also knows I've no fear of the truth. He's gone off with the Royal Nincompoop[12] who's the worst possible advertisement for his hereditary profession. A blowsy gusty day with a thin layer of sand on the page as I write. Luckily the *khamsins*[13] are over.

10pm All goes as well as it can in the circumstances. We must make it go better still before long. I grudge the minutes before we can finish these people opposite, for finish them we must.

20 July 1942 El Alamein *7.30am* A lovely morning, no wind blowing. The enemy is quiescent, licking his wounds and hoping for a lull to get his strength again. We'll see. The general layout of our dispersed encampment has become almost as 'solid' on its haphazardly chosen patch of sand as a village. Skyline's familiar now. The whole desert's familiar to a Bedouin, charted in his mind. He knows its possibilities and deficiencies and is designed to overcome them. Only to us is it strange, for we all fear the unknown. That's the trouble with new troops. People have for so long told of the mystery of the desert that a fear has grown. The Swiss never went into their glaciers till about 1880, afraid they were the abode of dragons. Proves one must face facts, enquire, unveil. Dare!

21 July 1942 El Alamein Freddie's[14] stretched on a row of four campstools in my lorry office, clamouring for soft women and developing a fixation over the back page girl of the *Egyptian Mail*. Deplorable spectacle! Today's papers show the public have been told that since we started to fight we've taken 6000 prisoners. At least another 6000 casualties since, as they indulged in fruitless attacks and counter attacks that we know were costly. A pretty good rebuff to the Axis who were confident they'd take over Egypt during the first week in July. 21 days ago we were a very dicky army with little or no apparent fight left. Now it's they who are dicky, and they're hanging on by their eyebrows for reinforcements. The worst is over. Remains to be seen which of us can gain enough offensive strength first.

9.50pm A half-moon glazes the sand and our barrage shakes the desert from 12 miles away. We're beginning a big action and the firing is heavy. If this assault succeeds, the wider plot will unfold and try the German-Wop medley opposite. Their morale is putrid, especially now the vision of the juicy brothels of Alex has been replaced by the gleam of Australian bayonets. For a victorious army almost within marching distance of its goal they're singularly ready to throw in the sponge. I felt the Wop would behave badly once we got him in the open, and if Italy packs up the Germans couldn't go on. If only the Yanks could open up a third front against Italy *now*. The bombardment continues – our bombers are taking an intensive hand now and the enemy have come a long way for what they're getting. This will blunt any edge they retained during the last three weeks fighting.

22 July 1942 El Alamein *8am* Curious that one can sleep so soundly when the fate of nations is at stake, except for a midnight report on our dusk assault. One wakes up chilly with less than two blankets and this is Egyptian summer. Some day the world will discover this glorious coast, ripe for development. Musso's vision is correct, only he never sees beyond Italians. The myopia of Nationalism and National Finance! News 85% good at the moment but it's going to be a day of hard fighting.

12 noon Exasperatingly slowly reports come in to Army HQ, but must fight against curiosity. Fever must take its course, and we're not yet at the moment of crisis. Depends whether the enemy can keep his distended front intact or it begins to crack. His Italians must be a sore trial – quite unpredictable. Must be immensely difficult for him to gauge the strength and success of our attacks against such uncertain material. Am attempting to project myself into the mind of his higher command and think we'd not care for it at all.

3.30pm The battle will flare up again soon. It's been swaying to and fro all day, but on the whole the advantage lies with us. It's for us to make the most of it. Poor P2. Wish I could take some of the pain from him. The issue is so long drawn – we'll probably not know till tomorrow evening. Certainly we aren't to be awarded a walkover.

6.30pm Disappointed but not disheartened. We haven't yet played all our trumps. Thud! Thud, double thud! Behind me there's barrage – fire and ceaseless rumble as our guns support an evening attack. For all the enemy has held today he's been well stirred up, and his losses must be heavy because for 24 hours we've had at him with little pause for rest.

Chink's reorganisation proposals for mobile divisions containing a mix of armour trucked infantry and guns were now agreed by Auchinleck. The next step was to seek Brooke's permission, but that would have to be via his like-minded emissary McCreery. McCreery's refusal to Auchinleck to change the traditional separation of cavalry and infantry divisions had just been witnessed by Chink.

23 July 1942 El Alamein *8 am* We hold some useful observation we didn't have before, and shook them again just before dark. Operations continue. On the balance of this month of P2's direct command we've plenty of assets to show. They aren't decisive but have a cumulative effect. Tomorrow I reach the perfectly ghastly age of 47. Freddie moans about being 42 and impotent! Blast him! I'd swop – but no. My 46th year was one of the most vivid of my life – great happiness and distress mingled.

24 July 1942 El Alamein *11 am* Our hopes of yesterday were unfulfilled. We attacked, and it's something to be able to attack heavily with an army which so recently has known defeat and humiliation. The road to victory isn't a speedy one, and I'd not care to be in command. Simply couldn't waste time and energy on some of the people I'd be expected to cajole into cooperation. If we were a disciplined organisation, yes. But we aren't, and I've no flair for persuasion in action – wasn't made of compromise material. Hear I'm likely to be DCGS Ops, Plans and Intelligence:[15] important enough and agreeably beyond the limelight. I'm a retiring violet and like to get my effect from the shade of big leaves. Haven't yet recovered from the bad time professionally in the last 18 months which left its mark on my assurance. Am content as to my influence on events, so long as P2 remains in ME. His good intentions were reduced by the men he found around him and by others he appointed. There his judgement is faulty. It may also be faulty in appointing me, save that I'm not in a position to be much danger. It's the job P1 [Wavell] would have given me too, so that's a counter check.

1.30pm This is the hot time, but though the sun is strong the air feels cool and invigorating. We on this coast are to be envied for our climate, not pitied. Never seen men look healthier than the troops – hard to recognise them as Europeans. Much of their work they do naked, and the newcomer looks painfully pink in comparison – definitely at a disadvantage till tanned and acclimatised. I know P2 hates the idea of going back to GHQ. He may have to do so soon if we aren't strong enough to get a move on our enemy and the thing settles down. Suppose I'd accompany him, though reluctantly.

5.30pm The brave days are over. It's stable now and there's little scope for my bright ideas. I'd be more use in Cairo – leave this business to the heroes and go where in the shortest time I can reorganize and remake an army fit for this sort of war. Gather I'm reckoned to be an 'intellectual' rather than a 'hero'. Delighted, if so – am tired of practical men. Ritchie would be regarded as 100% practical, and what should he have done to himself having thrown away not only half an army but a golden chance of licking the Boche to blazes? Could one go back to England, go into the club, call officially at the War Office, seek new employment? Wouldn't one be dead of shame and grief? Want never to be seen again, never to wear one's name again? I would. When not used again, even through no mistake of mine, I couldn't bear the idea of facing my old world. Suppose Ritchie won't mind.

8.15pm Guns are at it again. Undoubtedly the Boche is up to something. Neither of us have done much today but it'd have to be pretty good before he makes any real advance.

25 July 1942 El Alamein *8 am* Am studying P2. He's undoubtedly a leader, but what are leaders and what sort of people like being led? People who only do the right thing because they're told seem to me infantile. People who know how to make the infantile do the right thing have a nursery maid genius, but the maid knows she's dealing with minds which lack the facts and mental strength. Adults shouldn't lack either. To be leadable seems to me to betray mental weakness. To lead the leadable exploits a weakness which one possesses in some degree oneself. Oh, the high degree of stupidity and slow-mindedness in this war game from top to bottom!

P2 is fundamentally simple-minded, with very nearly an inferiority complex. He's always been badly off and envied and despised those in high places by birth. Auburn-haired, determined, ambitious, he's studied how to deal with men as it's only through men that one rises. He's not unintelligent – a good soldier, with physical courage and a certain moral courage. His technique with people is simple, flattering, insincere, effective. Good-looking himself, he's susceptible to good looks in others. He's gradually realizing that a man like Ritchie, for all the presence and bluster, may be morally and intellectually feeble. I was brought here because of a lack of certainty in himself. He likes me and suspects me at the same time, and will use me to his advantage without giving me the power, position or reward which might make people say I influenced him. He'll keep me where I am as a tame brain. I possess an awkward set of values so am hard to handle, since I can neither be led nor driven, nor bribed by future promises. Dispassionate and critical, I'm an awkward customer. Every man has his price, if one can find out what it is, and he doesn't know mine which makes it difficult for him. He's a little frightened that I may lead him. He's by far the best soldier and biggest man out here, and yet!

So stupid, this war business. Men like war – it gets them away from women they've ceased to love, working at dull things, thinking, earning money. We don't think. No books, no new ideas, no past, no future, plenty of food, stimulating dogfight excitement. A purely animal atavistic existence with a sugaring of sentimental glory and a claque of war correspondents.

26 July 1942 El Alamein *8pm* A rotten day. A mass of work, mainly trying to convince people with vested interests that they must sacrifice those to principle, and it's blown sand continuously. Everything thickly covered in gritty layers – not hot like a *khamsin*, but breathing and eating sand all day.

27 July 1942 El Alamein *8am* We've been at them again since early morning, and the air is full of great thumps and rumbles as our mass of artillery beats up the enemy positions. Began well and the real business is to keep it going – he isn't far from cracking point. Can we apply sufficient pressure to snap the taut band of his resistance? Today will show. Sand has ceased to blow, thank goodness. A sponge-over daily helps and at last clothes can go to Cairo to be washed. Privacy's completely lacking. Freddie comes up today to take over Brigadier General Staff. As I thought, I'll be Deputy CGS controlling Operations, Plans and Intelligence, a new departure and most interesting. The job will suit me, and is all I want to keep me out of mischief. Can ensure from its 'eminence' that things don't go avoidably wrong, though it mayn't be possible to do anything spectacular. The time seems past for that for obvious reasons.

1.15pm Guns still hammering away, but no news of his breaking. Fighting here is becoming reminiscent of the middle years of the first world war. Still, we've made Rommel think differently – a solid gain. Pity we can't get him going back more definitely. I anticipate our marking time for a while and am superfluous as there's quite enough staff to keep things ticking over. P2 would be better back at GHQME. We've turned a broken and retreating army into something that not only stops its enemy but attacks in turn, and that's an achievement and a re-establishment. Now it's time to turn to the wider picture. Must stop now to write an Appreciation,[16] for the time has come to reconsider our policy and plans.

6.30pm Have been writing red-hot for two hours, and can only come to one conclusion which I regret because I'm not a prudent person. On the other hand, a counsellor must be objective and impersonal. Hope I'm both. The difficulty is to give proper value to all the factors, or to know what the factors are. The moon is fat and full, rising to the roll of cannon. Today began well, and a good idea has foundered. Mightn't have done if my first advice had been taken, but can't expect everyone to back one's hunches all the time. Have had a fair run for my money, particularly at first when the show was critical and P2 had no-one else to discuss with. Now he has two Corps commanders – Lt' Generals – and they're obviously experienced people so I must listen to advice being accepted that I wouldn't give myself. Compromise solutions are typical and usually they're the devil. In the last week they've killed a lot of good fellows.

10pm P2 has just scrambled out of the lorry office into the moonlight saying "Bedtime – blast the PM!" We do lead a queer life. Tiny battles and highest policy all muddled up as only Lewis Carroll could imagine it.

The complex Appreciation which he began writing at 14.45 that afternoon would, postwar, become a key document in the controversial claims between 'First' and 'Second' Alamein. (A myth has also grown that it was written by Auchinleck because Chink's signature is often omitted.) Emphasising current Eighth Army casualties, losses, shortages and lack of training, the conclusion was inescapable. Auchinleck should call off the offensive until mid-September, because the Eighth Army would not be ready to face an attack from Rommel, most likely to come from the south, until then. Prior training should exclusively be for a break-through in the north, near Alamein, and a decisive battle at the nearby ridge of Alam Halfa.

Handing the finished document to Auchinleck next morning, a 'sharp altercation' ensued on the grounds that it was insufficiently aggressive. Chink told him to read it again and list the military mistakes. Auchinleck conceded, and initialled the Appreciation. Both men were only too aware that an unexpected Cairo visit from the CIGS,[17] General Sir Alan Brooke, was imminent, but Chink did not yet know that his bête noir Churchill would be accompanying him.

29 July 1942 El Alamein *8am* Want a photo taken inside the lorry from which this campaign is run – the map on the wall's quite interesting. Am getting restless sitting here, but think P2 feels that so long as he stays heroically stemming the invader he'll not be bothered by post-mortems or chickens coming home to roost. That's only putting off the evil day because he hasn't heard the last of it yet. Proper people will soon forget that he stemmed the tide of defeat, whereas Tobruk won't be forgotten. I'd sooner go back to Cairo and get on with the 101 big problems cropping up there. Thank the Lord he's just gone off to the sea in his jeep – I declined the offer of a ride. He's getting very odd. Begin to wonder if he's tiring but think it's only a reaction after the last months of fighting and strain, added to the shock of the earlier collapse. He's inclined to be fatalistic about himself. If the enemy don't get to the Caucasus we'll be able to turn on the heat here good and proper and make them sorry they came so far. That is, if they've not gone back already.

30 July 1942 El Alamein *5pm* We're quickly approaching the great turning point in this war. The Germans have Rostoff and are over the Don. They can go east to Stalingrad or south to the Caucasus, and the situation is serious. Interesting to note the unanimity of the Western press, rubbing in the gravity. Surely if things were so desperate we'd pipe down? Germans seem pretty elated, so now's the time for the third front. We haven't shifted the enemy here yet, nor he us, but his tone is different and his confidence rudely shaken. Everything's changed for the better. After a quiet day, a bombardment to start the sunset going. He often attacks about this time, getting the benefit of the sun behind him and the night to sort out. When the panic was on, the people who Levanted[18] from Cairo and Alex were a pretty good cross-section of the wealthy classes – lucky they had the Lebanon.

31 July 1942 El Alamein *10.30am* Lunch today with a journalist avid for copy. Won't tell him anything – he's too dangerous a gossip. Damaging back-biting which ordinary chaps indulge in normally does little harm, but in high places it's devastating. Human beings are really cannibals. We've given up eating each other physically but feed on each other in every other way. Doesn't matter when people are busy – books, love, clothes, garden, business, cars, good times, bad times, money – but when everything concentrates on military personalities the feeding becomes particularly foul. News today not helpful. We just hadn't the resources to reap the benefit of earlier efforts and for the moment it's stalemate. Such a pity, but am not one to hammer at locked doors. We'll have to defer the wiping up for a little. P2 will forget this humiliation as quickly as he can once it over. It's a bitter business for him and failure is a thing one tries to forget. Feel in my bones that his time is numbered.

1 August 1942 El Alamein *4.15pm*   Great as the enemy's success on this front has been, its effects are already wearing off because it wasn't 100% complete. He can't afford half-measures, that's his predicament. The joke is that at the end of June Mussolini and some very important Germans came to Libya for the drive into Cairo. They've since gone home – annoyed. Bad luck on Musso, he's due for a break. A spate of gloating propaganda – 'We've come to strike off your chains!'– got released too, but the chap who only has not to lose must win, provided he wills it hard enough.

Next week P2's fate will be in the balance. Had we won our battles at the end of July all would have been well, but through this and that hierarchical failure in the bluer-blooded stratas[19] – very thin this blue blood – we didn't pull off what in a more efficient party would have been a cinch. My bright ideas got clouded over by consultation with bigger noises, and robbed of some of their glitter. That wasn't so in the earlier stages when things were going too fast for consultation. We may see some changes. Can't find time to go round the front, have made every kind of plan, helped tidy up a good deal of rubbishy ideas and put up with considerable crustiness from P2. He's not nowadays an ideal companion, though charming and great fun when the crisis was on. Now this moody reaction. Ritchie, we hear, is fishing in Scotland.

2 August 1942 El Alamein P2 will probably take me to Cairo[20] with him and maybe I'll not come back. There's a packet to do at GHQ. It's filthy hot there but am sufficiently nomadic to welcome the change. Inertia, not getting anywhere is death. Worked this afternoon on Rommel's system of command. Brain suddenly becoming active again.

3 August 1942 El Alamein *8am*   P2 and I go in today and probably come out again tomorrow.

4 August 1942 Continental Hotel, Cairo *2.15pm*   Luxurious evening yesterday. Flew in to meet the very-very-great man[21] and other importants. Not me – got away in time to have a hair-cut and wonderful bath. Dined at the hotel cabaret with friends. A girl danced with the loveliest figure and very few clothes – white body and barely veiled breasts. Pharaoh 1[22] arrives this afternoon and tomorrow we go back to the sand with our distinguished visitors.[23] People straight from England look so pink and how they sweat – didn't realise how we've salted down. A conference at 3pm and everyone edgy because of the visitors and P2's long absence from his real job.

Am worried about him. He's desperately irritable; rather unreasonable too and difficult to deal with. Hope they make him get back to his proper role. Too much to be done here and Corbett's[24] equally on edge. Very uncomfortable. Am booked to motor most of tomorrow in the same car as the very top soldier of our Imperial army.[25] A joke, considering he's been patently side-tracking me[26] for so long. Met him this morning. He looks fit and active, though feeling the heat, and seems to be carrying his burden well. But P2 looks like death – he's lost weight, aged, stoops, snaps and growls and is rapidly losing his two great assets: the power of making men want to work for him and his level-headed grasp of essentials. If he allows this to go on it will be serious. Must bathe and change. Another night in civilization.

5 August 1942 *In flight 6am* A cloudy morning and we're flying low across the delta. The Nile is just behind, that was the last canal, and here's the sand. No more real life till the green in Cyrenaica. P2 seems happier. Had a short talk before we took off and think the very great have been sympathetic. Didn't see P1. Hope he'll come up before he goes back but all depends on what the very great decide about P2. Met friends at the Mahomet Ali last night and drank quite a lot without damage – wonderful what living in the open air does. Bed midnight, up at 4.30am. Behind me the sky is gold, in front still night and overhead a large squadron of aircraft – for a moment thought they were enemy.

2.15pm A tremendous do. The old man has been round El A. making V signs with his fingers and giving away cigars to Aussies. They had a late breakfast about 11am and have now gone back. P2 in a much better humour, so suppose all is well. Can't help feeling that my time in the sand will end fairly soon. The great WC looked well but he's an old man. Brooke too has aged – now shrunken and potbellied but quick, alert and intelligent. Must be the hell of a job and don't envy him a bit. Apparently he's prepared to give us a free hand as to reorganisation[27] from remarks dropped by P2.

6 August 1942 El Alamein *7.15am* The month creeps on and it's not to the advantage of either party that time should pass in idleness. Too much at stake and decisions are needed. One certain rule in war: directly one feels strong enough it's essential to attack. P1 comes today and I'll take him over the front.

6.10pm A busy day planning, arranging, talking. P1 turned up, looking well. He never says much but walked about with his arm in mine. Dour yet charming person. He made one or two shrewd comments about the campaign, went round the front and returned covered in dust. I may be going into Cairo tomorrow.

7 August 1942 *In flight.* Force-landed with engine trouble at Wadi Natrun – hot and no breeze. Will stay tonight and return to the Chief tomorrow who's much more reasonable. Am changing jobs to Ops, Plans Intelligence under title of Vice Chief of the General Staff, to distinguish me from my opposite number John Harding.[28] Will spend part of my time in the desert and part in Cairo. Things moved so fast in early July that even now I can't piece together cause and effect. It's all an anxious blur.

1.15pm Got off in another 'dizzie'[29] almost equally decrepit and made spectacular landing at Heliopolis. Meeting CIGS[30] at 3.30pm for business.

Saturday 8 August 1942 Continental Hotel, Cairo *7.15am* Spent yesterday afternoon in GHQ entirely overrun by august 'Foreigners'. P1 in his old chair (accommodation only of course), and top quarter of the army council littered all over the place suffering badly from the heat and ceaseless batterings of our inexorable Fuhrer, the great W. Incredible man – monstrous! He wears his surroundings to shreds. The trouble is that those on whom he works (don't think he works *with* anyone), his victims and utensils in high places, are left in little mood for calm consideration by the time he's finished with them and merely pass on their agitation and fatigue.

Of course the great man is right to trust nobody. He won't take no for an answer, even hard facts are insufficient. Bricks must be made, straw or no straw. Told that the straw is still standing in the field, he affects unreason and demands the brick complete. No argument! Or so they tell me. Agreed that it's a time for ruthless treatment, but pitching demands to the limit of achievement without there being the means only causes fruitless action. Suppose only he can judge. It's an intolerable burden for a man to bear, even one who's trained himself to greatness. Interesting to watch, but a grim experience for those in close contact with this devouring furnace – literally a furnace, since he consumes about 24 cigars a day.

GHQ pretty well cluttered up with extraneous activities and its dignified calm has wilted. No wonder P2 keeps to his fastness in the sand. I came in on a mission and met the head soldier [Brooke] tattered and nerve-wrecked from a stormy 24 hours with the Leviathan.[31] Failed dismally to get my ideas across [about reorganisation, already submitted in writing] and got no clarity at all, just an ill-tempered, negative reaction. The logic of events will force action along the desired lines – there's no other way. Got the impression there was no great desire to change. There's a rumour that Death has put his hand on Strafer Gott, our best younger leader and one of the very few in whom I believe. It's these unforeseen events which shape our destinies, and against which we and our enemies struggle as much as against each other.

9 August 1942 Continental Hotel, Cairo *7.15am* It's wrong to discuss casualties but Gott's death has saddened me. It's almost as great a loss as Dick's capture – loss to the army, for he was a great man and would have gone far.[32] Think the enemy shot down his aircraft over the landing place I was held up at too; might well have been travelling with him. So much luck in this game. We relatively senior soldiers fly almost as often as senior airmen nowadays. Spent yesterday getting in touch with the wider picture of the ME as there'll be plenty to do here. Should go eastwards to Persia to see things for myself soon. Mustn't all of us get too engrossed in the western issue.

2.10pm P2 will probably go into limbo. He hasn't told me personally, but he's in from the WD with all his kit, and I doubt he's going back. More will develop in the next few hours. Haven't begun to make my new job effective and will make no changes in existing machinery until I know the form. The downfall of the great means the collapse of all who work with them. No use guessing.

7.30pm Back early from the office. Nothing public yet, but the brokers are busy in the background, P1 among them. Poor P2. Wish I could help him, but these are things which must be borne alone. The appearance of Gen. Alexander is interesting. He passed through from Burma six weeks ago and he's back again. Everything's so uncertain that it's useless to settle down to planning. Pretty sure I'll not go back to the WD, and no certainty the present General Staff setup will suit anyone who relieves P2. Always thought it was unwieldy, so am prepared to be squeezed out.

10 August 1942 Continental Hotel, Cairo *3pm* Just been lunching with P2 and the truth is out. They've produced a P3 – Alexander, no less, recent defender of brother Reg. That should make the Middle East safe for mediocre aristocracy, and all the snob parties P2 outraged will come into their own again. Obviously no place for me. Gather P3 and Reg did <u>not</u> see eye to eye.

P2 has been tentatively offered a pig in a poke.[33] If he took it I'd go with him, for it would be almost as odd a venture as stopping Rommel, but am pretty certain he won't look at it. Don't discuss my personal business with a big man in distress, but know he thinks that any intimate of his will get short shrift from the new dispensation. Obviously he can't initiate any more policy, and Alex won't want to either. By Heavens, there'll be sighs of relief from the old gang when they know. It's a pretty complete anticlimax to the last two months. Ironically have just seen a paper written by WSC in October '41 which begins, 'Renown awaits the General who succeeds in restoring artillery to the position from which it has been ousted by the Heavy tank.' Exactly what was successfully done at El A! Only there's no 'renown' – at least nothing in advance of history. Will hang about to see what's going to happen. Still very hot so sleep under a sheet without pyjamas and note with amusement that the lady in the room opposite does too. At least she walks about her room stark naked in the mornings – quite a good body! Wonder how much she thinks she reveals of herself, and whether she cares?

10.40pm Tom Corbett got his *congé*[34] this afternoon. Poor little man, he was stunned. Mine comes next but hasn't been administered yet because I'm on a slightly lower level. Only hope they'll get on with it. I'll ask for a month's leave in my present rank and permission to go at the end. Really rather pathetic in the office this evening. By 7pm nothing doing, shutters up, everyone gone.

The 'pig in a poke' was the offer to Auchinleck of the Iraq-Persia Command. This would split the Middle East command into two, giving Egypt, Palestine and Syria to Alexander. Auchinleck's portion might put him under the ultimate command of Wavell as Commander-in-Chief India, therefore subordinate. But, far more importantly, the air command would not be split between Auchinleck and Alexander, so Cairo would have priority. If, as was fully expected, the Germans went for the Caucasus and its oil by pushing down through northern Persia, no air cover could be got in time because London's permission would first be necessary. This was such a dangerous flaw that Chink advised refusal, despite the likelihood of being appointed Auchinleck's CGS which implied the rank of Lieutenant General and a knighthood.

11 August 1942 Continental Hotel, Cairo *7.15am* My claim to fame will be that I attended the only occasion on which an Axis drive[35] was stopped in full career. Rather a negative claim, for already those 'laurels' have faded and been forgotten in the march of events. Today am one of a committee to carve up this demesne,[36] a little like Solomon's threat to the baby, and incidental to it my own employment will go. An amazing transaction to take part in straight from the battle. P2 [is] engrossed in his own affairs, and I won't know my own fate until Pharaoh 3 gets into the saddle. Possibly before, if the High Ups accept the recommendations of this committee, because automatically I'll become 'for disposal'.

2.15pm Spent part of this morning with poor P2 who's being got at by WSC to accept the unsound command. We read through the correspondence together, trying to see how much double-dealing it contained. Don't think he'll accept. He made it clear that the new brooms would never employ me as

I'm too much of a revolutionary. It's all so wrong. P2 has made mistakes but so has every other commander in this war, and he's a very big man with lots of fight left in him. He's lost confidence in himself, naturally, and doubts if he's fit to go on. Grievous to see him checked in mid-career, all his power and energy dammed up. Meanwhile of course the war goes on – men die. Heard from Estelle, settling down in barbarous Kakul[37] in an army hut without furniture to run a canteen. She has guts. She says Wavell suddenly declared a 10-year-old passion! No wonder he tried to get me to Delhi – what an unbelievable world. After all, he's sixty.

4pm Have grossly idled these last few days and don't intend to go to GHQ till 5pm, when I sit on the Committee of Disruption or Partition.[38]

12 August 1942 Continental Hotel, Cairo *7.15am* After a day's thought I've been forced to the conclusion that I *must* advise P2 against accepting the pig in a poke, not because of his personal feelings, but for the inherent fundamental unsoundness of the proposition which sprang from Winston's fertile but semi-senile brain. I'll have to add a rider to the committee's proceedings, disassociating myself from the proposition on strategical grounds. That will be the end of me, but I can't have it said afterwards than nobody at the time protested. Even the top soldier[39] who's being dragged about minus his advisors behind this juggernaut is mute! So there we are. The Western desert is calm and, unless P3 makes a muck of things, will so continue. Don't doubt the dynamism, will to win or demagogic ability of the man, but do doubt his strategic judgement. P2 is fairly up against it, for if he accepts my views he'll have to say no. It's more than ever clear to me that my military career is drawing to a close. Yesterday I didn't see this quite so distinctly and thought it would be a question of them getting rid of me. Now I see that it is for me to say what I think, which can only result in me getting rid of them.

13 August 1942 Continental Hotel, Cairo *7.45am* August is half over, and the war goes on, but still the enemy have not got Stalingrad and the Caucasus barrier hasn't yet been penetrated. We don't know what Stalin's got behind it, and nor does the Boche. Meanwhile our desert adversary[40] is rumbling, and I wouldn't worry if P2 and self were out there. Sounds conceited but it isn't, for we had the measure of him. The new management goes in today, very green to this sort of warfare. P2 and Ritchie started from scratch this time last year. Now P3, with less ability, and a new man, Monty,[41] face Rommel. Monty's able in a narrow way. He was one of my teachers at the Staff College. Poor Freddy.[42] The new chap will be glad of his insight (if he's any himself!) but they all naturally want their own creatures to help them. Changes are inevitable. Dined with P2 at Mena last night. So odd – quite like old times in spite of the threat 100 miles away – and we discussed the cigar eater's new offer. More than ever I think it's more patriotic to refuse, and have said so plainly. Don't know what's in his mind. Just possible I may have to go off with Casey[43] to the City of the Caliphs[44] to meet the cigar man and discuss the whole business of ME but hope not. Don't like the interminable third degree, single point, cross-examination-plus-argument in which he indulges.

2.30pm Today I formally advised Auchinleck against accepting. He challenged me – 'And yet when I first asked you, you said it was my duty to accept!' Replied that I'd only changed my mind after much consideration, concluding that the whole proposition was most dangerous. P2: 'You realise, of course, that your advice, if I take it, ruins you too, for I'd have taken you with me.' Myself: 'I'm honoured, but that makes not the least difference to my advice.' Left him with a note on my reasons and suppose he'll act accordingly. Wish fate hadn't cast me for the role of his only advisor. It's altogether too serious a business, and the cigar smoker will be furious. P2 went on to say that in due course Wavell would find me employment, to which I replied that [I] didn't want it. 'Oh yes,' he said, 'everyone wants to go on. Even I would like to go on.' Pathetic. Propose to write to him tonight asking that before his successor comes he accept my papers and lets me go on a month's leave to South Africa in my present job. If he agrees, I'll lecture there.

4pm In a few minutes will climb into my enormous car and roll off to GHQ looking a typical unpleasant general, a trifle on the lean side. If only people realised how transient this appearance is. But they don't know P2 has fallen yet, and P3 is lying low. Have only seen him once myself – a joke to think we were at the Staff College together, he in his second year, I in my first. He isn't really very intelligent. Wonder whom he's sent for as CGS? Possibly Oliver Leese,[45] also a guardsman. P2 knows, but I've not asked him.

14 August 1942 Continental Hotel, Cairo *7.15am* P2 hands over on 16 August. Gather, though he hasn't said so, that he's refused to accept Churchill's offer with good sense. After this experience I've come to the conclusion that it's absolutely necessary to divorce the military direction of this war from the political. Politicians, particularly those raised in our school of politics, work on intuition, emotions, hunches, personalities and, ultimately, compromises. They don't deal in hard irreducible facts and, however well read, they aren't trained to assess the problems of war. There's no doubt that we need a supreme director for our military effort – a military man with sufficient political support, a true Commander-in-Chief or Minister of Defence. Don't care what he's called, but he must primarily understand the nature of war, and know how to work through facts to deduction, and then to organisation and action. Our present system appears to be guided entirely by expediency.

WSC is a consummate politician, but he's also an artist and a dramatiser, and highly narcotised by expensive cigars. Doubt the clarity of his brain. I grant him a supreme capacity for saying the right thing and admire his energy and power to bridge the enormous gulf between the UK and the Americans. All these are invaluable assets. But when it comes to directing strategy he fails utterly. He knows nothing of the machinery of modern war, and his tendency to interfere is disastrous. One can't conduct a war with one eye cocked on the Conservative party! Politics and military strategy, including supply, don't go together. There must be a new approach. If necessary, the Commons should hand over some of its constitutional responsibility to a non-politician responsible for defence to coordinate the three great services and supply. Lord knows who such a man would be or how he'd get his advice, but without him our work is wobbly, weak and almost undirected. Pity Smuts isn't a younger man. Possibly the worst of our society today is how the propaganda people – the advertisers – can force an entirely false personality on the public. It's even more dangerous when the individual being boosted is no mean publicity agent himself.

2.40pm No news. P2 was busy this morning. Corbett and I paced his office, both doomed, neither knowing when the doom will fall. Quaint, after the urgent days of the last two months, not having anything to do. Asked Frank Theron to dinner tonight.

Midnight Heaven knows what's going to happen next. This morning the only real live adjutant general from the War Office, who'd just been to India, said he'd heard Wavell tell Brooke that he wanted me at Army HQs India, and I was to do nothing till both turn up here on Monday. An order! Don't want to go to India, as it will mean a show down with Estelle, and all Delhi-Simla will be in on it so I'll only get sacked again.

15 August 1942 Continental Hotel, Cairo *7.20 am.* Theron was interesting last night about the cigar eater, and talked about the amount of brandy he consumed. Not so good this whole business. P2 finishes today and at 3.30pm I officially meet Alex and get my *congé.* He's decided, I'm told, to put in as CGS a man who's so stupid as to take my breath away. But Alex and this man are personal friends in the same social group, so that is that. It lets me out in fairness, for this new man is a good deal my junior.[46]

4.15pm Had my Hail and Farewell interview. We fenced about for a little. He'd seen Estelle in Karachi and tried to talk about Reg.[47] Finally got him to face facts and said I was sure he'd be happier to be rid of me. Alex replied with avidity, 'Oh, but you're going to a command at Home.' Countered by saying

I was told Wavell wanted me in India. Upshot is, I'm graciously permitted to sit about in Cairo until I know my fate. My final comment was how they used me was all one, since what I'd like would be to go altogether. Apparently that's unthinkable.

16 August 1942 Continental Hotel, Cairo *7.15am* What a strange year this has been. Completely unsettled right from the beginning, and then to be employed in something stable and to lose it again. To see the Auk wrecked after he saved Egypt, and to see how it's all been done. It's quite enough. No longer wish to have anything to do with these people or a society in which such completely insane things can occur. There's nobody I want to serve now. Not Wavell, not the High authorities at Home, not WSC. The feet of clay are too apparent. The Auk had only to make one mistake for them to destroy him. They'd put him in here to make room for APW in India, but never intended that an Indian Army general should remain in such a British preserve as ME Command. Tried to warn him but don't believe he ever realised. Now he knows.

9.30pm Phew it's hot. In half-an-hour I'll go off to Mena to see my late chief, if he's still there.

11.30am Went out to Mena but P2 has gone – they've taken 'my Auk' away. He's so vulnerable, so painfully young at 58, much younger than I because all this business of high command – dealing with men, the pomp, patronage, responsibility – thrilled him. He was great, but with the greatness of a magnified second lieutenant who'd achieved his dreams. Suppose all good soldiers are like that, that's why they can't understand bad soldiers like me who don't think of success like the boy-at-heart. He walked off without raising a finger to help me. To some extent that's the measure of his distress.

Monday 17 August 1942 Continental Hotel, Cairo *8am* Still awaiting military execution, only the Gestapo haven't turned up yet. Have made up my mind that if there's any option I'll go to England. However I'm employed there, Eve and I can live together.

7.45pm Just been photographed again. At least they'll be evidence that for a little I was a general. Then walked to the Turf Club and had a couple of drinks.

18 August 1942 Continental Hotel, Cairo *7.15am* My judges have been in Russia and how amused the Russkis must have been at the circus.[48] Winston with his childish V signs made with two fingers, his old doctor to ration the cheroots and brandy, plus one-eyed Archie[49] and the CIGS! Suppose they'll turn up here tomorrow, then I'll know my orders. Don't want to explain that while I'm absolutely certain Germany can't win this war – she's got to make

NO mistakes which is impossible, whereas we MUST win it however many we make – I've lost faith in the combination of party politics, international politics and military strategy as applied by WSC. Don't want to have to fight my way out from under P1's kindness either, and if he asks for me it's so easy for the CIGS to agree. All this is conjecture.

10.30am Before it got too hot I walked in the morning squalor of Cairo. How much worse is the oriental city behind this occidental *façade* – that beastliness of the Middle Ages, a world of stenches, dirt and corruption. Difficult to see how men's minds escaped, yet beside it was the glory of architecture, music and poetry. Now we've a world from which squalor could be excluded – technically possible – but the gutter in men's minds keeps us in the rubbish. Damn the selfish rogues who hold back the journey of humanity. The Egyptian parliament's currently debating morals and some wish to send all women, including Europeans, back into traditional dress – literally back to the Koran. There's a strong reaction against Europe here: people are tired of the endless struggle conducted over their heads by races for whom they don't give a damn. Certain locals would mind, those who'd backed our star, but not the man who tills the Nile mud. He'll remain the most real thing in Egypt. P2 has vanished utterly, nor have they announced the change of command.

12.30pm Suppose WSC is thinking out how to break the news while the claque's still applauding his genius in meeting Honest Joe [Stalin] face to face. After years of vilifying Bolshevism, that's the time to launch a major announcement, and the loss of a great man will go unnoticed. Dick O'Connor, Gott, Auchinleck – all great losses. Hate to think it, but suspect P1, by taking a hand in P2's undoing, has made his own tenure secure. P2 was moved to make room for him, so how awkward if he asked for India back.

12.45pm Orderly has just brought the written message that I'm to return to the UK. 'In consequence of the reorganisation of the General Staff you will relinquish the appointment of DCGS. A passage by air will be arranged at an early date and on arrival you are requested to report in writing to the Military Secretary. In due course you will receive instructions as to future employment.' Well, that is that.

19 August 1942 Continental Hotel, Cairo Packing and waiting. The papers cover Winston's visit to the Front, but hardly a word about P2. Presumably the Press have orders to go slow on him. They haven't announced the change of command but it's an open secret. Am told he's still in Cairo, so he's done what I suspected he would when I was of no further use to him – partnership dissolved. Perhaps I misjudge him – what *does* it matter? If I've not gone by

Saturday am asked to a Press lunch, and if asked to speak I'll do a 'Friends, Romans and Countrymen' oration praising him.

20 August 1942 Continental Hotel, Cairo Today the papers are full of the change of command and by inference writing down P2. They haven't even published the full text of his farewell order of the day to Eighth Army, a fine terse piece of writing. To know behind it all – to have been his brain in battle and to an extent his heart, to know the new men and all the difficulties they're going to meet. To know the history of these campaigns backwards and to know more about this war here than anyone taking part – yet to be powerless, unnoticed, disembodied. A little like being dead and watching one's family making a balls of things. But it does give a dirty taste, and all the truth isn't out.

Am told the Pretoria authorities don't want me to lecture down there – it was 'unwise' to mention Tobruk – but the obvious reason is that they've all agreed not to emphasise past operations or the Auk's part in them. He hasn't appeared in any photos of the PM's visit and all mention of him has been suppressed. May as well get out of here.

21 August 1942 Continental Hotel, Cairo *12.30pm* Sent off my heavy baggage on its three-month journey, or at least gave it to the Military Forwarding officer and took a receipt. Release in one way. Last night slept badly, thinking how beastly it'll be to arrive alone in England and wait for my future to be decided in a grimy London hotel.

7.00pm Have been to GHQ and they say that 'Priorities standing unchanged' I fly to England by Liberator on Sunday 23rd – the anniversary of my first battle at Mons. That gives me tomorrow to wind up my affairs and get vetted for high flying, then off.

22 August 1942 Continental Hotel, Cairo Saw the Paymaster and the Bank and paid a few bills. Tonight dine with friends at the Mohammed Ali to pay my debts for kindness and sympathy. Think I'll be kept employed in some way, though none worth anything. Hope for lots of work to do quickly, otherwise I'll die of claustrophobia on those petty overcrowded little islands. Got my movement order with details about oxygen, parachutes, dinghies. Instructions include thick clothes – haven't got any – and talk about frostbite! Journey's about four days.

4.15pm Brother Reg is in town, passing through to Simla and still Governor of Burma in exile. Will see him after weighing in my luggage as he goes east at 4am tomorrow. Well, he's kept his job, more than I have. Clever lad, Reg.

Sunday 23 August 1942 Continental Hotel, Cairo *7.10am* Reg very tired after days of air travel and interesting about the people he'd met at home. Gave me several names of younger Conservatories who might be interested in me. Says Duff Cooper[50] hasn't forgotten that I outlined modern war to him in 1935! Warned me against Hore Belisha[51] etc who he says attack Winston, but unconstitutionally, and agrees that the military direction of the war must be taken from W's hands. Recommended the RAC[52], so plan to get fixed up there, go to the WO and see James Grigg,[53] Secretary of State for War. He was a friend in India so want to tell him about P2, not me. Then to the Military Secretary and Leo Amery.[54] Feel fine and offensive now, utterly free to do my duty as I see it.

11.30am Just told my plane isn't going till tomorrow evening so another 24 hours in Cairo. Damn. If I'm to go, prefer to go at once. Vetted thoroughly by the RAF doctor for high altitude flying today – heart, lungs, blood pressure. 'You're a most remarkably fit man. Do you smoke?' 'No.' 'Take a lot of exercise?' 'Not as much as I'd like.' 'Drink much?' 'Not less than three a day, not more than five.' 'Well, you're as fit as anyone I've ever seen. Sound everywhere.' Good to know.

7.15pm Went to the Gezira Club for tea and it might have been 1938. Terrace crowded, cricket, and the only thing wrong being the men and women in uniform. A curious atmosphere for wartime, but it's hopeless to expect our stock to walk about grinding its teeth and talking war shop if it can go and play. All the same, difficult to believe they realise the enemy's only 100 miles away.

24 August 1942 Continental Hotel, Cairo *7.15am* Didn't expect to be still here this morning. From the news today I see that old trimmer Jumbo Wilson's[55] accepted the new Iraq-Persia Command I advised P2 to refuse. From now onwards our tireless premier has split the Middle East between two generals with one airman superimposed, an arrangement I know won't work. Jumbo's so anxious not to be retired (he's over 62) that he'll accept anything, regardless of principle. Oh hell, it's a crooked game in the borderland between politics and soldiering. Saw the old elephant at Gezira yesterday watching the cricket, backed by his usual claque of black-buttoned riflemen. So the Middle East cut off at Syria will be made safe for cavalry and guardsmen, while the ME in Iraq will blossom for riflemen. No place for me! At 5pm today will be handed over to Imperial Airways for the oxygen-sustained flight which will land me in England in a few hours. Not a comfortable transaction, never the less.

Military historian John Lee: Clear nowadays of the febrile wartime atmosphere, how decisive does Auchinleck's achievement in the summer of 1942 appear today?

And were the dice loaded against Chink all along, as he believed? Let's look at this new evidence. In a letter of 22nd January 1941 to Richard O'Connor, Chink chided his old chief for not making him his new Brigadier General Staff when he had had the chance. O'Connor replied, assuring Chink that he had specifically asked for him as BGS but "he wasn't allowed to have me". So someone at the War Office was already 'marking his card' and seeing that he did not get the sort of posting that suited his talent.

Similarly, General Wavell had told him on 26 December 1940 that he did not want to keep Chink at the Staff College, Haifa and was looking for a chance to employ him in close connection with operations, so he could be of "more use than just giving encouragement and advice". Both Wavell and O'Connor were already aware of the major contribution Chink had just made to the rout of the Italian army in Cyrenaica.

Chink could have had no better job than that of BGS to O'Connor. It would have obviated the critique that, as commandant of a staff college, he lacked practical experience in desert warfare. Even on 9 December 1941 he was thinking of all the good he could have done if he had been a Chief of Staff in Middle East Command. His advanced ideas on organisation and training of 8th Army would have had the time to be absorbed.

Wavell did bring Chink to GHQ in April 1941, but only as a temporary BGS. Chink was shocked to learn that he had appointed Arthur Smith as his Chief of Staff. Chink was to be used for various special missions, such as warning Freyberg, the general defending the island of Crete, that Ultra decrypts had revealed the German attack plans in detail, but that his troops could not make any deployment that would reveal that the British knew as much. And Chink was duly sent back to Haifa when Wavell was replaced by Claude Auchinleck.

Once again Chink found himself being consulted informally, this time by Auchinleck, for his advice and approval (see 17 July, 27 and 28 December 1941). On 18 December 1941 he was told by the new Military Secretary for the Middle East that everyone was 'aching' to promote him, but that he was 'too brilliantly disturbing' to lesser talents around him. He was even told that there was a 'first class' job waiting for him but that the War Office would not authorise it.

In February 1942, on recall from Haifa to GHQ Cairo, Auchinleck unburdened himself to Chink about how he admired his 'brilliant mind with twice the vision and speed of anyone else'. But went on to say that that very brilliance 'made Chink's seniors feel small', and made it difficult to appoint him to a senior role.

Auchinleck, however, made a point of discussing all his plans of campaign and high strategy with Chink. (See 10th February 1942.) Chink's subsequent tours of inspection of 8th Army deployed in the desert would result in the classic paper,

'*Command and Control in Modern Warfare*', *and further entries in early March 1942 show how Auchinleck continued to discuss ideas and plans with him.*

After absence on staff college matters in South Africa, Chink was back at GHQ Cairo by 12 May 1942. Auchinleck then appointed him to Neil Ritchie's old job as Deputy Chief of Staff, presenting the War Office with a fait accompli. *Chink had no input into Ritchie's defence schemes at Gazala, but would never have approved the scattered deployment into fixed defences that invited the enemy to over run them one by one. His clear advice at all times continued to be, 'Keep your forces concentrated for battle. Don't lay yourself open to being defeated piecemeal.' (13 February 1942).*

The Battle of Gazala, 26 May–21 June 1942, was a catastrophic defeat for Ritchie's Eighth Army, ending in the collapse of resistance at Tobruk and the surrender of its garrison, although from as far away as GHQ Cairo, Chink had detected serious errors made by Rommel in the opening attack. Someone with a grip on Eighth Army and an understanding of modern warfare could have won that battle decisively. Instead, British armoured formations again attacked in a badly co-ordinated fashion, head on into German anti-tank defence screens, resulting in the whole Allied army being put to flight.

On 25 June Auchinleck intervened personally to dismiss Ritchie and take direct command of Eighth Army. He took Chink with him at his side for the remainder of the fight. Over lunch in Cairo before leaving for army headquarters, they decided that a stand at Mersa Matruh was pointless, and that the whole army had to get back to the El Alamein – Qattara Depression line where Rommel could – and, indeed, would – be halted. See Chink's letter to Auchinleck of 16 October 1958.

Chink had long been one of the few senior staff officers privy to the Ultra decrypts of German secret military messages. By now the Eighth Army was receiving daily intelligence, just one day behind real time, of the state, strength and position of German and Italian units, their shortages of fuel and ammunition and often the intent of their commander Rommel, including any last minute delays to his plans. Chink had never been afraid of Rommel and thought him a bit of a reckless 'chancer' (as did many of the 'Desert Fox's' German colleagues). Now he had the opportunity to help Auchinleck direct the fighting power of Eighth Army to stop Rommel's stampede towards Cairo, Alexandria and the Nile Delta.

As a result, decrees went out that British artillery would be massed under centralised control, that the army would hold forward defences lightly and remain as mobile as possible, and that they would try, without obvious success, to oblige the 'separatist' elements in British armoured formations to co-operate more closely with the infantry and artillery. Rommel gave his Panzer Armee Afrika none of the rest they so desperately needed and threw them into the attack on 1 July 1942. This and attacks over the next few days were comprehensively defeated. By 4 July Rommel admitted that he could make no further progress towards his final victory.

Having thrown Rommel onto the defensive, and under continuous pressure from Churchill in London to attack and destroy the enemy, Auchinleck and Chink began using the Ultra decrypts to pinpoint where Italian divisions were in the line, delivering a series of attacks on them to great effect. Soon Rommel's famed Afrika Korps panzer divisions were racing to shore up a crumbling line.

Not all the British attacks went as well as hoped, however. The use of Ultra decrypts often left only a fleeting window of opportunity, and attacks had to be put together very quickly, using the troops to hand. It has to be said that Eighth Army staffs were not used to this sort of rapid response; they were not good at it. Here it is plain, the speed of Chink's brain and his concept of battle often ran far ahead of his fellow officers. His position in the hierarchy of command was anomalous, and that must have caused anger and upset to many.

Certain now that Rommel's drive towards the Nile was ended, Auchinleck decided to regularise Chink's position on 24 July 1942 by making him Deputy Chief of Staff for Operations, Plans and Intelligence. This was a job, at last, for which he was ideally suited.

One of Chink's final acts for Auchinleck was to write a formal staff 'Appreciation' on 27 July 1942 of the military situation in which, meticulous as ever, he examined all potential outcomes. In conclusion, he predicted that Rommel, once suitably reinforced, would make one last great attack to break the Alamein line, and that would probably be along the southern face of the Alam Halfa Ridge sometime in August, before Eighth Army was ready to resume the offensive. Suitable defence measures were put in place as a result.

It can be seen, therefore, that Montgomery's first victory in North Africa would be entirely due to Chink's foresight. But during the upheaval of the change of command in August, the fact that the Appreciation also covered 'worst case' scenarios requiring further retreat would be used effectively and deliberately against both men. The stigma would bring down Auchinleck and Chink together in 1942, and it wrongly overshadows the reputation of First Alamein today.

England:
October 1942–March 1944

With the rank of Brigadier in the Home Forces, Chink was posted to 160th Infantry Brigade in 53rd Division, part of 12th Corps in Kent preparing for the invasion in France. Numbly he reported to his HQ, an Elizabethan manor-house named Provender near Sittingbourne in Kent, and the parochial setting confirmed his black mood. If Haifa had felt like a backwater, Provender was solitary confinement and his sentence interminable.

But Eve reached England in mid-September, and the direct train service to London would enable them to spend his regular 24 hours off duty together. They lunched openly at the Berkeley Hotel then, careless of gossip, and booked a room above for the afternoon; in due course she would rent a London flat. He was touched by her extreme possessiveness, unaware of the effect that Estelle's hospitable fence-mending had always had on his career. In isolation his low spirits persisted.

To Basil Liddell Hart

Home Forces.
160 Infantry Brigade,
5 October 1942

My dear Liddell Hart

I wrote to you shortly after my return from Cairo but the letter was bitter and rather compromising (to you, not myself) so I never posted it. Perhaps too it was small-minded. But I was and am angry, for Auchinleck has been treated shamefully and the state has not profited at all by the transaction. We have merely lost the one general capable (like Rommel) of learning from his misfortunes, and besides the one staff officer who would see that he didn't forget his lessons. A super victory of the forces of reaction.

I'm now commanding three very useful battalions of Welshmen and were it not for a feeling of complete frustration I'd be content. I have yet to learn whether I can deal with simple soldiers, and suspect I'll be unable to take myself sufficiently seriously to be the sort of paternal demigod, petty Jehovah, a brigadier should be. I find our home army fit, keen and obsolete – ready and anxious for the cenotaphic[1] experience of a second front (Anglo-Dieppe model). We have a lovely army of lemmings prepared at any sacrifice to go down to coastal dissolution! I doubt whether we've learned anything in the higher fields of warfare, and the men who might help us – O'Connor, Gott, Auchinleck – have been set aside by fate.

I have much which will amuse you, confirm your predictions and depress you all in one. I remain, however, optimistic as to the Axis' inability to gain their war aims which gives us a technical victory. But as to our achieving a real victory over the dark forces within ourselves which created this totally unnecessary war, I'm less optimistic. Reaction is strongly entrenched. But we'll talk about this.

Yours ever, Chink

13 October 1942 HQ, 160th (SW) Infantry Brigade, Kent *9.30pm* Damn cold. Liddell Hart seems to think he can regild my blazons. Interesting – he's not far from the mark about Auchinleck in one or two respects, but won't tell him too much until I can see him personally. It does justify me in thinking it won't be quite so easy for the authorities to lay my ghost – unless I'm prepared to exorcise myself. It's just so empty sitting about here, pretending to be interested in the little world of this Brigade which anyone reasonably efficient could run so much more enthusiastically. I'll find things to do gradually as I get to know what's going on, only what a waste of time.

14 October 1942 HQ, 160th (SW) Infantry Brigade, Kent Brother Victor suggests I switch to the Air. The idea's amusing – damn dangerous but worthwhile, even if have to sacrifice myself to Boy Browning's[2] ambition. If I've got to be with troops in my present rank these airborne 'Boys' are much more fun and am quite prepared to hop out on a parachute. But perhaps I'll be translated to London? All the brave chaps want to get away from the War Office. Smuts is in town – wish I could see him. Means that North Africa's in the limelight. It would be grand if he could stay here as Minister of Defence, he'd be much more objective and trustworthy than Winston. The war is taking a new phase. This Nazi talk of consolidation and defence is interesting. Bet the army's told Hitler that he can't go on attacking forever, and this policy is the compromise. More than ever it's going to pay us to put the heat on Italy and clear North Africa. If we own North Africa the enemy has to prepare for attacks everywhere from the North Cape to Crete, and Italy becomes open to intensive bombing. Don't believe Hitler can defend Europe unless he can close the Med at Gib and Suez. Must get down to 'home guard training problems'. Dear me! Wish I could get a kick out of all this, but don't seem able to.

To Basil Liddell Hart   HQ, 160th (SW) Infantry Brigade
                                                                      14 October 1942

My dear Liddell Hart

I was cross when I wrote that deleted letter and said one or two things about Winston's interference in strategy and personal selections which might be inadvisable to send through the post. I've learned to trust no official acquaintance, so thought it wiser to keep my comments for our next meeting. That was all. I'm persona non grata because of certain advice I'm known to have given Auchinleck.

As for the poor Auk, he's a queer cuss. Likes picking one's brain but hates the idea of acknowledging his indebtedness, particularly to me. As you know, he got his real start through the Chatfield Committee, which accepted plans for modernising India's peacetime forces for which I was largely responsible. Incidentally, the higher ranks of the Indian Army never forgave me.

Last December Auk told me formally that 'Brilliant as I was', he could never use me 'unless I learned to suffer fools gladly.' His actual words! I laughed and said that the Nation couldn't afford the luxury of fools in high places and I proposed to be as intolerant as ever. For that stupid answer I languished at GHQ ME from December '41 to May '42, doing odd jobs which included a tour of Eighth Army that disclosed many of the defects which finally produced disaster. The Auk attached little weight to my recommendations, one of which was Ritchie's removal.

However the tide suddenly turned and simultaneously Archie Wavell wanted me for his DCGS in India and the Auk required me as DCGS at GHQ ME. He had an Indian Army CGS, Corbett, who was never up to the job. Anyhow, from May on I became DCGS and he leaned on me with increasing heaviness. It was all too late. We buttoned up the command in Iraq-Persia in case we had to fight on two fronts, but then the Axis attacked in Cyrenaica. I'd failed to persuade the Auk to go 'forward' before Tobruk was lost, so we at GHQ were reduced to advising Ritchie from a distance. A distressing and unnecessary episode. At least the Chief did make up his mind to take over Eighth Army, I went with him and between us we made the best of a bad job. All that story I'll tell you some day, as well as what happened when Winston appeared. Obviously when Alex took over, anyone closely associated with the Auk was *de trop* so I was sent back.

Of course the army in England has grown up independently of the army of the ME, and anyone coming home is an intruder, difficult to absorb. I ask you to do nothing on my behalf, for really I don't think there's much to be done. Professionally I don't particularly care. I'm a little tired of the army. But looking at it objectively, I think the authorities are wrong to appoint a man of 47 to the command of a brigade – 40 is plenty old enough. There seems to be no organisation for using the abilities of passed-over officers who are still relatively young mentally. I'd have thought it better to use people like myself on a panel for planning or organising or training, rather than employ us in jobs which in wartime should be treated as senior regimental appointment. Things seem to be regrettably like a continuation of peacetime.

You mention the Auk's caustic remarks. He had a difficult task as an Indian Army officer who only knew the British service superficially, and I doubt that Smith or Ritchie were sound advisors. Both are snobs, and both depressingly orthodox! Typical regular officers, in fact. Odd that both having failed signally are still 'well' employed. With their advice, it must have been hard for the Auk to form balanced judgements at short notice.

Please don't 'return to the charge' on my account. On the other hand, I've much to tell you of interest and if you are ever in town I can usually get away.

I am gradually getting used to England, though I still feel very much of a Rip van Winkle.

Yours very sincerely, Chink

Friday 16 October 1942 HQ, 160th (SW) Infantry Brigade, Kent Lunch arranged with Boy Browning on 19th. I'll get built into this mediocrity if I let myself get buried here. Better to go 'Buchaneering', and about time I took a risk or two. May lead me to London: the War Office or Combined Operations HQ. If the war plugs on, my turn may come to go to it again, and if it's possible to do reasonable work intend to do so.

My Brigade Major Rex Cohen[3] took me to his little house off Eaton Square. They're rolling in money and his wife is small and charming – so pretty at 33, and tragically can't have a child. Woolton,[4] an intimate friend, was at lunch too, and talked most interestingly. Impressed by him, and would like to meet him again. Excellent lunch and first class sherry, burgundy and brandy. Got both feet into the trough and wallowed – am getting greedy. Provender living is austere.

Rex Cohen, an urbane and sophisticated wartime soldier, liked Chink personally and gave him sensible advice. The army was more like a business than a profession, he pointed out once. The arts of compromise, obliquity, mutual flattery and concession simply had to be deployed, because the only leverage was through a propitiated hierarchy. Chink's smile was wry. 'Of course, the extent to which Rex is right is the extent to which I am inevitably doomed to failure.'

Eighteen-year-old Lieut E.E. Dorman-Smith of 1st Battalion, the Northumberland Fusiliers at the start of the war in 1914. 'It's really easy to go on once you've seen the enemy's trenches shelled a bit and got your blood up.'

Basil Liddell Hart, like-minded military journalist at *The Times*, the *Morning Post* and the *Daily Telegraph* in the 1930s.

Chink's two military heroes confer in 1940. 'Archie' Wavell, Commander-in-Chief Middle East (right), and 'Dick' O'Connor, commanding the Western Desert Force (left).

Claude ('the Auk') Auchinleck, appointed in midsummer 1941 to take over Wavell's command as C in C Middle East. Having thought alike about modernisation when colleagues earlier at AHQ in India, Chink welcomed the move.

Prime Minister Sir Winston Churchill, the personification of power.

Confrontation in the desert. Churchill faces Auchinleck and Chink at TAC HQ on 5 August 1942. 'It was a little like being caged with a gorilla.'

No meeting of minds. The CIGS Brooke and Chink had recoiled from one another's military outlook and personality at the War Office in 1935; seven years later nothing had changed. 'I mistrusted the influence of Dorman-Smith on the Auk', Brooke subsequently pointed out.

Cap defiantly tilted, Chink poses for a Cairo photographer on 17 August 1942. 'Then I walked to the Turf Club', he wrote later in the same recalcitrant mood, 'and had a couple of drinks.'

Chink postwar, with Eve (lower left)

Ernest Hemingway, whose novel *Across the River and Into the Trees* testifies to his sadness at Chink's lost army career.

Home – Bellamont Forest, Cootehill, County Cavan, Ireland.

20 October 1942 HQ, 160th (SW) Infantry Brigade, Kent Must preside over a General Court Martial of a wretched creature with 17 cheque cases against him and some funny work with £70. There are all sorts of archaic charges for offences 'unbecoming to an officer and a gentleman', when the offender probably never in his wildest dreams thought he'd ever be an officer, let alone a gentleman. A dud cheque carries little stigma in private life, but in the regular army it's a tremendous crime. This fallacy that every temporary officer must conform to the regular model persists, applying a whole host of false standards, and the army council would be better advised to keep a private fund of £100,000 for settling groggy accounts. Every time we court-martial a non regular, we as likely as not lose a good leader of men for a paltry sum. He might in battle do the enemy many hundreds of pounds of damage.

Friday and Saturday ahead: that damn court-martial. Sunday: fun and games with the home guard. Next week: CinC Home Forces comes down on Wednesday, Divisional exercise on Thursday and Friday. Corps exercise the following week.

21 October 1942 HQ, 160th (SW) Infantry Brigade, Kent Liddell Hart in his letter today says 'It does seem absurd that you should, at this stage, go back to training an infantry brigade. Now that the overhaul of our planning organisation is under consideration there may be better opportunities.' Don't know what he means by that.

22 October 1942 HQ, 160th (SW) Infantry Brigade, Kent *8.45am* Off to look at my diciest battalion and generally gain impressions. Not sure if the CO's fit to keep but don't believe in blind sackings, especially on the strength of oblique hints.

8.45pm Turned out some mobile columns supposedly in 'readiness' during the hours of darkness – not 100% a good show. Also spent today going round two units. One a temporary accretion, the other a part of my army. The soldier of the King leads a pretty Spartan life with huts possessing the bare minimum, a small degree better than being under canvas. The men look fit enough, but it's a dull life year in, year out, with no fighting or excitement. They're wonderfully well behaved on the whole.

24 October 1942 HQ, 160th (SW) Infantry Brigade, Kent My poor victim was disposed of in a day's trial. We were all kind to him, poor devil. Anyway, the offence against military law laid him open to the direst punishment. Hell! It's all damnable and need not have happened. The judge advocate who ran the court was a barrister and charming. The prosecutor was a barrister, the defending officer a solicitor, and in the face of so much legal talent the President – myself – a cypher who merely signed on the indicated spot. One of the worst parts of my punishment to limbo is that I have to live with the 'much-younger-than-I', who being bred in England are practically illiterate. The atmosphere of this Brigade mess – it's like having to feed in the nursery. Don't know how I'll tolerate this kindergarten life. Never to speak to anyone of my own outlook or status is as bad as being in the ranks.

9.30pm We seem to be stoking up the air effort in the WD, and Russia is gumming up with seasonable mud so there'll be a pause before the winter fighting on snow begins.

25 October 1942 HQ, 160th (SW) Infantry Brigade, Kent A wet but warm October day, and I'll spend it puddling round the men's billets of two battalions, seeing how they live. Their discipline is really excellent – an occasional and understandable absence without leave, that's all. Am getting to like this old house. If I must do this period of probation before the powers-that-be decide I'm nice to know, I'll make it tolerable by keeping busy.

28 October 1942 HQ, 160th (SW) Infantry Brigade, Kent Visited by a nice young Scots doctor who commands a field ambulance nearby. Very liberal-minded, and we talked war aims and the political postwar target. We've got to offer hope to people. He feels as I do – interesting how many men do. They accept Winston's war leadership, but are far from sure of his political leadership or judgment in military matters. They certainly don't expect him to lead them into the new world. Nothing more from Eighth Army. Monty is trying to wear down the German reserves while eating his way through the enemy's defence system. A slow, costly business and casualties will be heavy. The question is whether having got the enemy's main forces pegged at El Alamein, we can develop a movement from elsewhere. Wireless tells us nothing, quite properly.

30 October 1942 HQ, 160th (SW) Infantry Brigade, Kent Last night Liddell Hart phoned from Westmoreland,[5] wanting me to dine in London next Sunday. Can't, so compromised with tea on Monday at the Park Lane Hotel where he always stays. He was mysterious, and there was the underlying suggestion of power to help. All I want now is to find adequate work, however inconspicuous. The army's had few casualties yet and isn't short of Brigadiers, though woefully short of brains. Brigadiers need to be a cross between a good farmer and a good overseer – fine qualities in a limited field in which excessive imagination or creative ability is a handicap. Not that I don't like the work – part of me is perfectly happy to be with men again and to exercise authority. There's something in male nature which makes this necessary, only I know such luxury is *not* for me. Two depressing days of military study held by my Divisional Commander. Things are far from right with the new-model Division – there's ignorance about so much that's vital to success, indicating a faulty set of planners at the top. Entirely wrong organisation at the War Office and lack of intelligent direction.

Eighth Army gets on so slowly that it's hard to believe it moves at all. Hope Monty hasn't stuck for good. At GHQ ME the changeover did *not* increase the brain power, and doubt there's any notable increase of ability at HQ Eighth Army. This non-success is worrying – interesting to hear what Liddell Hart has to say.

2 November 1942 HQ, 160th (SW) Infantry Brigade, Kent Liddell Hart now wants me to lunch with him on Wednesday to meet Stafford Cripps.[6] Hastily accepted. Have always wanted to meet Cripps, so here goes.

The second battle of Alamein, under Montgomery, had begun on 23 October. At midnight on 4 November the BBC was finally able to announce that 'The Axis Forces in the Western Desert after 12 days and nights of ceaseless attacks by our land and air forces are now in full retreat.' Chink's immediate reaction is unknown.

7 November 1942 HQ, 160th (SW) Infantry Brigade, Kent The Divisional Commander has been bitten by my new tactical system which renders all current tactics obsolete. Gave him two hours of it this morning, and he's gone off a convert to get on with its propagation. It's a new doctrine with unlimited possibilities: simple, clear cut, potent. Don't see how anyone can argue against it. Only if it founders officially will I try to carry it forward politically. On Tuesday hope to sell it to the rest of the Division on behalf of the commander – but everyone will be so excited at the Mediterranean news[7] that nobody will listen. Must get cracking on Liddell Hart's challenge about higher organisation, and write something on 'Staff Officers' for the PUS[8] Air Ministry.

8 November 1942 HQ, 160th (SW) Infantry Brigade, Kent Landing in French North Africa, and at last the correct strategy begins with pressure from both ends. This is the proper way to wage world war – not being led into stupid direct assaults on Germany in Europe until we've stretched her everywhere to the limit and make her prop up a collapsing Italy. Trusting to the submarine war to weaken us is a policy of despair. If we can re-establish control of the Mediterranean and eliminate Italy as an Axis, German occupied Europe is exposed to sea, land and air attack. And there's still Russia – after failure this year, what can the Germans hope for there in 1943? Cripps the other day thought the Germans could carry on even if we occupied Italy, but his arguments overlooked psychology. Will make a date with Hore-Belisha who's already used something I said to him in his weekly war commentary in the *News of the World*. Pretty scrappy – odd how these people live on scraps.

Am told to allot one hour weekly here for the Padres to hold religious discussions with the troops as *a training parade*! This in addition to the poppy-cock compulsorily pumped into them via their reluctant officers by the Army Bulletin of Contemporary Affairs, whose pamphlets might have been written by the Conservative Central Office. Smacks of 'The Squire and His Relation Keep Us in our Proper Station.' Don't like it.

10 November 1942　HQ, 160th (SW) Infantry Brigade, Kent　A bright and lovely day. Automatically had a very British reflex when found myself trying the lawn with my heel to reassure myself that it wasn't too hard for hunting. Fox hunting will have to die. Wonder if it'll die hard, for it's in our hearts despite the snobbery and ostentation that often go with it. Do wish I was on horseback. Instead will exert all my store of tact and persuasion to get across the new tactical conception to the not very bright group of men who make up the 'high command' of this not very important Division. By the end of the morning hope to have persuaded them that it's all their own idea. If, as I think, the forces left in these islands are merely waiting to cross the channel when the German resistance in France has been softened, these tactics will enable us to move twice as fast and hit more than twice as hard without any radical change in organisation and equipment. Part of me is in a hurry to get it across, while part says that unless the war clamps into stalemate it will be ended long before my views are accepted. Hope for the sake of the young men who've been soldiering in England since before the war began that it does come soon. They're getting restive and I don't blame them. Natural that they should.

13 November 1942　HQ, 160th (SW) Infantry Brigade, Kent　Suppose I should mix more with people locally in Kent, yet they bore me for I can't see myself going back to county conversation. Do I know a.b.c.d.? No! Yes! No! Am expected to know everybody, however negative, and never find I know anybody, so conversation wilts. Bits of ideas float about the back of my brain where I keep my military conscience. Feel I ought to be walking about my brigade probing into its training and registering an interest I don't really feel. Oh well. Must do some work in this odd little job.

14 November 1942 HQ, 160th (SW) Infantry Brigade, Kent Today the Corps commander, lately commandant of the Staff College, visits Provender and will be introduced to the new tactical concept. It's almost complete in my mind – can visualise it from the smallest to the biggest detail – and it's simple, water-tight and revolutionary. That last word is the trouble. It all seems so easy once one sees the whole process, but I've little hope of success because it would need an enthusiastic, powerful personality to get the concept accepted by the War Office.

16 November 1942 HQ, 160th (SW) Infantry Brigade, Kent Last night dined at the Woolpack at Chilham, a pretty little place 5 miles SW of Canterbury, with my wealthy BM and a Battalion commander. A good party, consuming each four sherries, our share of a bottle of Burgundy, Port, Brandy and a nightcap of whiskey. Total cost plus taxi £1.2.6 a head, not bad for wartime. With the alcohol circulating we laughed and talked a lot. Good drink too, as no sign of a hangover today.

Yesterday had the Corps commander, Monty Stopford,[9] for two hours and tried to sell him my idea. He's biting, I think. Hope he does, for the previous idiom was responsible for our defeat at Gazala, given Ritchie's inertia under Rommel's energy. I'm merely a functionary, but someone must think. Am told the War Office is forming a directorate of research and better late than never, but how very late. Everyone who comes back to England has to start at the bottom of the class, but by Christmas I will have been in limbo for nearly 4 months. Surely more useful employment will show?

17 November 1942 HQ, 160th (SW) Infantry Brigade, Kent Can't guess what Archie Rowlands[10] wants of me. [Rowlands turned out to be the PUS Air Ministry whom Liddell Hart had advised him to contact.] He's in close touch with James Grigg. Does he mean to warn me to lie low? Does he have a good job in his hands? Certainly he was very affable – 'Eric', indeed. Three years ago I'd not have tolerated that familiarity.[11] Am either learning sense or losing my guts.

19 November 1942 HQ, 160th (SW) Infantry Brigade, Kent The powers that be won't have anything to do with the potential of my ideas. Largely, I think, because they see the changes they'll bring and they don't wish to face the effort of adjustment. Easier to crush the whole idea and go on as we are now – temptingly simple to say No, without any need to give considered reasons. The 5th Column, like the Kingdom of Heaven, is within us. They even ordered that I was to be relieved of the duty of opening a Corps debate on 'defence' because I was clearly sold on my system which they declare Heretical. They propose to transfer the burden to my stupid but utterly orthodox brother infantry brigadier. That was too much. I said I was prepared to put across orthodox views, if that suited them. It's like Galileo being forced to recant his solar system heresy while murmuring *Eppur si muove*.[12] I'm impelled to say 'But we <u>could</u> move.' Orthodoxy, safety first orthodoxy. Safe to the men on top, but danger, death and disaster to the men below. The orthodoxy of Singapore. The orthodoxy of Hong Kong. The orthodoxy of Burma. The orthodoxy of where next? Must think out a new line of attack, but carefully. Now certain people have sabotaged this effort, there are many who will delight in its reverse.

20 November 1942 HQ, 160th (SW) Infantry Brigade, Kent An angry morning thinking of the fatal stupidity of men in power, and how deadly dangerous it all is. The super-Panjandrum Minister of Defence – Auk-Sacker[13] —is coming to visit one of my battalions by train and am to meet the bloated creature at midday. An element of comedy in this encounter, and what it's all about, don't know. Must wait for WSC to arrive. Blast and damn!!

5.30pm Today lunched in an extra-lavish private saloon with the PM, the CinC Home Forces, General Ismay, and my Divisional commander. Discussed old battles, the fall of Tobruk. El Alamein etc and the present campaigns. He said of Darlan[14] 'If the devil himself wanted to shoot at Hitler I'd encourage him to do so. One can't be too pernickety in war.' Was most amusing over his talks with Stalin, and put down a couple of whiskies and soda and a very good brandy.[15] That makes ex and present Cabinet ministers since my return eight, counting Grigg and Amery. Tomorrow will be tedious, restless, unimportant, with little sleep. A little like soldiering, so amusing obliquely as one doesn't altogether slough one's skin, even with a kick in the pants to help.

21 November 1942 HQ, 160th (SW) Infantry Brigade, Kent Got back from the 4th Welsh dance before 11.30pm, having behaved myself admirably. Even danced, as there was a host of young women and the chaps weren't doing much about them. Every female I met seemed to belong of the same family – nice looking, and one or two amusing, but this soap shortage made participation in the Paul Jones a trifle risky! Can't *do* with a smelly woman. Left early – just as it was warming up in effect.

To Basil Liddell Hart   HQ, 160th (SW) Infantry Brigade
                                                      24 November 1942

My dear L.H.

W.C. came to Faversham to see some of my people and afterwards (at his invitation) to my surprise I lunched on his train. The quartet was Ross, my divisional commander, who sat mute, WC who was in good form (two large W & S and brandy), Paget, costive and pedantic, and myself, volatile and vulnerable as usual.

WC talked about the summer campaign in Libya. The loss of Tobruk still rankled. Apparently he was staying at Washington at the time (where all the proportions of war are permanently distorted) and felt his position vis-à-vis the Yanks. He also felt that Auk should have taken over from Ritchie during the Gazala battle!

We discussed the course of events in Tobruk because he couldn't understand how a place which had survived the 1941 summer could fall so rapidly. I tried to show how different were circumstances and he took that point, but went on to the lack of fight shown by the garrison. I again tried to explain Klopper – his background and the atmosphere in Tobruk – to show that in Klopper's mind once the Germans had reached the harbour and water supply, further resistance was hopeless. His concern with destroying petrol, transport and ammunition may prove more objective than detractors allow. But in any case, Tobruk was not defensible. Paget was terribly pompous about <u>surrender</u>, showing the PM as a desperado, and tried to tell me off for attempting to explain. I think he's jealous of Alex and Monty, and anxious to win his spurs in the field. Heaven forbid.

The Times, in its review of the year, has been less than fair to Auchinleck, at places almost brutally unfair. I've had 5 months to see something of the Home Army's theory and practice, and not only is it 1917–18 vintage, but there's a drag from 1916. We're no further forward in mobility or flexibility. Tough infantry, their minds befogged by ABC and padres, will move to their death insufficiently supported, as on the Somme. Our small army has to get onto the continent, and once there has to beat up the Germans in France before more Germans come from elsewhere. The time factor is <u>everything</u>. We'd only move ponderously, in phalanx at foot pace, for with our force which is 'all Front' we can't protect our L of C.[16] The true theory for a small expert army – flexible and of high morale, the theory of Alexander, Cromwell, Frederick, Wellington – is waiting to be grasped. Exasperating, isn't it! Incidentally of Nelson, too, for in the 'combined ops' command <u>properly developed</u> we have the real instrument. High Command should be at war school learning a new technique flat out.

We want an army as good as the Japs at seaborne attack, better than the Boche at airborne attack and air supply, and with a mobile armoured force of 10 Divisions even better than the Afrika Corps. We must specialise for an objective campaign – not one but <u>three</u> new model armies. No 1 – Home defence. No 2 – Aquatic (seaborne attack to limited depth). No 3 – Armoured. All airborne and air supported for deep rapid action in the heart of hostile territory. Plenty for 1943!

Best luck, Chink

30 November 1942 HQ, 160th (SW) Infantry Brigade, Kent We're having a difficult time with Eighth Army. But Monty is an intelligent and far-seeing tactician and he knows perfectly well what he's up against. I'd plug along in the north with my right on the sea because though it's the strongest part of the Axis front it's also the tenderest, and if it goes the Axis forces are lost good and proper. The time will come when Rommel will have to move all his reserves, particularly his armour, to the Daba area, and then perhaps Monty will strike much further south and hard. That's the proper thing to do because there will only be Italians. If Monty does that, all should go well. He's still in the opening phases of his campaign. But if he just continues to hammer at one place he will fail.

1 December 1942 HQ, 160th (SW) Infantry Brigade, Kent My office is warm, having an electric heater, and the anteroom – the old smoking room of Provender – has a magnificent wood fire. We've masses of wood cut and stored, so almost independent of coal. Chilly getting up this morning – colder indoors than out. Breakfasted alone as my staff still in bed, this being the Sabbath.

9 December 1942 HQ, 160th (SW) Infantry Brigade, Kent Must make a decision about that awkward Battalion commander – all indications point to the need for change and new blood in the unit. Having been so unfairly treated myself for so long, hate the idea of being unfair to him and it isn't easy to be certain it's his fault.

16 December 1942 HQ, 160th (SW) Infantry Brigade, Kent A sinking feeling that I'm tilting against windmills. This little study-office seems so remote from the places where things get done, and I feel helpless against the mass of well-meaning complacent quasi-efficient inertia which towers above me. However, David disposed of both Goliath and Uriah the Hittite by unexpected application of the usual methods, so perhaps there's more hope than is apparent at 9am on a December morning in Kent.

17 December 1942 HQ, 160th (SW) Infantry Brigade, Kent Rex returned from town last night where he'd been to see Woolton on this question of the army. Woolton gave him a hearing and said he could do very little to help since Winston and Brooke were inseparable and W believes all the CIGS says – or words to that effect. We really are in a difficult position.

21 December 1942 HQ, 160th (SW) Infantry Brigade, Kent Back from an exercise dragging on since Thursday. From the professional point of view depressing in the extreme. Am delighted it wasn't real war. Had it been, my Brigade would have been to all intents and purposes destroyed, with very little real result on the campaign. As exercises go, extremely comfortable. Slept for two nights in a viscount's mansion and for two more in a charming panelled room in the manor of Ulcombe, but this semi-camping out in other people's guest-rooms is disturbing.

Lunched and dined with the desiccated cream of local society, and in between umpired my own actions, all on paper. Do find talking insincerities to people unnerving. My partner at lunch, wife of a former ambassador, was a formidable affected creature and though I made headway against the strong tide of her inanity I was exhausted when it ended. Bless their hearts, they're still so certain of themselves and they little knew how much I felt that their necks needed wringing – except for old Goschen,[17] he's charming. There don't seem to be any young or middle-aged people about, just the 60s and 70s waiting for the war to end or death to come. Hanging on desperately to the final laps of a departing way of life which consists of their families and relations. Isolated, dreary, and bleak, particularly for ageing women.

30 December 1942 HQ, 160th (SW) Infantry Brigade, Kent A conference at Tunbridge Wells today confirmed my long-held view that there's got to be a radical change in the training of higher command. It's by far the most important task for the army this spring. High command's had no fighting worth a hoot and won't send itself to a school instead – far too busy training the troops to train itself. With my head already full of clearly the proper answers, had to listen to a contemporary holding very high rank gibber and grope towards the proper way by error and trial. All my work planning 'Tara' in Pretoria was a waste of time. Looked at the military audience today – beef and no brain whatever. Guts, endurance, wasteful stupidity, yet an inevitable steamroller strength. Might as well argue with a steamroller as with most of the faces there. A letter today from Papa written in June that's since been to Cairo and back. He congratulated me on my 'well-earned promotion' – ironical!

1 January 1943 HQ, 160th (SW) Infantry Brigade, Kent Spent today working out the way an invasion of the continent might go. Can hardly see it happening, as to make a success of it we'd need a revolution in our military outlook. Cogitated and wrote, but don't think got very far. I'd do better to accept things as they are and get on with the business of managing this Brigade, but am so sure the whole system has got to be rethought – re-organised and totally changed – that it's almost impossible to interest myself in work which I believe to be 80% a waste of time. It's hell to be born with the rebel's badge. Wonder what Archie Rowlands will do or say with the stuff I've written him. I see it more clearly every hour and always this frustration. Am deadly scared of what may happen if they ever attempt to use us as *they* see us being used. The Somme, Passchendaele, the disasters of March 1918 – they can repeat themselves, and so narrow is our margin of superiority that we can't afford defeat. We must organise, plan, and design for absolute victory – not just for stumbling, fumbling half-success. How on earth to improve?

7 January 1943 HQ, 160th (SW) Infantry Brigade, Kent Yesterday my residual military conscience took me to the training area where one of my battalions had been for two days. Poor devils. They were blue and stiff with cold but cheerful, busy and tough. Don't believe they noticed their misery but it would be difficult to conceive of a bleaker prospect: snow, slush and a chill half-fog. If they'll stick to that they're right enough, but in 1914–18 we survived four such winters and were none the worse.

At 10am my three COs, their adjutants and quartermasters, are coming for a post-mortem on the disastrous exercise we did two weeks ago. Makes a peg for me to hang my own instruction on. It was what will happen when we go to war, so we may as well face up to it. We'll thrash about in detail and I'll learn about them. I want their own initiative and enterprise to go into the common effort unimpaired as part of a team, not because they blindly believe I demand it. That form of leadership is easily mistaken for weakness. It's a matter of making one's juniors believe and one's seniors respect. Seniors seldom believe in juniors!

8 January 1943 HQ, 160th (SW) Infantry Brigade, Kent Rex dined with Dermot Morrogh, *Times* Leader writer, last night. Think am on the way to righting the 'wrong' done to the Auk, for phoned Rex when I heard of his guest.

9 January 1943 HQ, 160th (SW) Infantry Brigade, Kent Today out in the country with my COs and their staffs, trying to show them how things may happen in war. Need their intelligent cooperation. Rapier or bludgeon? We bludgeon along clumsily, but with a falling birthrate this nation can't fight as if its manpower's inexhaustible. Our army's still built from the bottom upwards, and left like an Egyptian mud house unfinished at the top except for lots of space for rubbish. I've got to teach a doctrine which I don't believe in, and train an organisation obsolete for modern war. What is one's duty – to go blindly on with the mob, or to say 'No, this is wrong', and take what comes? I struggle with this question all the time. On the recent exercise I knew my master was sending this Brigade to useless destruction in an inefficient way – knew it scientifically – yet still did as I was told. Hope I'm bright and convincing and effective today, but in my heart of hearts will feel uneasy and bloody.

22 January 1943 HQ, 160th (SW) Infantry Brigade, Kent Imminent postmortem on that exercise is already spoilt by my Divisional commander's mental failure to apply certain elementary tactical principles. None of these men have any military education. He never went to a staff college and isn't even fit for a Brigade in peacetime since he can't educate his officers. His staff is raw, untrained and uncertain, so the blind lead the blind. His new plan is messy with the seeds of failure in it, and even if it succeeds we'll get a lot of good men killed. There'll be no surprise or deception. These are the chaps they promote because they're 'nice fellows', loyal to their class – 'Blood brothers of the bum-warmer'. Doesn't matter so long as we aren't *really* fighting.

25 January 1943 HQ, 160th (SW) Infantry Brigade, Kent Last night noisy, but nothing fell our way. Enemy attack came in from the south, so watched the string of planes being taken on in turn as they threaded across the sky towards central London. We'd stuff over us, but no incendiaries or bombs. From the hall door could see a great line of flares.

31 January 1943 HQ, 160th (SW) Infantry Brigade, Kent Rex was comic last night, practically telling me that my mental characteristics, by their forcefulness and because they brought to slower men threats of disruption, made it impossible for those in power to let me go further. Just what Auk said. Thought I'd been so docile down here, seen nobody, said little, disturbed nothing, lain low! A cold note of doom in Goering's speech.[18] He's seen disaster before and sees it coming again. Not a word about America or North Africa, little even about the Atlantic campaign. Much about the Russian struggle making for Germany a spiritual claim as Europe's preserver against Bolshevism! It's usually in failure that one's ego tries to idealise one's stupidity. I know that too well. I do it myself, and Goering's speech might have been made by me.

4 February 1943 HQ, 160th (SW) Infantry Brigade, Kent Back from half an hour flying about in a baby aircraft – one of those little chaps which take off and land almost anywhere. The Gunners own them, and they represent the beginning of the army possessing its own aircraft. They're aerial observation posts – one sees very well. This part of country is so wooded and broken up by farmhouses and villages that we can hide 30,000 troops and all their vehicles easily. Hardly saw any.

In January Eve's pregnancy had been confirmed, and Chink spent his next leave with her at their new flat in London where they had the privacy to rejoice together. 'I'm still glorying in the happiness', he wrote incredulously on return, but within the month she suffered a miscarriage.

27 February 1943 HQ, 160th (SW) Infantry Brigade, Kent Must sit for an hour to watch two units boxing each other – how quickly one moves into a totally different world. Hope against hope makes me think that some day I may be usefully employed again, but the dry dust of evidence shows my time in the army is over. There's a darkness on me tonight. It will drive me to a final challenge of the system which rejects me, but I'll wait until this exercise is ended. Intend to ask the Military Secretary flat out what they really have against me. My record as DMT in India was good. No-one could deny the success of Haifa. Archie Wavell must have approved of the period when I was at his GHQ in 1941 or he'd not have asked for me as DCGS in India in 1942, and they can't complain about my record as Auk's DCGS. So what's it all about?

28 February 1943 HQ, 160th (SW) Infantry Brigade, Kent A fine frosty morning and am less morbid, though my sleep was full of anger. A good article by Boney Fuller[19] in the *Sunday Pictorial* about our North African strategy, and the faulty combination between the Cairo or Egyptian front and the Algerian front which he says made it possible for Rommel to get to Tunisia pretty well intact. He's right, and of course we're at the same old game. Am visiting one more Battalion for a 'pep talk' and then to check over my reactions to the forthcoming dummy war – get my views straight. Doesn't seem much use, but one never knows. The wind may change again.

3 March 1943 HQ, 160th (SW) Infantry Brigade, Kent That bad February week, which seems so long ago already, tore me to pieces. I hated the passing of the small entity, though I accepted the inevitable once it was so, and began to rebuild in my mind.

21 March 1943 HQ, 160th (SW) Infantry Brigade, Kent Rex says I ought to try to work from inside affairs outwards, not from outside inwards (not his simile). Know what he means. To a businessman it's common sense to think that way, but to go inside to gain leverage runs the risk of being absorbed and never wanting to push out again. According to him, with people who may affect my promotion I create this feeling of 'outsiderness', because in conversation I not only look at what I have to do, but at the operation as a whole. I make them feel my opinion of that is as important as their opinion of me in my subordinate role. They find that disconcerting, for the big picture is no concern of mine.

The residual impression is that I take a dilettante view – principle in a vacuum – in preference to a practical man's view of the importance of his own task. If I restricted my outlook to my own field they'd accept me, appreciate my capacity, and *in their own interests* advance me. The fact that I remain dualistic is a fatal hindrance to progressing my own ideas. Since I consistently aim at changing the whole, they can't see how I'm playing my own part sincerely, though I do seem to do so. I'm my own enemy! In his view I'll never be any good anywhere until I abandon sticking scrupulously to a set of rules of my own making. He accused me of masochism! Sees no virtue in my contention that a professional soldier should state the truth as he sees it. His argument's an echo of the Auk's 'Suffer fools gladly' refrain, or Wavell's 'Chink is his own worst enemy.' In the depths still is the old demon of ambition which suffers extreme annoyance at seeing contemporaries being advanced because they have the characteristics I lack, rather than the abilities I possess. In wartime one is a prisoner of the war – very neatly trapped.

26 March 1943 HQ, 160th (SW) Infantry Brigade, Kent Apparently the Divisional Commander has reported that I'm fit to command a Division (nice of him), and that's gone forward to Corps and Army HQ. Am not supposed to know that. Seems to be a general atmosphere among the gossips that things are bound to move, if only because I'm too 'overpowering a personality' to be left in a subordinate position under a Divisional commander who isn't too sure of himself. Heaven help us all if that's how things are run.

Wrote at length to Basil LH – we're boys together – saying he, least of all, can escape responsibility, being the recognized authority, if he doesn't act on his own paper criticising military organisation. His facts: 27 Formation HQs (Brigade, Division, Corps, Army) control 53 fighting units – the equivalent of one commanding General or Brigadier, not to mention the senior staff officers, to every two Lieutenant Colonels in command of infantry or tank units. He must force political or press action. He ought to go direct to Winston for change or an enquiry – wonder if he has the guts, but doubt it. He's data enough, for I'm not his only correspondent and I can't be the only senior officer who feels the army is unfit for future tasks.

2 April 1943 HQ, 160th (SW) Infantry Brigade, Kent Just returned from one of the stupidest meetings it's been my fate to attend. Kaput my Divisional general, as far as I'm concerned. No man who is as bad a psychologist is fit to lead men. A decent stick but hopelessly stupid.

3 April 1943 HQ, 160th (SW) Infantry Brigade, Kent Today my Divisional commander asked for my view on the lessons of Spartan [the recent exercise].[20] Know he gets uncomfortable if I raise the discussion to any level higher than 'what exists for him to deal with', for he prides himself on being a practical man. All the little things that are wrong are inherent in the big things, and hardly worthwhile discussing them if the big things go unchanged. Analysis always leads back to the instructions given by higher command since my arrival. The consistent unreality of the work we've done is unnerving. Facile, superficial, short-term and so it will remain. Useless to talk but right for me to continue thinking critically, so long as there's the remotest possibility of having wider scope. Am preparing models and maps for my three Battalion commanders, who are good. A little like Napoleon playing with soldiers on Elba, but he did stage a comeback and he nearly got away with it. Plug along, because I must.

4 April 1943 HQ, 160th (SW) Infantry Brigade, Kent Got an embossed 'Mentioned in Despatches' certificate for December 1941 – pretty ironic bit of shoddery [*sic*] really. Rex was shocked to see something issued 'By the King's orders' torn up, but I felt better after that gesture.

Wrote to Archie Rowlands supporting the paper by Liddell Hart – a good effort demanding utmost support as the facts can't be disputed. It's as if a platoon of 50 soldiers had to be run by 25 NCOs. Poor old taxpayers. Poor old soldiers. No wonder our orders were late and bad, and Spartan was a muddle from beginning to end. The greatest number of Formation HQs that could possibly be required is 13! But that's the way we run things. We laugh at the Mexican army for being all officers, but at least it's not all HQs.

5 April 1943 HQ, 160th (SW) Infantry Brigade, Kent My circle includes the officers in this little HQ and the three COs, their adjutants and seconds in command. Occasional very brief contacts with the Divisional commander and his staff. Except for him and the three COs, none is professional or knows about the service world. Clubland, the War Office or serried phalanx of higher HQ are beyond my contact.

7 April 1943   HQ, 160th (SW) Infantry Brigade, Kent   Have been looking at myself objectively, trying to find out whether in my run of professional misfortune since I went to India I've been particularly stupid and brought disaster on myself. Undoubtedly had I lain low then, said nothing, been nice to the Little-Tin-Gods-on-Wheels and their deplorable wives, kept mute till I could get away, I'd not have aroused opposition and dislike. It might have been discreet, but also dishonest. Then to MidEast, accepted by Archie W 'with gratitude', Haifa, BGS to Arthur Smith. Gave them the key to Sidi Barrani – Dick O'C admits as much – then a red-hot report on his campaign, and I carried a lot of that awkward time when the MidEast was going to pot in the summer of '41. Then '42 and the stupidly conceived Liaison job, the report on the Western Desert, the Commanders school and finally DCGS and El Alamein. There's nothing they can get against me, except that I don't – nor ever shall – suffer fools gladly.

8 April 1943   HQ, 160th (SW) Infantry Brigade, Kent   It appears that this Division was adjudged by the critics to have been outstanding on Spartan – despite its fighting being designed by me. They can't very well ignore that. Monty does at least appear to have done something effective[21] – he may have caught Rommel really napping.

19 April 1943   HQ, 160th (SW) Infantry Brigade, Kent   So odd to feel I'm being punished without any word being said, as only our class can punish anyone it considers a disturbing influence. Liddell Hart in the *Express* has an article emphasising Rommel's successes, and suggesting obliquely that though Monty's army is capable of striking hammer blows, it's not sufficiently well-organised in mobility and flexibility to outwit a man as capable as R. It's a point the authorities will find it difficult to refute, and borne out by the lessons of Spartan.

21 April 1943   HQ, 160th (SW) Infantry Brigade, Kent   Monty is attacking again[22] but now from a difficult start line into dense and hilly country which won't suit Eighth Army. In fact Rommel has at last made his *Querencia*,[23] the lair he's chosen as a final battlefield, and we're going to have a lot of casualties before we get him out. I feel Monty will go on hitting as long as he can. Alex should take command in the field, and certainly move parts of Eighth Army into 1st Army.

1 May 1943 HQ, 160th (SW) Infantry Brigade, Kent War Office telephoned – Mil Sec in person, no less. The VIP – an old friend from India who wears two monocles,[24] is in England and demands to see me. Arranged to meet him in person there on Monday at 11am. He may only want to say 'How do you do, and Estelle is very well, Eric!' He may want me to go back with him to India, or he may be taking over something big here and want to give me the once over. I'll come in on that one, but the first is more likely. Why is he over here? Most intriguing. I asked whether the VIP's name began with 'A' or the other end of the alphabet. 'It's not 'A'.'

3 May 1943 HQ, 160th (SW) Infantry Brigade, Kent Archie Wavell feels his army is 'rather too orthodox'. Can't believe anything I'd be let do would help. Tried that business once from Delhi and warned the army-in-India for two long years as to what was coming to them, with conspicuous ill-success. Only way to get that sort of thing across – the way to fight Japs and lick them – would be to train a formation specially for the job, and then prove it in action. Will wait till he's gone and then drop a personal line to the MS saying I couldn't flatly refuse the FM[25] but have already had a good whack of the Orient and suffered professionally, so would rather take my chance with the chaps here.

12 May 1943 HQ, 160th (SW) Infantry Brigade, Kent Visit from a dull official from GHQ because the poor man's tired of being in an office so drives down from town with a pleasing-looking ATS[26] driver. Visit spoilt my day – would otherwise have been out in the country with troops. My mind's busy with a set of new ideas about training, without quite seeing the road to the end I'd like to attain. Always torn between the immediate interest of getting this little party as good as I can, and the wider, more diffused frustration from realising how little I'm allowed to do. The Army's a shrinking service affording less and less opportunity to men at the top layers, so it's inevitable that only those 'well in' can expect full employment. Should cease to think about anything other than the immediate task, but that's not easy.

According to Hitler, Rommel's health broke nearly a year ago after his July defeats in Egypt. I'm pretty sure disappointment almost broke his heart in addition. An interesting side-light on history if we can prove the Auk was the man who quenched Rommel's spirit, leaving Monty nothing but a shell to tackle. Looks like it.

25 May 1943 HQ, 160th (SW) Infantry Brigade, Kent Liddell Hart's letter says, 'I now hear a suggestion that APW may not return to India and that the Auk may succeed him.' Not impossible, and it would be a good thing. APW is too used now to making war in a small way with restricted resources, and has got into the habit of averting defeat, rather than organising victory. He knows much more of the philosophy and history of war than Auk will ever know, but Auk is more of a man of action, and if he goes to Delhi he'll purge some of the GHQ limpets pretty drastically. He won't ask for me because he only calls me in when he's in extremis, so that lets me out from going east. Basil L.H. also quotes General Fuller the other day. 'Is that the Dorman-Smith I once gave 1000 marks out of 1000 to in the Staff College entrance exam? There was a row over it. I was told that it had never been done before and was never to be done again. I answered, "Next time I will give him 2000, for his answers were 100% better than any I could have provided to my own questions".' So have established one all-time record, at least.

10 June 1943 HQ, 160th (SW) Infantry Brigade, Kent War Office interview[27] this afternoon, banal to a degree. Came out of it badly in my own eyes because felt I couldn't argue with the well-meaning, obtuse yes-man opposite. He was unconcerned with the mess the top boys have made of my brigade HQs, and thought it better to accept the situation rather than make a gesture which wouldn't get anyone anywhere. Didn't care a damn what his officers thought of himself or of me, and by the time he'd finished burbling on, neither did I. Efficiency wilts and dies away. 'Off the record', he said, he'd formed the impression that I neither cared nor heeded what I did because I felt my career had ended, which in his opinion was far from the case, only the authorities were chary of promoting me because I was felt to be 'agin the government'. Irishmen often were, etc. Out it all came. If I'd be a good little boy the authorities would relent. If not, I'd become a sub-area commander. Didn't attempt to impress my point of view as can compete with everything in life except the conventional English mind. Now can't make any resigning gestures without being accused of running out.

12 June 1943 HQ, 160th (SW) Infantry Brigade, Kent Army reform is the one thing which any soldier will be wise to leave strictly alone. It's only possible for a Dictator. There are many ways of fighting wars, and the old clumsy bludgeoning way is best understood by the majority. Men of integrity and intelligence and insight should not be soldiers, such men must go for the free callings – science, medicine, art.

3 July 1943 HQ, 160th (SW) Infantry Brigade, Kent In the carriage[28] opposite me were a couple I tried hard not to dislike. He aged 60 say, actively fat, well dressed, wealthy, full of selfish acquisitive aggressive movements. She dark, brown-eyed, tall, Spanish type, shaded moustache on upper lip, male conscious, well off, age say 28. Large diamond naval crown and a quite lovely Kerry blue terrier. These two monopolized one side of the 1st class compartment emphatically, with the KB on the seat too. My side was eventually filled by a shy RNLR officer and his shyer little wife. But the plutocrats were glorious – everything I'm fighting against, even though brown eyes was prepared to recognise that I might, from my appearance, be a tolerable member of a world they consider suitable to live in.

5 July 1943 HQ, 160th (SW) Infantry Brigade, Kent Late last night Liddell Hart rang up from the Park Lane, hoping for a meeting. He said my last letter was composed entirely of epigrams, which seemed to please him. He himself totally lacks spontaneity or any light. Asked him what else he expected of an Irishman and discovered that he never realised I was Irish. So much for Anglicization!

16 July 1943 HQ, 160th (SW) Infantry Brigade, Kent A distinguished visitor. Men are queer fish! Am sure a new commander of his status would make a much better impression if only he could bring himself to turn up in one car minus motor-cyclist outriders, and without corps and divisional commanders. It's ridiculous that so imposing a cortege should arrive merely to look at me, and made me bloody-minded. The great one has an impressive aura. He carries his personality outside his body instead of inside it, but looks old, grey and fatly-podgy. We paced about the garden. He's humourless and unoriginal but with plenty of determination, and I doubt he'd mind being unpleasant. He went off followed in strict seniority by the other 'big shots'. Afraid I created a poor impression, but can't do with pomposity.

19 July 1943 HQ, 160th (SW) Infantry Brigade, Kent Travelled back solo with a red girl and a black dog, she so smart that the very powder on her said 'Rendezvous', and not a long rendezvous as the grip was too small. She read my paper eventually but lived way off in her thoughts, till just before Chatham she put things in order. Hair to my eyes quite excellent (Venetian red) and some more powder, then to lean out of the window till a man came in – a commander RN – fell over dog and into her arms and then almost sat on her in a corner of the carriage. He shook all over and she was sweet with him. Her rather spoilt face softened and grew gentle in anticipation – phew! I almost howled. He was so in love with her and she, I think, responds. Left them at Sittingbourne, climbed into my car and drove back to this empty-hearted place to feel the grey hours clamping down on my neck with cold dragging weights.

The Daily Mail *dates the beginning of the end of the war to this time last year at El Alamein. Truth is coming out. They link El Alamein and Voronezh[29] as the two turning points. Oh well, a little patience. But it's hard to be patient when one might be contributing to the ending.*

Such speculation tantalised Chink in his isolation, especially when it became known that the Grand Council of Fascism had met in Rome as early as 24 June to pass a vote of no confidence against Mussolini, and that his old enemy had put up no struggle. Mussolini was now out of the war: a turning point indeed, and Chink realised that Hitler had done nothing to prevent the Fascist collapse.

9 August 1943 HQ, 160th (SW) Infantry Brigade, Kent Hard to believe that Mussolini and the Fascist regime in Italy collapsed abjectly over two weeks ago, yet Italy and ourselves are still at war. Evidently we're no longer fighting Fascism there, so interesting to consider what and whom we're fighting. Relations between Germany and Italy are no longer political but strategical, concerning *Wehrmacht* policy. Italy is on the same standing as Rumania now the Musso–Hitler link has disappeared.

19 August 1943 HQ, 160th (SW) Infantry Brigade, Kent Local WASP[30] girls and girls of the village teamed up for a dance, beer bar etc. Went down to see the fun. One girl amused me. A strapping land worker about 22 with the figure of Juno – would look marvellous stripped. The chaps were after that magnificent anatomy like flies after jam. She'd like bed but she'd never let herself go. Talked to her, and then to the WAAF[31] officer, about 28 and typical head girl, captain of hockey team: nice but totally unreal. Having done my bit went off to bed and read. Not seen a paper for 24 hours, a serious matter. The war's reached a curious pass. Superior reserve powers of the UN are beginning to tell – we're able to apply more brute force – and apply it more brutally – than the enemy. Warfare's become a business, with science behind it and government geared to it. Technique is established, and the art of war (best seen in lean years) yields to the business of war. Until it ends in German collapse there'll be no more major developments in technique and no great calls on Generalship. Soldiers will die necessarily and unnecessarily, but the results are in the end inevitable. We now plan and design offensively with the necessary means for successful execution.

31 August 1943 HQ, 160th (SW) Infantry Brigade, Kent Caught the train comfortably and dozed like a bloated pug-dog for most of the way. There was a very tired grey woman in the carriage with a boy about 11 and a younger girl.

She seemed so alone, poor person, and so utterly clogged with children. She'd given up all hope in life. Not yet 40, I estimate. Hair anyhow, hands gone to pot. So sad. The children were restless and noisy and uncontrolled. I was so sorry for her. Returning, found a letter from Liddell Hart. One quote: 'Your letter with reflections on the real background to the present situation impressed me so much that I had a copy typed for Herbert Morrison[32] and several others. Important to make people scratch their heads and think.' Amusing if it falls into Winston's hands.

1 September 1943 HQ, 160th (SW) Infantry Brigade, Kent Tomorrow is a day of prayer, so the vicar's snorting around for soldiers to swell his congregation. Won't go myself – such undertakings seem so hypocritical. Equally am making it voluntary for the soldiers. Glad it's not peacetime. A brigade commandant who didn't go to church would soon get spoken to.

To Basil Liddell Hart   160th Infantry Brigade
Sittingbourne, Kent
17 September 1943

My dear Basil,
    Until Mankind finds some way to abolish war, the main form of international political action, warfare must be regarded as the most permanent occupation of mankind next to prostitution, religion and agriculture. There is no such thing as peace unless and until the whole of mankind agrees to put warfare on the same plane as cannibalism, which is where it belongs.
    Best luck, Chink

19 September 1943 HQ, 160th (SW) Infantry Brigade, Kent Saw my Divisional commander last night and congratulated him on beginning his second year, and myself on starting mine in disgrace under his command. That began a conversation in which he maintained that the powers-that-be were playing fair and intended to give me a division. Told him the truth as I see it and he was pretty shattered. Don't believe he realised that so many high up were gunning. Told him that if he were wise he'd put my successor in now for the good of the service. He finds me an embarrassing subordinate, but oh no! He couldn't do that. It would be unfair on me. Answered that I was so used to unfairness that a little more wouldn't hurt and left him with the barb in his mind.

23 September 1943 HQ, 160th (SW) Infantry Brigade, Kent What does it matter how Archie Wavell or Auchinleck think or feel about me? If it's in the interest of the state that I should be employed, it's their duty to do so without fear or favour. But if, as it seems, it doesn't matter either way, why waste time thinking about it? My most dignified behaviour is to do without complaint whatever I'm ordered and seek no favour from anyone.

28 September 1943 *Exercise 'Paul'.* Woken at 3am to go out into the rain in pitch dark driving a jeep to watch some night attacks. Returned at 7am wet to the skin. The Downs were swept by a drenching sea-mist which swamped everything and sent a miniature torrent through my lower HQ. Today's been full, ending at 10.30pm, then full of ration rum to bed. Tomorrow has one of our big attacks. Lots of artillery shooting and a real good splash. Spent hours today buttoning up the action of guns, tanks, infantry, engineers etc with a lot of decent chaps, brigadiers and Lt Cols, who only half knew their jobs. Begin to think there really is something in experience, if only as an alternative to intelligence. The poor soldiers have been dismally wet, but that's part of their training. Eventually changed my wet clothes, stripped, washed and shaved in my blacked-out car. Pretty Spartan! This intensely real yet utterly artificial life absorbs us all. Eating everything I can, but still hungry. The air up here is wine. Two flights in a Moth aircraft over the chaps who never know from which angle their brigadier is looking at them. They're all trying hard to do their stuff. If only some top boy would come and look.

29 September 1943   HQ, 160th (SW) Infantry Brigade, Kent   Rise at 3am tomorrow and stumble about in inky blackness to watch a night attack. Won't be much to watch, just muttered 'sexual acts' in Welsh [swearing]. Exercise going well, I think. People are settling down and beginning to feel they'd better do things properly because if they don't they're bound to be found out in the next ten days. Shot everything we had today and glad nobody was hurt. Time enough yet, for we're three more days with live ammo. No more rain but nights really cold with a ground frost. Poor soldiers haven't the stimulus of a live enemy with lethal means, so they feel the discomfort. They're good and patient and most try hard to be right. Have been busy stage-managing, producing, playwriting – a sort of Noel Coward of the South Downs, Actor, Audience, Author!

My Divisional commander tells me with pride that he's spent all night at a party and not been to bed. Fatuity! Yet he gets a division because he's 'safe', though everyone knows he's bone idle. Liddell Hart has potted a group of my letters into the foreword for his book. I must edit it. To bed without undressing – the hardy warrior.

13 October 1943   HQ, 160th (SW) Infantry Brigade, Kent   A note of appreciation on the exercise from the gunner Brigadier. "Paul' was to us a success unqualified, and is classed as one of the few really worthwhile exercises by all who took part. Thank you very much for the realism.' Gratifying, seeing we had a lot of artillery behind us and training is dull for gunners who don't get shot at back.

Been to a nearby drill hall to see the set up for tomorrow's talk on 'Paul'. First class and must strive to do it justice, so after dinner will make notes. So much I could say – outstanding fact being that we can't make 1943 warfare with a 1918 army. Winston's refusal to introduce any major social reform was inevitable – he's leader of the Conservative Party, after all. It shows we can expect few fundamental changes until the Tory majority is appreciably reduced.

14 October 1943 HQ, 160th (SW) Infantry Brigade, Kent The almost top Panjandrum of parsons from Army HQ has elected to see the Division, and invited himself to tea on 4 November. Let him come in office hours by appointment. These parsons are getting too big for their boots and am certainly not playing, so have written accordingly. It may raise a rumpus as higher authorities seem to rely on some C of E God-of-battles to rescue them from their tactical errors on the day of fate. From the fuss the lesser padres are making, he might be an Archangel. Snobs and lickspittles! Enlivens the tedium to have a smack at somebody.

22 October 1943 HQ, 160th (SW) Infantry Brigade, Kent A letter from Estelle, difficult to understand. It's tolerant, humorous, affectionate, yet she seems to realise that even if Eve never existed I'd have difficulties in facing her again because of all these failures. Poor E doesn't realise that we couldn't survive together thrown only on our own resources. We'd both go bats if we hadn't soldiering to support us.

Basil L.H accepts most of my prunings from the foreword culled from my letters, and protests at my 'mock-modesty'. 'I would reiterate my emphatic disapproval of the way you look at yourself. It is not justified, even as a reflection of others' views. For in my own range of contacts I have found that you have quite a number of warm supporters in fairly influential circles, and when the facts become better known I am sure there will be a big swing of opinion, as in Boney's[33] case. As for the critics, it's amusing to find that they usually express their view in the same one-word opinion they used of him, 'Unbalanced'. I remind them of what George II said of Wolfe when people complained that he was mad: that if this was so 'I wish that Wolfe would bite a few of my other Generals.'

11 November 1943 HQ, 160th (SW) Infantry Brigade, Kent Armistice Day. Since lunch – a good solid military meal, everything stewed to nothingness – heard as reliable gossip that Ritchie is coming to be two above me. That makes a nice layer of inefficients. Told my informant that, if true, he would be very uncomfortable with me under his command and that I'd better go somewhere else. Clearly that's necessary in view of what happened in July 1942. That remark's being passed to Ross[34] tonight, and I hope he'll get cracking. To think they'd re-employ the man who by his stupidity and ignorance lost us Tobruk in command of anything real defeats me. Still, there it is. Another letter from LH, in which he writes, 'What a brilliant satirist you would make if only you would exchange the sword for the pen, and, incidentally, learn punctuation.' With this new shadow over my fortunes, am unlikely to get much choice as to which I use.

14 November 1943 HQ, 160th (SW) Infantry Brigade, Kent Am told the War Office has accepted one of my ideas as 'a good practical suggestion to help the infantryman'. Damn their condescension but thank goodness for the poor soldiers' sake. Wrote back to ask that the system be named after the Battalion which first tried it out at my direction. They'll probably call it the AlanBrooke system or something equally fatuous. Think I've saved a lot of lives. Had also submitted it through the usual channels and understand it was turned down as useless by Corps HQ – probably because it came from me. Have heard nothing from Ross about Ritchie[35] but believe the two have already met, so Ritchie will realise that I'm going to be in his command. A most anomalous situation.

Later that day Chink learned that he was to be moved, but with no indication as to where or when. He at once put in a formal request to depart quickly, stating that his contempt for the incoming Corps Commander would be impossible to cover up since his men knew of their previous confrontation. 'So unfortunate for General Ritchie …', he politely worded his ultimatum.

16 November 1943 HQ, 160th (SW) Infantry Brigade, Kent Rang a friend at Divisional HQs to ask if he knew what was going on. Gathered that things are happening about me, though when asked if I should begin packing he replied, 'Oh, not yet'. Ross hasn't answered – he's a bit of a moral coward but may be trying to work a wangle and get a friend of his in my place. I may misjudge him. It's almost exciting – better to be kicked downhill than to stagnate, but to most people nothing's more stupid than to offer oneself for crucifixion on a point of principle. Suspect this really is the end of my military life. The Big Shots won't put up with rebellion twice, though I dressed up my offer to withdraw as made in Ritchie's interests. Ran in the cold winter air for four miles – as keen as a knife and crystal clear.

Eve had changed her surname to Dorman-Smith by deed poll at his urging, although neither had yet applied for a divorce. He spent all his leaves with her in isolation, and she stoked his anger.

To Basil Liddell Hart   HQ, 160th (SW) Infantry Brigade
                                                        17 November 1943

My dear Basil

I'm waiting to go off to Divisional HQ's to learn my fate from Ross. I gather that the authorities agree that Ritchie and I should not be in the same Corps.

I don't expect to be in this job for long now, and don't expect another job. A good thing in many ways because I've been stagnant for 14 months. Too long at my age. So, my dear Basil, goodnight. I may even be an honourable and free civilian when next I write to you.

Yours ever, Chink

PS [4 days later] I've just heard that I will be ordered to leave this brigade this coming week and to await instructions as to my future (if any). I'll keep you posted, but c/o Royal Automobile Club will always find me.

19 November 1943 HQ, 160th (SW) Infantry Brigade, Kent Apparently some senior military secretary official is visiting Divisional HQ and seems to be in the know. If I don't hear anything tomorrow will write to Ross suggesting that as I'm to go I'd better get on with it, so not here when Ritchie visits the Bde. That should stir the frog puddle into activity. A dull train journey brightened by the patter of the daughter of people who shared my first class compartment. She was a pretty, under-developed, brown-eyed little mouse who asked what I did. So I said, 'Soldier.' And had I been fighting? 'A little.' 'How long had I been a soldier?' 'Much longer than you've been alive.' 'Oh, that is impossible. I'm nineteen.' Rather sweet. All very pure, but at Chatham had to change trains and partners. My time here and my work here draws to an end. There may be developments tomorrow. Apparently some senior military secretary official is visiting Divisional HQ and he seems to be in the know.

On 21 November the Assistant Military Secretary phoned Chink to say that on General Sir Bernard Paget's[36] direct order he was to vacate command of 160th Brigade, and stay on leave of absence until further notice.

The obedient weeks accumulated without word, however, and his sense of outrage mounted. Even his foreword to Liddell Hart's book The Strategy of the Indirect Approach *had to appear anonymously because Auchinleck's despatches were not yet out. On 19 December he wrote to the Under-Secretary of State for War to say that it was increasingly difficult to explain his suspension from duty and requested a formal explanation 'to protect myself from possible calumny'. On 28 January 1944 he was told to relinquish the temporary rank of Brigadier and was placed on half-pay. He was back down to Colonel's rank. Asking for a court of enquiry, he turned down the offer of a junior command in Northern Ireland on the grounds that it was 'a noncombatant vacuum'. On 3 March, despairing at the silence, he notified Wemyss,[37] the Military Secretary, that he was prepared to serve in any rank anywhere as long as it was on active service.*

On 30 March a telegram was delivered, and he tore it open. 'You have been selected for the apt of commander of an infantry brigade Allied Armies in Italy.' He had been taken at his word. He was to serve on the beachhead of Anzio, which he and Liddell Hart agreed was turning out to be the worst shambles in the war.

But optimism took over. His 160th Infantry Brigade experience would stand him in good stead, and after command of an operational brigade came a division if he was recommended by the four monthly report.[38] He promised Eve to write, made a will bequeathing his military papers to Liddell Hart, and attended a command 'séance' with the Military Secretary on behalf of the Army Council for a 'let's kiss and be friends' settlement. 'I feel', he wrote hastily in advance of that meeting, 'I will not be very forthcoming.'

To Basil Liddell Hart   c/o RAC, Pall Mall
                                                                            2 April 1944

My dear Basil,

I'm off to Italy to command an infantry Brigade, as yet unspecified. I'm uncertain whether the enemy or our own side object the most to the appointment. At any rate Alex put up a sustained if unsuccessful resistance. But they had to do something with me and this with its Uriah the Hittite[39] flavour is the best way out. I'm glad, even though it means that for the third time this war I must begin again at the bottom of the class, and under a new cloud – or perhaps the same old cloud intensified.

I don't look forward to the military situation: both Anzio and Cassino are extremely unpleasant and dangerous operations. However it's something to be going off to the war again at my mature age, and to a job which should be done by someone ten years my junior. I certainly am sustained by no illusions as to what I'm defending – Churchill, Grigg, Brooke, the City of London, individual enterprise, the rights to unlimited profits, Toryism and all – uninspiring things, even when contrasted with Continental Fascism. I think I'm fighting for the right to fight, and I only hope to survive so that I can return to wage war against the Anglican Junker system which promotes Ritchies and generally regards only its own interests.

I feel much as Raleigh must have felt when he set off on his final voyage, glad to be free of prison and loyalty and hope! Meanwhile, continue to be a signpost to good sense and a warning to tactical road-hogs. They also serve who only point and wait.

Yours ever, Chink

Anzio:
April–August 1944

The beachhead landings at Anzio had gone wrong from the start, due to a last-minute shortage of landing craft which severely limited the number of troops that could be got ashore. A combined Allied operation, it depended on swiftness and surprise for success, and the odds had been reduced further by the over-caution of Major General Lucas, commanding US VI Corps. On landing in January Lucas had ordered a halt for the first 24 hours, which gave the enemy in the surrounding hills enough time to pin them down under fire. 1st Division, containing 3rd Infantry Brigade, had drawn the worst of the fighting with the majority of casualties, as Chink was about to discover. 'Shambles' would prove to be an understatement.

21 April 1944 *Algiers* A good trip out, and my fellow passenger was Harold Nicolson MP – good company and a curious type: a National-Labour representative but not genuine Labour in mind.[1] He loaned me *The Structure of Morale* by T.S. MacCurdy, interesting psychological stuff. Jumbo Wilson[2] was at conference with the French, so haven't seen him yet. Found the officers shop and bought a raincoat, thick pair of boots, two blankets, two flannel shirts, and another beret. Better equipped. It's raining and quite cold. Algiers is very noisy, smelly, crowded and war-dingy. Pictures of Giraud[3] with his face stamped over by the Cross of Lorraine.

On arrival next day at Anzio, via Naples, Chink was met by the commander of 1st Division, and with misgivings recognised him. It was Ronald Penney,[4] pompous contemporary in his old Staff College syndicate of four whom he had often ridiculed during their hours together, and given little thought to since. Obviously Penney had not forgotten. 'I did not want you at first,' Penney said flatly when Chink mentioned his War Office briefing about low morale, adding 'and I do not want you now.' Fortunately Penney was said to be about to go on sick leave, so Chink put the awkward encounter behind him.

24 April 1944 HQ, 3rd Infantry Brigade CMF[5], Anzio Written in a mess tent in an olive grove, my new HQ, in the act of taking over. The outgoing Brigadier seems a first class fellow – 40 years old, good-looking, bold. He goes to another division to command a Brigade – part of the juggle to fit me into this army group. Attractive here except for lethal noises, and this afternoon went out along the coast road where the country was green and sweet, swelling with flowers. Away in the distance the hills are remote. Undoubtedly we're good colonists. Winston need not worry – 'What we have we hold'. There's even a brick fireplace in the mess tent! But perhaps we're better 'Stay-Putters' than 'Move-Onners'? 9pm news – damn, the Boche are jamming it and nothing coming over except German.

It seems both familiar and strange, because we live much the same life as 160th Brigade do in Kent, but the faces and names are new. I like the look of these young men, but haven't yet seen my troops. Two loud cannon shake the tent.

25 April 1944 HQ, 3rd Infantry Brigade CMF, Anzio Rained last night and my dugout leaked abominably so my few things had to be dried out to be worn today. Went about yesterday and saw the chaps. Tomorrow my predecessor leaves me to carry on. The enemy were tasteless enough to shoot at us, but did no damage except to my nerves. Don't think the current war has any parallel to the life we live here. It's more like 1915 at Ypres. Soldiers are learning to go to ground again, with dugouts of every sort appearing. My predecessor, Jimmy James of the KSLI,[6] is certainly not 'politically minded'. Unprompted he expressed doubts about the attitude of ordinary men towards the future and the present – says the further 'left' the better, as far as they're concerned. Tragic that after the stirring words of 1940 our national leaders are so tongue-tied. This command's had a hard and disappointing time[7] but has passed its reaction period and its tail is going up again markedly. Will know more after seeing the front-line chaps – must be like drowned rats today. One of my battalion commanders is 'Bunny' Careless,[8] Cassells' Controller of the Household, who told me he'd recently met Elaine's[9] husband, Kenneth Bols,[10] in Naples.

Am gradually collecting equipment – a camp bed today. A black market of barter exists between us and the Yanks, so prize bits get swapped. Money won't buy anything here – no place for millionaires. News reception bad, what with atmospherics and jamming. The Great Monty doesn't cut so much ice out here as the press would have us think. I still believe that steady, sympathetic, firm, efficient handling is the real recipe for high command. Showmanship may have some justification in the forward area, far less behind. Wouldn't be in his shoes, anyway.

26 April 1944 HQ, 3rd Infantry Brigade CMF, Anzio Take over full command today with Jimmy James' departure. Feel a little strange to them all because they've just had a pretty terrific battle experience which I haven't shared and my transition from peace to war has been so quick I've hardly adjusted. That will pass when we've had enough fighting together, unlikely to be long delayed. The beachhead is the most interesting place professionally. Newspapers give an accurate map, more or less, of its dimensions but, big as it is, it's astonishing how little spare room we have for our Anglo-American venture. Can't escape the impression that it's a somewhat Pilgrim Father affair, with the harbour full of assorted Mayflowers. The natives on our frontier are definitely hostile, and we're colonising under difficulties. We've even got a hen-run and a dozen good laying hens. The enemy does some day gunning and night bombing but surprising how much he misses, considering how thick on the ground we are.

Found a young Italian liaison officer living in the Mess, a nice fellow, remarkably at home and speaks fair English. Wonder what he makes of us? We've been here nearly three months, so the chaps who landed first are old inhabitants. They had stiff fighting when they started to advance as Boche counter-attacks were determined affairs, so their war experience will stand them in good stead. Interesting professional problems about the proper modern way of exploiting a sea-borne surprise. Undoubtedly difficult with our present obsolete arm, unless simultaneous group landings with very large forces and possibly carrier air support.

The outgoing brigadier is going off to say goodbye to Bde HQ personnel. He seems a particularly fine type: pleasant, efficient. They moved me into the Brigade to put me under a divisional commander of my own seniority, which was rough on him. They've just given three cheers and I feel a bit of a beast. He's undoubtedly got them with him. The other two brigadiers are junior, one a St James's guardsman not up to the calibre of this chap, but it's unfair to make first impressions too definite. Haven't seen any of the Great Ones. Jumbo [Wilson] was away from Algiers, and I only had a night in Naples so didn't see Alex or his CGS. Suppose if I fill the bill here, I'll qualify eventually for something better, but the average brigadier is many years younger with recent experience. Trust my probation period won't be unduly prolonged for am sure to get attached to these nice fellows.

Spent both mornings in the front area near enemy positions. We're almost as close to each other as we were in France in 1916. Extraordinary situation at this stage of the war, dictated mostly by the ground. Part of our front lies in a system of deep water courses running like branches of a tree to the sea, and while we're in one, the enemy's in the next. Awkward trappy country, but elsewhere more open. Anzio Archie, an enemy high velocity gun, is putting stuff down near us and also firing into the harbour down the road. The night is warming up and our guns are getting noisy. The Anzio barrage is a spectacular party, and my sandbag hut is neither deep nor solid. A big fire is burning somewhere and the flaks[11] still coming down. Damnation, a new lot. The young moon is bringing them in.

27 April 1944 HQ, 3rd Infantry Brigade CMF, Anzio Busy all today. Went round part of our front, spoke to chaps and attended a demonstration. This country is really delightful in the spring weather – flowers everywhere, nightingales singing in daylight. Everything lovely except man, who is vile. Rome is really so close – a short jeep journey, were it not for the enemy. My predecessor went off today by air, so am now in the saddle. All I ask is that they'll leave my battalions under me so I can get to know them. There's a tendency to split things up here, which I dislike. Must go carefully, for my immediate master is inevitably going to feel things a little different. He's not an infantryman and somewhat my junior. So awkward when people break up the seniority business. News is desperately hard to come by – miss the papers badly, even the French ones in Algiers. Opinions are as important as the bare news.

Can't believe I was in England seven days ago – that life seems intensely remote. Have moved into the Brigadier's dugout, more commodious with a solid roof against rain and splinters; even a mirror taken from a deserted cottage. Most local inhabitants have gone by now, but even in the earlier fighting one guardsman managed to get VD, and a corporal of Sappers got engaged to the station master's daughter at a little halt almost in the front line! We feed early – it's 6.45pm but won't go over just yet as the enemy's doing a spot of shelling round here. It doesn't deter the evening football match – when soldiers play football they forget everything else.

8.15pm. Still chilly in my dugout, so I'll go back to the mess where there's a fire. Tomorrow must have a military haircut. Civilian locks too long for this life.

30 April 1944 HQ, 3rd Infantry Brigade CMF, Anzio *10pm.* It's a dull war here, now that our first effort's failed. Dull yet dangerous, much like 1915 in Flanders. Position warfare, with the forward companies close to the enemy. In places each side can hear the other talking. Men in the line live all day in sandbagged trenches and ravines, and there's only patrolling at night and shelling and mortar fire by day to relieve the monotony. Behind the front, between that and the sea, lives a great concourse of troops engaged in vital but non-fighting business, and the whole area is subject to spasmodic shellfire and air attacks. These do little damage, particularly air attacks which meet terrific flak. It's a wearing life to soldiers used to the greater freedom and security of North African warfare, but they'll have to settle down to a life much like their fathers underwent. Certainly the enemy isn't having it all his own way. In spite of having plenty of room behind him in which to disperse, he suffers from our constant fire and this is a noisy place with all the large calibre artillery at our disposal. The countryside is varied with great stretches of scrubby woodland but is steadily accumulating damage, and if we're here for long will look like Ypres.

I like the people with whom I work. My brother brigadiers are Erskine of the Scots Guards and Gore of the Rifle Brigade, both much younger, of course, while my COs are younger still. Am a bit of a Methuselah! Can't help that, and they'll keep me at this job while I'm on approval. I'm glad of the battle experience at this level, particularly in a place where my last war experience will stand me in good stead for most people here have absolutely no experience of it.

Meanwhile the world waits for the second front[12] to begin, and much will depend on its early phases. Even if it settles down to a great series of Anzio salients, it will force the enemy into battle on all possible fronts, and he's less well equipped for a slogging match. A great American army is yet to enter into action and the Americans can, with experience, be formidable fighters. Provided we can establish ourselves and form, say, four or five 'Anzios', we'll give the Germans a lot to think about. Presumably the Russians and everyone else will push at the same time. Glad I'm not a War Office or Home Forces general, or even behind the front in Italy. This is the proper place to be, no doubt about it. Last year was desperately lowering to my morale, but that's all over now. This is the return route to normality. Whatever becomes of me professionally, haven't shirked my duty anywhere.

1 May 1944 HQ, 3rd Infantry Brigade CMF, Anzio This afternoon visited the Brigadier on my flank and after a bit he said, 'I used to be at GHQ Cairo. Are you any relation to Chink Dorman-Smith?' So I admitted it and he couldn't believe it, saying I looked years younger – left him muttering.

We live among the battle noises, rumbles, whangs and great C-rumps, then the little Vickers and Spandaus go stammer, stammer, stutter, stutter, and the Long Tom's blasting roar drowns everything. Will spend tomorrow night up with the Battalions seeing what goes on in the darkness when Troglodytes come alive on both sides. It's ridiculously like the last war. Feel as if the years between had miraculously peeled away and I was a 20 year old back in 1915. Totally different from desert fighting, even from Tunisian and Sicilian fighting, and the chaps undoubtedly feel strange in the new stresses. Psychologically they find it more trying than movement warfare. They've got to learn to dig like beavers and wire like spiders, a whole host of new tricks of this siege trade. Above all, they've got to learn to endure.

It was tragic that we didn't make more of the initial landing here and, as the papers made clear, the enemy ripostes were skilful, savage and determined. Our casualties and his were far from light, but as things are we're a thorn in his backside. He has to keep us sealed off from Rome and that's a costly business, even though the country favours defensive fighting. He must realise that our extra-powerful artillery is as potent for offence as for defence, and he seems resigned to our free sea movement with all that implies. I really believe that if we keep at him wherever we can get our teeth into him, he'll crack. It's he who now feels we've the better weapons and more of them. In some ways wish I had my 160th Bde people here. They'd learn quickly after the first shock of battle, and they'd be fresh and keen and hard. They'll do Monty proud.[13]

2 May 1944 HQ, 3rd Infantry Brigade CMF, Anzio *8.15pm.* A busy day. Wandered round the foremost HQ and later a conference. Have got my policy fixed for defending what I've to look after, but it will take about a week to get things shipshape. As in the last war, daily shelling and mortaring takes its steady toll, slight but cumulative. Am going forward soon to see our rations go up the line. Want to see how things really get done, though people know the routine so well that there's seldom a hitch.

A lovely day, hot and calm, and am already sunburnt and blistered, if short of exercise. Had a bright little Mess dugout made with Pin-up girls from *Esquire*, and we're getting it running more *comme-il-faut*. My dugout is now re-floored with bricks, and hope to have its ceiling whitewashed. Also having a proper Control dugout made from which I'll be able to coordinate everything on my front – except the enemy. We've just chased away a Boche recce plane with gunfire; we'll have more. They drop canisters of anti-personnel bombs into the back areas, and occasionally on their own people – not un-amusing.

3 May 1944 HQ, 3rd Infantry Brigade CMF, Anzio Our forward positions lie in a network of precipitous water-worn ravines, deep in places and confusing. They give a fictitious feeling of security, but are absolute traps unless one holds the top. They're uncanny places – might almost expect to meet a tiger.

Have been wondering so much about the chaps, and wish I could get into their minds. So many must be vague as to exactly why they're in Italy fighting Germans. Both Russians and Germans can see clearly that they must defend their native soil. To the Russians that must be profoundly inspiring, and even the Germans know their backs are towards the Fatherland. The issue isn't so clear-cut for the average uneducated working-class Briton or Yank. It's not easy to make him realise that Germany's main war aim is our elimination as a world power. Bless his heart, he does his stuff to the best of his ability, but I wonder how much he really believes in it all. It's a problem how to enthuse him. He's a poor hater.

4 May 1944 HQ, 3rd Infantry Brigade CMF, Anzio An old acquaintance has just taken command of this division, Ginger Hawksworth.[14] We were fellow teachers at Camberley. He's for so long modelled himself on Monty that it's ridiculous. Same incisive tone, same repetitions of the telling phrase, same mannerisms. Find it hard to keep a straight face. He's become his own ideal of a general. Suppose that sort of success-hunter has to merge his individuality into a model, but how grim. He has no poetry or warmth and keeps saying, like Monty, that he's 'out to win the war'. But he knows his job well enough. Am wearing shirt and shorts – more comfortable; battledress was getting stuffy. We can really dig down now. Before, we had to build obvious breast-work defences, difficult to make bullet-proof. At our HQs in the wood we've a little cat who joined us and hunts mice in our dugouts. She came to the dinner in the Officers Mess tonight and behaved beautifully.

5 May 1944 HQ, 3rd Infantry Brigade CMF, Anzio A cheerful morning in the front posts, where being a cunning old bird I chose the enemy's breakfast-time. That spot has an unenviable reputation as being too close to the Boche, but is better organised for defensive action now because I'm making the earth fly everywhere. That's partly because of the mud, and even more because due to that mixture of ignorance, idleness and solid unimaginative courage – our contribution to warfare – we'd done little to make ourselves strong on the ground. I believe the British soldier, officer and man, thinks of battle life in the order:

a) The next meal, including a meal between meals.
b) Somewhere dry and reasonably safe to sleep.
c) His fighting positions.

He's incurably a civilian and amateur. This life in close contact will do the men good. They don't know how to make bullet-proof posts, or realise that the art of survival lies in digging while suppressing the enemy by sniping and mortaring. They're learning to really watch him, and becoming callous to casualties. Think they've recovered from the shattering early battles, and morale I judge to be good. Journalists have made an unnecessary tall story over the dangers and discomforts here. We enjoy lovely weather, and inhabit holes and caves with plenty to eat. Dull, perhaps. There's nowhere for troops out of the line to go, but we've hot showers and a mobile cinema. Anzio and Nettuno, formerly pleasant little towns, are far from inviting now.

We all wonder when the great Monty will issue a flamboyant order and launch his armada[15] across the Channel. We battle-scarred veterans are cynical. Monty is not looked on as a superman. The glamour of El Alamein II has faded, and he's no other adequate laurels to show. No question but it will be a terrific struggle. I picture Basil Liddell Hart with a seat in the back-row of the stalls, absorbing red-hot history and muttering, 'I told you so.'

The enemy chucks propaganda leaflets at us a lot. He's entitled to crow over Cassino,[16] a theme he combines with variations about the bombing of London and the social life of Yankee 'blockbusters' in the houses of olde England. This he illustrates with seductive sketches of American soldiers busy with luscious nudes. The men like the nudes and pin them up, having cut out the Yankee! But what's the use of propaganda on a people like us, with almost no imagination? Very little indeed. The sad little graves left over from battle in the positions we now hold are better propaganda. Am getting a move on the things I want done, and hope people won't think I'm chucking my weight about just because I'm ambitious or anxious to retrieve my fallen fortunes. That's not the case. I'm happy to be here and content to have my career go the hard way. Count myself fortunate that I've been allowed to run myself in, and am full of admiration for the company commanders who carry the real heat. The ones I've seen so far are very good types, bless their hearts.

6 May 1944 HQ, 3rd Infantry Brigade CMF, Anzio Met Vaughan Thomas[17] of the BBC in my sector. Thanks to a new communication trench he was able to go into one of my forward positions and this afternoon we had a chat. He knows Reg a little. Seems a decent little chap, an educated Welshman, but I was glad he was able to go into the *avant-poste*. He'll get a better idea of the real soldier there than elsewhere in the beachhead. Think he was secretly relieved to find that I'm taking work in the forward area seriously, which when finished will really save lives. Want to get our sniping effort better organised, too. People may think I'm regarding our length of stay through pessimistic eyes; not so. I'm no Monty, but don't believe in putting off the defence work of today because of the hypothetical advance of tomorrow. Back to the old game of snipe and shovel, a type of warfare believed to be obsolete.

Now dress for dinner, which means getting out of shirt and shorts into battle-dress. Necessary because mosquitoes become pretty enterprising as it gets towards sundown. An incredible full moon, too bright for comfort. In daylight most of the beachhead is blanketed in smoke, purposely created from chemical generators. We're well supplied, except for whiskey. Cyprus brandy's a bad substitute. Light reading much appreciated by the mess, but I miss the *New Statesman* and *The Times*. Important to keep men here well informed about outside thought.

8 May 1944 HQ, 3rd Infantry Brigade CMF, Anzio Am as busy as Uriah and enjoying a situation which has so distinguished a precedent, only the Army Council has no Bathsheba[18] (and wouldn't know what to do if it had!). All the little complaints of England have left me. Sleep so well that last night two 150 HE[19] shells burst within 50 yards of my shack and I didn't waken. Killed a viper this morning up the line.

9 May 1944 HQ, 3rd Infantry Brigade CMF, Anzio At 5.30am they tell me the night's reports. At 6am I rise. Breakfast at 7am. Operation room till 7.30 or 8am, then up to the line to visit my subsidiary HQs and probably a company or two in the forward positions. Our communication trenches aren't all they should be, but 7 days ago they hardly existed. Back to HQs by 12.30 for quick lunch and planning for the afternoon. Sometimes go up again in the evening, for there are places can only go to in the half light. Dinner 7.30pm and more work in the operations room. Bed about 11.30. A longish day with plenty to think about. Without a doubt the place to live here is underground.

An American group of guns close-by make a damnable din. This HQ isn't well situated, but so fully established that I can hardly move it. New operations dugout is a fine affair, plenty of room and fair security. There the BM and I work and brood. He's a good lad, Harry Leask[20] of the Scots Fusiliers, a little man with a large moustache and strong right-wing sentiments. Very regular army and efficient, with plenty of staying power. He finds me trying, because my predecessor let him do things his own way. The staff captain is a nice chap with a sense of humour and very efficient. The intelligence officer's first rate: large, cheerful, keen. I've three good Battalion commanders: Hackett, a Sherwood Forester, Bunny Careless of the KSLI, and Webb Carter of the Duke of Wellingtons. All good, reliable men, with fair battle experience. I think we'll make a good team when we get to know each other. As usual with soldiers, the beachhead is full of canteen rumours of impending battles, and am almost the last to hear of them. We'll all be glad when something develops to shift the Hun from central Italy, but he's strongly posted and it will be at his leisure that he'll go, rather than at our insistence.

10 May 1944 HQ, 3rd Infantry Brigade CMF, Anzio 17 days in this discomfortable [sic] place. Can't believe we've long to wait for the next phase to begin. Still fear it's going to be a long, slow business. The enemy is definitely on the defensive, but with a purpose. Thanks to mistakes and delays on our side, we've given him too much time. We're also extremely strong, but this is no country for an offensive.

11 May 1944 HQ, 3rd Infantry Brigade CMF, Anzio Most interesting place to be professionally as no two sectors are alike in their characteristics. At certain points we aren't 200 yards from the enemy. Elsewhere there are wide open spaces where night and day patrols play Tig with each other. One sees more of the back of an army than anyone has since 1918, for the men who normally live well in the rear – supply units, workshops – have to be right forward to occupy their proper places in theoretical military geography. More visiting war correspondents today. In the forward posts I found a brace of cameramen complaining that we weren't doing anything worth filming! Have a resplendent jeep which mounts my blue flag, and a red panel with a white star. American Brigade commanders are 'one star' generals, and ours are only three star colonels. There's no doubt our allies are general-conscious.

9.30pm. No moon yet. A German high velocity long range battery is pumping shells into the back areas, and a heavy thump sounds close by, then the quick eager hunting whine of the questing shell. But they're going a long way over, towards Anzio. Both sides have a lot of artillery deployed, and being men with dangerous playthings insist on firing it off. If the result of physical defeat wasn't so serious, this would seem the most ridiculously foolish business that sober men could undertake. However I sincerely believe that we must dispose of the enemy or else he'll dispose of us, and that gives reality to the whole grim stupidity.

It's an odd business being a Brigadier of Infantry at war. One straddles two worlds – that of the high command, where planning becomes more abstract and there's often a great deal of wishful thinking and window dressing, and that of the fighting soldier, who deals in the actuality. The gulf between the two is immense – can't really be bridged. The Brigade is the last echelon of command which deals with real human beings, so it's an absorbingly interesting command and I'm glad I've had one at home and another abroad in battle. Shall also be glad when they decide, if ever they do, to give me work with greater scope. Problem is how to exercise command without constantly playing the part of being a commander. I still believe in Cooperative Command, planning in consultation with my Lieutenant Colonels when time permits discussion. Not too sure they do.

12 May 1944 HQ, 3rd Infantry Brigade CMF, Anzio *1500 hrs.* The Ball has properly begun on the main southern front. Daresay Kesselring[21] is guessing what we're doing and going to do, but can't really risk taking any troops away from in front of us. Hope to Heaven the battle goes well, as another defeat in Italy would be depressing, and if Alex were decisively checked the enemy might well feel free to turn on the beachhead again. I'd dearly love to enter Rome in triumph. It's become a symbol of this war, as did Baghdad and Jerusalem in the last one. Well, after many tribulations and failures we entered both.

The BBC padded [*sic*] on the battle news tonight: superfluous warning about difficulties ahead and a tangential description of a BBC hero's flight above the battlefield in a very secure Fortress. Suppose stronger medicine is still thought too much for the Home 'back'! Have forgotten the black days of December and the crookedness of the exalted. I seem to have been doing this always. If certain generals among my junior contemporaries come to see me, we converse with careful avoidance of old controversies. The bread is only buttered on one side, and pariahs may have rabies. Laughable.

13 May 1944 HQ, 3rd Infantry Brigade CMF, Anzio The main offensive goes slowly forward. Reports today speak of violent enemy counter-attacks. Hold lightly, use mortars liberally, counter-attack savagely with tanks and infantry. That's his game, and also what I taught 160th Bde in Kent to expect and do. Our fear of elasticity, or being caught on the wrong foot, precludes our doing the same. He's got a number of good divisions tied up against us, and can't have us walk onto Highway 6 or into Rome. Behind him broods the Colle Lasciali. 'The hills like beasts at their hunting lay/Chin upon paws to await their prey.'

How sinister those not far distant shapes look, yet they're the beautiful hills which shelter Frascati, Castel Gandolfo and lovely villages and vineyards. From them he can see all the beachhead – the ships lying off the little curving bay, the busy mine sweepers, the destroyers out beyond. He also sees the balloon barrage floating nostalgically over the Porto. To take the sharp edge off his observation, we make smoke continuously across our back areas, and the two little towns hide their poor pathetic scars in drifting veils of chemical clouds. My woodland site is summer green and sandy underfoot – it might be a Camberley plantation. But the trees play tricks with sound, and shell-burst are hidden.

To the north where the woods end the country opens, and from ground-level looks as if it fell flatly away till it begins to rise again. But this is deceptive. It's seamed with deep ravines and water-courses worn out of the soft sandstone by torrential rains. They're tortuous and dark, with tree-grown banks, and once down in them they're hard to climb. Shocking places to be trapped in, as our poor Guardsmen found in the early fighting. In the open country corn is growing, and homes of the sowers are shapeless heaps of white rubble. Beyond, the country looks intact with little towns of the Alban foothills shining in the sunlight. The enemy is fortunate in that his front backs onto all this safe country with Rome back of that, while we've only the sea behind us. The beachhead's existed for nearly four months and has an air of permanence and solidity. The front is adequately manned, and defences have a depth which Tobruk never had. Even so, we must be always vigilant, for if ever the front was penetrated a short advance would enable the enemy to control the port and its exits. Rest of our perimeter defence would be useless.

Talked for a little today to Ginger Hawkesworth, temporarily in command of us, who was at the Staff College with me. We spoke of Auk, Ritchie, Arthur Smith, and his views were mine. He realises that Ritchie owes his salvation to Brooke, and insists that he's too slow-minded for modern warfare. This is a new war, invasion of fortress Europe – a new war with new methods, or what *should* be new methods.

BBC News, through the warble of interference from Rome to jam it, says we're attacking from the coast to very near Cassino against a strongly garrisoned German sector. Behind the Gustav line[22] is the Adolf Hitler rear position, but this business is no route march to Rome. It must be a mixed army down there: Yanks, Canadians, British, Indians, French, Poles. They're all in it, and they seem to be fighting well and steadily pushing in the Gustav line. But we haven't taken any prisoners, and I feel the enemy is using the area between his original front and the strong Hitler line as a buffer zone to exhaust us before we can seriously threaten his rear position. In due course it will be our turn here to take him on, dependent on the success of the main battle to the south. I'd sooner be here at this stage of the war than anywhere else.

15 May 1944 HQ, 3rd Infantry Brigade CMF, Anzio Ordered a new road built through pretty jungly country a few days ago. Today walked on it – not quite finished – astonished at the capacity of modern road-making machinery. It's just a dirt track, but the bulldozers and graders have torn their way through matted jungle, virgin *maquis*,[23] over water-courses and round hillsides – wonderful job. It's going to make a great difference to our movements in future, avoiding a valley called locally 'the Oh God Wadi' (well deserved as the enemy shoots there too frequently), and it sets one thinking of other possibilities. This 'Burma Road' is a wide carriageway, and now I can jeep to every battalion HQ forward.

16 May 1944 HQ, 3rd Infantry Brigade CMF, Anzio *9.45pm*. If I survive this I'll be in a strong position to ventilate my wrath when I get home. Begin to wonder whether the main requisite of a British general isn't an extroverted heartiness, combined with a capacity to play the same game as the rest of the team with little critical sense. A capacity for analytical thought, or a tendency to search for novelty in warfare, is decidedly unnecessary. The sole recognised requisite is battle experience. No matter whether a good or a bad battle, having been there is the thing! For the rest, simple hail-fellow-well-met extroverted heartiness does the trick, hence our situation in Italy.

The generals, having mucked it at Salerno, Cassino, Sangro and Anzio, are now trying to crack the terrifically tough nut of the Liri valley.[24] We made much ground in a part which doesn't lead anywhere, but got on far less quickly towards Cassino where success would be damaging. Of course, this is a subsidiary front, a blind alley, but we have to go on attacking to pin down as much as possible of enemy effort. Am more than ever convinced that my views about modern military formations are sound. Here we're a desperately top-heavy organisation with a totally unscientific ratio of fighting units to directing HQs. We've far too many! I'd love to reorganise this beachhead army, regardless of the forces outside. Both sides here watch each other with one eye cocked on what's happening 70 miles to the south-east. Can't expect quick results.

17 May 1944 HQ, 3rd Infantry Brigade CMF, Anzio Released from the 'Gateway' for a few days, it's nice to walk about. Had a lovely hot bath, wallowed in it and came out scalded pink. My hair looks like a retriever's. Getting strongly sunburnt, which suggests the hardened warrior that I'm not. A long letter from Reg today. He's pretty bolshy about Burma, which he says will never get a square deal while Winston's in power. Only too correct. Tonight Cassino was evacuated and the Hun's going back everywhere to the Hitler line, as I think he intended though shaken by the speed and vehemence of our initial assault. The French have done very well indeed, and the Poles have fought like tigers. Meanwhile we sit here. The Hun is windy, inclined to shoot at shadows.

18 May 1944 HQ, 3rd Infantry Brigade CMF, Anzio The BBC is making much capital out of the outflanking of Cassino, trying to rub it in to the enemy. But the Germans have every reason to be proud of its defence. It certainly paid a dividend.

We're still watching events further south. At last we've got back to the warfare of 1918 which we seem to understand. All we lack are the rolling plains of Artois – a bit too knobbly here. Now we're squaring up to the Hitler Line, and then it was the Hindenburg Line. No difference otherwise.

19 May 1944   HQ, 3rd Infantry Brigade CMF, Anzio   Daresay the higher authorities are watching me like hawks on a hillside for the least sign of incapacity.[25] Almost an inspiriting thought, for they won't find it by fair means, though they may by foul. The real problem is whether I can inspire my units to make successful warfare in the way I want it done. In our army we lack any objective political idealism to unite all ranks towards a clear-cut common end. The alternative is a quasi-feudal exaltation of the personality of various high commanders. Hence the aristocrat-appeal of Alex and Mountbatten, or the plebeian alternative of Monty mountebanking. Don't believe either's desirable, but both seem necessary. Not that I'm against a properly worded order of the day, but showmanship is a two-edged weapon. Monty's stock isn't conspicuously high after the song and dance about Eighth Army taking Rome, followed by the failure on the Sangro.[26]

If the Germans can't hold us on the Hitler line, don't believe they can stop us south of Rome, unless they give Kesselring considerable reinforcements in land- troops and aircraft. Has he repeated Rommel's error of staying too long within striking distance of a potentially superior offence concentration? Monty will begin pretty soon now, and then this front will become a sideshow. Am keen to break out of this little island and join the rest of the army. Am tired of crouching about sniper-beset trenches, and the moment of release can't come too quickly. Will have dreary Dick McCreery[27] as my Corps Commander soon and expect short shrift from him. Got a brief note from Wemyss, the Military Secretary – nothing in it, but affable. Wrote to Reg and took up his point about wanting to see that Burma got a square deal after the war. Told him I was a socialist and intended, if I survived, to work flat out for socialism.

But at news that the breakout was to start within four days such speculations vanished. It was to begin with a 1500-piece artillery bombardment at 5:45 a.m. on 23 May, and orders came that 3rd Infantry Brigade had to be in place by the evening beforehand. With the object to distract the enemy and make him think the main attack was coming from the west flank of the beach-head, Chink deployed one battalion of the Webb Carter's Duke of Wellingtons (DWR) supported by two companies of Hackett's Sherwood Foresters to attack on the right, and Careless's KSLI on the left.

22 May 1944   HQ, 3rd Infantry Brigade CMF, Anzio   *5.30pm*   In 3 hours my troops begin an attack as the prelude to the grand sortie from the Anzio beachhead.

Needn't be a military expert to see what a threat it will be to Kesselring's army, deeply committed further to the south with only one reasonably secure road linking it to Rome. The main front goes well, so our turn is now. Chaps are fit, keen, full of fight, anxious to get underway, bless 'em. This afternoon terrific sticks of Bombers went over towards Rome which heartens them a lot. Luftwaffe has vanished, making war on land easier. The enemy fights under a great handicap with no air force to help him.

The little picture, infinitesimal but important, is that tonight my sector and my brigade open the Ball by assaulting a point opposite us. Have done all I can to make the attack a success. We've powerful support and made our plans in the greatest detail, but the enemy is strongly posted and his reaction, once he recovers from the shock, may be violent. Suspect he'll sling a lot of stuff about, but think the strength and direction of our assault will come as a surprise. Don't believe he knows how easily I can move about my sector.

At nineteen I commanded a company of the 5th Fusiliers in just such another undertaking, and hope this will have better success. Hate what I know of fights of this nature. Hate giving orders for something I won't participate in, for my work is done – till something new happens. While the infantry companies are waiting on the hidden ravine slopes for the moment to assault, I'll just have risen from a good dinner in a dugout way back. Had hoped to be much nearer but the new HQs aren't ready, so must squat like a secure old spider at the middle of my network. Having heard all I can, I'll go to bed in pyjamas, which feels wrong but is right – except the contrast between my life and that of the men is too acute. Short of leading the assault, though, there's nothing I can do till the next phase develops. The enemy is going to get a terrific jerk, putting down bombardment wire is no small affair and it will fall where it hurts. Can feel the tension of 2,500 men waiting for 8.30pm, with 2½ hours to go. But it's quiet now, and the deployment up my communication trenches is going well. Thank goodness I insisted on them being made when I took this sector over.

Just been speaking to the CO of my assaulting battalion who at this moment is sitting in a deep ravine only 400 x [*sic*] from the unsuspecting Hun. Unreal to be able to bridge that physical and psychological gap by word of mouth, yet to be so differently situated. A little like telephoning from the *Café de Paris* to a chap in the condemned cell at Dartmoor. Officers of the assaulting companies must in their turn think my Battlefront HQs is a harbour of security compared to their own exposed situation an hour or two from now.

8pm We've dined well, under a camouflage net in a sandy pit dug in the open, which we use instead of the more confined dugout when the weather's good. A perfectly normal dinner: people formally dressed, flowers on the table, jokes, laughter, and not a word of the battle about to begin in 18 minutes from now. The liaison boys have fixed up a spare loudspeaker R/T[28] set in the mess dugout on which they can hear all that comes over the air from the scene of carnage. They will listen to that and BBC or Rome music simultaneously, the ghouls. Nothing doing up front – it's unprecedently quiet. Either the enemy is contemplating a spoiling attack himself, in which case we'll get ours in first, or he's thinking of going away. Here at these HQs it's a calm, beautiful evening among the trees. Nightingales sing. Fat clouds float overhead. What's about to happen seems unbelievable.

It has begun. The first guns have fired.

11.30pm. Attack made a perfect start and our guns shot magnificently, but there's been fierce hand-to-hand work. When darkness fell the situation was obviously more obscure. So far so good, though can't estimate our casualties. Information came in remarkably well considering, but my heart goes out to the brave chaps up there in the dark, erratically lit by Hun flares and our shell flashes. Will get called at 5am, so bed.

23 May 1944 HQ, 3rd Infantry Brigade CMF, Anzio *9.30am* Just back from the forward positions seeing COs, who all seem cheerful and confident. Hun stretcher-bearers out in strength on the front. Little sleep last night for my phone rang frequently. One assaulting company was very successful on the left – did its job completely and consolidated its gains. The right-hand company lost most of its officers early on, ran into a numerous body of Huns and fought with bomb and bayonet but didn't get its objective. Whistled in the survivors before daylight. On the whole our casualties haven't been serious, far less than expected. The troops have done a good job. Certainly the enemy's been well shaken up, and as I've got a lot of artillery at call and no restrictions on ammunition, he'll have a poor day today. The Big Sortie, of which my little operation ANT[29] was the prelude, began at 6.30am and seems to be going well, though little news yet. On the whole am not dissatisfied with the operation or the chaps. Anzio is going to be big news for a few days more and then I hope the two parts of the army will merge into one front, pressing on towards Rome. But there's many a slip. My first offensive battle – daresay more are coming.

That first day's fighting was intense. 1st Armoured Division lost 100 tanks, and casualties for 3rd Infantry Division would remain the highest single day figure for any US division in the war. In Chink's 3rd Infantry Brigade, the DWR had lost all officers by the time they were within 100 yards of the objective. Ten OR (Other Ranks) were killed, 8 officers and 89 men wounded, and 13 missing.[30]

24 May 1944 HQ, 3rd Infantry Brigade CMF, Anzio Looking forward to more extensive battling, confident the troops can both hold and attack if they aren't expected to be too clever. Kesselring is facing a difficult period, with grave danger of losing a large part of his force if he isn't lucky. It's been an auspicious beginning to the Allies 'Year of Offensives'.

Have got my third divisional commander, Charles Loewen,[31] in a month. A gunner, Canadian by birth, and a good, direct sort of chap. A lot junior but nice. Forthright, foul-mouthed, tactless, efficient, no yes-man. We'll get on. Of course they passed me over – am bottom of the class – but don't care. I like being close to our infantry. Have been in charge of this 'worst' sector for 3 weeks, and its been like going into the deep end all at once. All the better.

10.25pm A pitch-dark night and have been walking about outside my dugout in the flickering light of our gun flashes. Heard the story of our battle the night before last from the CO[32] of the assaulting battalion. An interesting tale, but nothing much I didn't already know. Elsewhere in the beachhead the Americans are making ground against strong resistance. Rumour says we have Cisterna[33] – good if true, but even without it they've advanced far. The French have made amazingly swift progress in the hills and are not very far from us now. The enemy wants to declare Rome an open city, but since all their traffic comes through it and their police control it, there's no hope of our accepting. Rome must take its chance with the rest.

Can hear the rumble of the Yanks' battle away in the East-north-East and sustained artillery fire. Suppose they're attacking as they work their way forward, but don't believe the Rome Line can be formidable – certainly nothing as strong as the Gustav or Hitler lines. Doesn't need much, I feel, to open up the Appian Way or link up with us via Littoria.[34] The enemy's unlikely to leave a large pocket of troops in the Pontine Marsh area, and won't move any troops from Italy to France.

We suppose Monty's party[35] will begin any day – say 1st June for luck. It will be heralded by a spate of 'orders-of-the-day' and sustained by Winstonian belches and a shrill chorus from the Yes-press. Don't mind much. The important matter is that it should have a fair measure of success from the outset. We can't afford failure.

25 May 1944 Temp. HQ, 3rd Infantry Brigade, Italy *5.30pm* We're settling a big account out here. Some of the repayment is slamming overhead. Today the Anzio beachhead lost its virginity and merged into the Fifth Army. Cisterna is ours and the Appian Way open. This morning two young Austrians deserted to us, and this afternoon I've been grilling the more intelligent one through an interpreter. Am sure the enemy is slipping away in front, and have got patrols going out at 7 o'clock tonight to his old positions. Kesselring has lost the battle for central Italy, and with it a goodish part of his army. Question is, can he now stand in front of Rome? Tremendous fuss going on with a hurried relief of my two forward units, so we may be needed for further offensives. Begin to believe that the Race to Rome is to start at any moment. We'll probably wade there through minefields.

6.30pm We're stinging the enemy up with our guns and plotting to give him hell at dusk, when I think he hopes to slip further away. Don't want to push at him because if possible he should be encouraged to stay opposite, so we or the Yanks can hook into his tail from the North East. The Yanks are now up in the Hills at Ceri and there's only 8 miles of mountain between them and Highway 6. If that were cut, enemy forces opposite Eighth Army would have only one roundabout road to lead them back to Rome. Our army would be nearer.

8.25pm A bold daylight patrol into the enemy's position found some Germans still there, though not many. Those there are have been pinpointed, and in a few moments will get a nudge from our 4.2 mortars. They don't like the big mortars, very lethal indeed. This is a moment when the ex-beachhead army feels it's finally on top of the Hun. There's an air of elation among officers and men. Victory is infectious, and its germs are in the air. Rome beckons us – Rome, home! Didn't think the old army in Italy had it in them. These chaps of mine have got their tails up again and will attack like tigers when let go. They've a lot of scores to settle. Any push which has had to sit and be mortared day in, day out in filthy grave-filled ravines not 200 yards from the enemy will relish a revenge in the open. This brigade is getting a new reputation for knowing both how to defend and how to attack, and we'll not be let stand idle. Besides, I must do something to justify the recently granted motto for the family, *Fortis in Bello*.

9.45pm Heaven help the Hun if he's trying to pull out from his foremost positions tonight. Our guns are slating his rearward routes and he'll make a highly precarious exit. My patrols are feeling well forward between the periods of shell and mortar fire so won't be surprised if we find his old haunts vacant by daylight. It's been an afternoon of order and counter-order. At one moment we were all to come out tomorrow, no sooner was than fixed up than cancelled, and now we're to stay where we are. The BBC is regaling the Great British Public with baloney about the relief of the Anzio beachhead by an armoured car from the southern front. Almost as if we were a starving post on the North West Frontier surrounded by thousands of fierce Pathans! Oh well, the Home front wants drama, and certainly Mark Clark[36] is not the chap to withhold it. Drama they'll get, so long as his name is in the story.

Loewen, new Divisional commander, just phoned to change all our plans for the third time in 5 hours. Order – counter order = Disorder. Gather Mark Clark is having Victory brainwaves. Am also told that the traffic from the south along the Appian Way has to be seen to be believed. All the coast is in our hands from here to Naples. We're free to reinforce this place overland, and it now becomes the offensive left-flank of a great encircling movement which may cut off the enemy still trying to hold Eighth Army in the Liri Valley.[37] Tomorrow should be interesting for everyone.

26 May 1944 Temp. HQ, 3rd Infantry Brigade, Italy Now got a fellow equal whom I taught at Camberley. Doesn't know whether to call me 'Chink' or 'Sir'. Ridiculous situation. One does get one's nose rubbed in it. Wind shows no sign of changing. Mustn't expect improvement.

27 May 1944 Temp. HQ, 3rd Infantry Brigade, Italy All fighting troops greatly amused by Vaughan Thomas's[38] valedictory description of the 'Hell of Anzio'. Grotesque over-statement, except for base-boys who had to encounter occasional shellfire. So is pseudo-history made. Everyone anxious to press on to Rome, invitingly close. People there should be able to hear the battle clearly. Read PM's speech on foreign policy – still astounded. Power politics without ideals or principles.

28 May 1944 Temp. HQ, 3rd Infantry Brigade, Italy The war has gone grumbling off northwards, leaving us high and dry in stranded inactivity. Good, because it gives weary men time to wash, sleep and recover from the strain of the last three weeks. Got belated opportunity of talking to men and officers today.

29 May 1944 Temp. HQ, 3rd Infantry Brigade, Italy This afternoon visited the German positions opposite. Found them skilfully sited, well dug and very clean – as professional as ours are amateur. Well constructed with a sound governing plan, while ours, like Topsy, 'just growed' untidily. But one has to be careful in German positions voluntarily abandoned, because liable to be mined and booby-trapped. Met some nasty sights and nastier smells: human decay sticks in the nostrils for weeks. Whole area is a testament to the violence of our artillery and mortar fire.

30 May 1944 Temp. HQ, 3rd Infantry Brigade, Italy Four days now left alone, and much refreshed. Took an hour off and went to see American guns bombarding the lovely slopes of the Colle Laziali.[39] Countryside in a deplorable mess. Once an outstanding example of Fascist agricultural resettlement, the pretentious hamlet of Apulia is a heap of rubble. Mussolini should be *walked* from Rome in full fascist regalia to see the result of his rule. Everywhere are battered tanks and smashed lorries. The wastage of war is terrific. In long stalks of still-green wheat are worse things: mines and unburied dead who stink to Heaven in mute comment on the stupidity of mankind. This was the richest part of tourist Italy, along one of the best-known autostrada where four years ago sleek cars of the well-to-do streamed to seaside villas.

Wrote to the Auk yesterday, and hope it reaches him by June 26th, second anniversary of our flight to Eighth Army. Told him some of the odd things which have happened to me. Wonder whether and how he'll reply. A boy just back from a course in North Africa tells me the atmosphere there is entirely un-warlike. It and the Middle East have become sleepy hollows in the trough of war. Can visualise old Jumbo slumbering like Chronos in the garden of the Hesperides among the oranges. '

1 June 1944 Temp. HQ, 3rd Infantry Brigade, Italy *10pm* Writing in my car in a shell-pitted cornfield by a battered farmhouse surrounded by noisy artillery after another day of order and counter-order. My HQs has temporarily come to rest exposed in the open, and our safe dug-outs are far behind. Have only to walk 10 yards to my skyline to see the Hun-held crest less than 3 miles away. At sunset stood on a hill looking across the battlefield towards Rome.

At short notice, orders came to take Aquebona Ridge on 3 June, which lay behind a ravine ahead. Intelligence revealed that it was held by crack units of 4th German Parachute Division, and Chink selected lunchtime as the enemy's weak spot. No attack would be expected, and it might coincide with a relief change-over.

2 June 1944 Temp. HQ, 3rd Infantry Brigade, Italy A day of rushed decisions and hasty plans, only to be expected when the Hun looks groggy. Have been flat-out planning and ordering, and now my wireless and very capable staff are tying up the ends – preparing confirmatory orders, coordinating fire plans, overseeing troop movements and generally buttoning up every detail. Will take a last look at progress. All this in order that tomorrow, after midday, some men will die. We all defy it, Allies and enemy alike, by dreams of women we love and hope to love. In that, as in the unburied corruption, all soldiers are equal. In one more quality, too: cheerfulness.

My 'very capable staff are tying up the ends....' Chink's view of the brigadier's classic role had by this late stage of the war – unknown to him – been superseded by the 'hands-on' encouragement of younger men who often broke precedent to bring reassurance. During the first breakout Chink's three COs had missed the previous morale-boosting presence of his predecessor Jimmy James, and tension this time was high. At the height of battle for Aquebona Ridge a hot exchange would take place between Careless and Chink over the RT. Careless could be overheard urging him to come forward and Chink refusing; it was still taught at Sandhurst and the Staff College that to ensure safe control for all three battalions the brigadier must stay out of harm's way. The outcome of the battle for Aquebona Ridge would be successful; the outcome of the RT call the reverse.

4 June 1944 Temp. HQ, 3rd Infantry Brigade, Italy Yesterday my brigade assaulted key position 'Rome Line', carefully prepared by the enemy with strong defence posts held by crack troops. Attack successful beyond all expectation, and large numbers of enemy killed, but they fought hard; already flyblown. Our casualties surprisingly light. Troops now prepared to assault anything or anybody. The attack was the neatest feat of arms, and great credit must go to the fine battalions, in particular to the newly arrived youngsters drafted from home. All feel we've repaid with interest the distressing events after first landing, and the subsequent costly holding on. During the night what remained of those in front of us vanished. Not yet found.

Later By the fantastic variations of fortune which war provides, I and my flock are at the moment sitting at ease in a pretty part of the golden *campagna* not 15 miles from Rome. So there's time to write more. I was given the job of cracking the Hun position at very short notice, and after an afternoon of feverish preparation and a night move into assault positions, we attacked at midday under cover of an intense Hurricane artillery bombardment. A deep dugout in each section post, and deep connecting trenches: formidable propositions and more seriously held by a tough battalion of Para-boys. A model operation, and its planning and launching has given me great confidence in my tactical ability – reads a bit like Monty, that. More importantly, it's given the troops a tremendous boost. Considering how soft life in the beachhead made them, they endured the sharp test well. Captured a tidy lot of prisoners, having killed a considerable number more.

I've walked over the captured ground, studying the enemy's methods. As usual, intelligently conceived and industriously executed. Fortunately he'd neither anti-tank nor anti-personnel mines, and little wire. I feel he thought we couldn't get tanks through the deep wadi onto his position, but we did. Anyone else in this army would be regarded as having won his spurs, but the kudos will go to our new Divisional commander [Loewen], considerably my junior. Am glad I've done my duty, justified myself professionally, and carried out two attacks with considerably less loss than expected. Now all three battalions have had a cut at the enemy they'll be more skilful next time.

Tonight the war seems to have ended. No sound anywhere – no artillery, no aircraft. Shall sleep well tonight in my tent under a full moon. My reactions to dead soldiers are not those of my youth, when death was abhorrent. Now I understand things more clearly I regard the dead and the not-yet-dead in much the same way. The first have had it and do not know it, the second are going to have it and know it. So I can't go as far as Rupert Brooke – there are no 'Rich dead', neither are there poor dead. They're just dead men. Very still, lumpy and tumbled, and covered with foul shining bluebottles.

By one young para-trooper lay a photo of Adolf Hitler. No British soldier has looked on Winston Spencer Churchill as his last devoted fetish. I gather the little people are beginning to protest at his concept of a world managed by four big shots: USA, USSR, China and UK. It's like him and his Tories to see only power politics, with large armed forces to absorb unemployment, and financial concessions to the working classes – so long as they remain working classes. Nothing to hamper or disturb the City or private enterprise.

Leask, the BM,[40] is not yet back from a staff conference at Div HQ. Will stay up to hear his news. Last night they asked me to nominate one of my units to take part in the victory parade through the Holy City. In an abandoned German gun position today found a German illustrated magazine for their troops in Italy. Beautifully got up, no expense spared, and on the big back page one of the nicest nude photos I've seen in a long time. Fancy us putting a nude in one of our military magazines!

Leask just in, says everyone is rushing madly into Rome where the Hun seems to have blown the Tiber bridges. Our troops are going to be kept out, except that battalion of my brigade representing Great Britain. Good joke that one, but hate the idea of good troops getting softened up in the pubs and brothels. Aren't we going on with this war? I want to get into Germany without delay.

5 June 1944 Temp. HQ, 3rd Infantry Brigade, Italy This morning's BBC news says the Yanks are in Rome. Poor Rome, full of undignified juveniles with vociferous enthusiasms, jaws ruminating over spearmint, pocketbooks bulging with cash. An army of Philistines.

We move up ten miles today, and to everyone's consternation I've refused troop-lifting transport. My infantry must relearn marching, so they'll go in the hot June sun across the cornfields, instead of rattling in a pretty uncomfortable lorry along roads in a perpetual fog of choking dust – worse than Alamein. No sound today. The infantryman's apt to take the absence of enemy aircraft for granted, but we who knew the early days in the WD, when even the Italian Airforce was an ever-present menace, note the difference.

6 June 1944 Temp. HQ, 3rd Infantry Brigade, Italy Went into Rome this afternoon to meet the Yank general about my unit under his command for the parade. We British have had it – we're a no-rate power. Afterwards had tea at the smart but deserted Grand Hotel, which I last visited in 1919.[41] As I entered, a US subaltern stopped a pretty girl, shook her hand and began to force a date on her. Inside, our allies were busy drinking everything in sight. Most have shaved heads, a Dartmoor or Newgate[42] clip which combined with their hideous green uniforms makes them appear utterly villainous.[43] Am sure they've hearts of gold, but they seem to lack self-respect or discipline. Our chaps look very well in shirt and shorts, with golden sunburned limbs.

Outside, hammer and sickle symbols crowned several suburban buildings. One was a Red recruiting office, with civilians lounging outside. Ridiculous number of Italian officers and men still carry arms in uniform. Women appear smart in their thin, unbrassiered summer frocks which clearly reveal their high firm breasts. One young mother was giving her child her breast, quite bare – these physical things catch one's throat. Rome itself looks dingy. The Borghese Gardens are dried up and unplanted. Maps on the wall of the Victor Emmanuel memorial showing the growth of Roman and Fascist empires have lost their freshness: memorials of disaster. Took our Italian Liaison officer with me, a young *tenente* law student. His family live in Bari so we control his fate if he misbehaves, but he's not bad. As soon as W. Spencer Caesar satisfies himself that no disaster is imminent, expect he'll come to Rome; could hardly resist. All the big shots will come tumbling and rumbling in, fortifying their egos with the dust of antiquity. Depressing thought.

My mind is beginning to turn over the lessons of our recent experiences. Nothing new. It remains astonishing that soldiers, whose lives and reputations depend on their memory, forget so easily. They've never made the facts of their trade a second nature. One thing is clear. The detachment at Anzio made a useful contribution to the general offensive, but the whole business was seriously mishandled. Ignorance, misplaced optimism and inertia. History won't flatter the conceivers, planners or senior executants.

Have also been analysing our recent actions. In view of their success am delighted by the small casualties, particularly in killed. But we've been fortunate in disposing of lots of artillery and heavy mortars, and of being able to begin our assault without a previous advance under fire. That was important.

Initial events at Anzio had damaged the men's confidence, not helped by two further months spent holding unsuitable tactical positions with the inevitable steady drain of casualties. It remains for us to ensure that morale doesn't fall back again. Think I know the right things to do, but not sure my fortuitously appointed masters are clear.

Here we are 24 months after Ritchie's disasters and Auk's subsequent retrieval of them and at last – after incredible delays and disappointments – we're in Rome, with our enemy in disorganised retreat. It's been a long, dusty, dirty road. I deeply regret that the men who began it – Dick O'Connor, Strafer Gott, Jock Campbell and others like the Auk who carried on their work – are not here to see its climax. Instead, new men who never knew the bitter years in the Desert lead the army across Jordan into the promised land and will reap the reward. I seem doomed to be a spectator from the periphery. But because the road from Cairo to Rome is a connected journey, I'm glad to have seen the end of it, as I saw its beginning and depressing middle part. For that reason I'd rather be here than in the French landings, though I miss the brigade I trained in England. Well, it's seven months since I left them, so they've forgotten me by now which is as it should be. But I have not forgotten them.

A peculiarity of command, as distinct from staff work, is that one is seldom busy. Given efficient and experienced staff officers (and I'm fortunate, for Henry Leask is first rate), given also good battalion commanders who know how to get on with the job, one has remarkably little to do. In fact, one must avoid doing anything except give decisions and keep a quiet eye on their execution. So I've plenty of time for thought – too much, perhaps.

7 June 1944 Temp. HQ, 3rd Infantry Brigade, Italy Tomorrow one battalion of mine goes into Rome as part of its garrison. Have arranged that they should march in with fixed bayonets, with detachments from my other units lining the street. Gave Charles Loewen the chance of taking the salute, so will watch them elsewhere as a 'civilian' for I'm his senior. It will be a brave show for the fighting men and put the 'contemptible' British into the limelight. No higher authority thought of arranging a proper setting for such a historic event, but I insisted. Am exasperated by rumours that for a time we fell out of the hunt to train and refit – hope it isn't true. Came here to fight, not to train, particularly when the Western front is going full blast. Have put Careless in for a DSO and my two other COs in for bars to their existing DSOs[44] for a really good show on 3 June. Hope they get them.

8 June 1944 Temp. HQ, 3rd Infantry Brigade, Italy The little show I staged caught on, and Mark Clark spoiled what should have been an all-English party by turning up with a bunch of thugs and taking the salute. Watched with our troops, fun to be among the Romans. Masses of Yank cars, many carrying the improbable blondes of Rome's poxy underworld who a few nights ago were in bed with Huns.

9 June 1944 Temp. HQ, 3rd Infantry Brigade, Italy Rome is swamped under a grey-green tide of gum-chewers. Also many signs of communist activity – an essential corrective to Fascism until re-establishment of democratic society is possible, but reckon Italians are too demoralised for that yet. Enemy still hurrying northwards, smashing communications as he goes.

10 June 1944 Temp. HQ, 3rd Infantry Brigade, Italy Idleness is most unsatisfactory, giving too much time for bitter thinking. Rome still *en fête* as no work to do. Unlimited capacity for discovery of cheap females, mostly neo-blonde, initially tinted to titillate Teutons. Adaptable creatures, albeit insanitary. The city has suffered far less than London – shows importance of a reputation for Holiness! News from France[45] still uncertain. Enemy seems to be counter-attacking and Rommel's dangerous if allowed to develop his own plans. He has a score to settle with Monty.

11 June 1944 Temp. HQ, 3rd Infantry Brigade, Italy Found a trio of wizened labourers sitting on the steps of a gutted farmhouse, eating scrounged ration biscuit. They have absolutely nothing. The younger woman – sixty? – lost her husband in the last war, one son's a prisoner in America, another a forced labourer in France. The soul-destroying misery brought on these poor people by world rulers screams for vengeance. Gorgeous noisy party in St Peter's today. Papacy coming down on the side of the big battalions, a trifle belatedly. Refrained from participation.

Many reports of ex-POWs turning up, some many months at large in Rome. Also authentic evidence of German mass reprisals. Hun attitude to Italians now is undisguised contempt, and they'll smash Italy as nowhere else, even Russia. It's essential that Germany doesn't escape heavy war damage by capitulating when invaded. Not a question of unconditional surrender, but unavoidable punishment on a nationwide scale. Until that's administered we can't begin to restore Europe.

12 June 1944 Temp. HQ, 3rd Infantry Brigade, Italy Dined in Rome last night at the Grand – masses of young asparagus, excellent steak, bottle of passable chianti. Yanks behaving badly with local strumpets during meal. Terrifying thought that there are 130 million of them.

Chaps like the BM, a regular, and the staff captain – efficient, public school – are completely Victorian in outlook. Authoritarian, class-ridden, genuine believers in the sacred rights of the individual to amass unlimited profit and a capitalist dynasty. They regard that as part of the laws of the universe. Both young men have exceptional qualities: physical courage, administrative capacity, probity, loyalties, elegance and high standards. Only they can't realise that the working classes are in fact their own poor relations. Political renaissance must develop in the factories of England, something they dread far more than battle with the enemy. This service is like perpetual immersion in a bad public school. I'm too much Mr Bultitude[46] for army life, unless I have the power to alter things. One selfish side of me cries out that since I know I can command in battle now, it's more than ever distressing never to have had a chance to do so until too late. God, I'm bitter about that. How fierce the desire to damage the swine who've wrecked me professionally, so keenly felt in this temporary inaction. The enemy is fighting stiffly about Vittorio[47] and using new troops from Germany and France as a rearguard. That shows how disorganised his older Italian divisions must be.

13 June 1944 Temp. HQ, 3rd Infantry Brigade, Italy Last night I and my three COs had our victory dinner in Rome. Cost me a packet, but worth it in terms of future unity in formation. First occasion we've been able to assemble the 'Command Club', and I do feel I'm getting the confidence of these men. Certainly they know I'll never put them into battle on the wrong foot.[48] Very good dinner indeed, impossible to get anything approaching it in London. Rome's beginning to settle down with something of a hangover as waiters and women in hotels show distinct signs of fatigue. Not easy to switch from grim correct Huns to the excited, vociferous heartiness of the Allies.

14 June 1944 Temp. HQ, 3rd Infantry Brigade, Italy The war has receded from us. Temporarily we live in a vacuum. A keynote of active service is ignorance of true events. One learns little of a watch by looking at the time and as part of war's clockwork we realise that all sorts of things go on inside the case, so tend to query the rightness of the hands. Flew for half an hour this morning in an 'Air' OP[49] looking down at our old battlefield.

15 June 1944 Temp. HQ, 3rd Infantry Brigade, Italy For the first time in two months sleep under a proper roof. It feels strange. Couldn't let this charming old *Castello*[50] go unoccupied, though it was gutted by 4th Paratroop Division who'd made it their HQs. They took all the electric fittings, smashed what they could and even broke the china when bolting after our attack. Feel entitled to the General's bedroom and bathroom – no running water. Views are magnificent, including a clear one of Rome. It's good for my officers to get their kit unpacked and sorted.

Must soon take Penney, the returned [from sick-leave] and legitimate Divisional Commander, over the battlefield of 3 June. Will have to do showman as best I can. Everyone who is anyone here is being knighted for the capture of Rome, so must have gone down big at home.

18 June 1944 Temp. HQ, 3rd Infantry Brigade, Italy Various High HQs are monopolising all the best hotels in Rome and there's uncertainty about getting a meal there at all. Instead, threw a little party for the COs at my *Castello*. The owner[51] came to see me yesterday and left behind a bottle of pre-war Scotch liqueur whiskey which we duly destroyed between the six of us before dinner. We're still out of the hunt but training busily. Am trying to develop 'normal' systems of tactical action to give us all a common outlook and save time in battle. Think I can go quite a long way with it. Can't reduce battle fighting to a drill, of course, but should aim to get a simplified standardised procedure. It suits troops to act with the minimum of thought, to react through habit rather than reason. We're generally moving north at fair speed, but so far have met no serious resistance. At the Rimini – Florence – Pisa line we'll meet the next heavy fighting

20 June 1944 Temp. HQ, 3rd Infantry Brigade, Italy My *Castello* continues to please me, though am bored by enforced passivity and feel the need for action. There's no limit to training necessary but we're fit to go on. The Military Secretary,[52] with whom I spent so much time in correspondence in the last six months, will be here soon but propose to leave him alone. More dignified to accept existing situation. By September I'll have done this sort of work for two years. That's a long time in warfare, because the authorities can't count the illegally enforced interval. Yesterday 24 months ago Tobruk was about to fall. Wonder if the frightened people who panicked in Cairo now think they were right to behave so undignifiedly?

21 June 1944 Temp. HQ, 3rd Infantry Brigade, Italy Tomorrow I escort Penney around the forward positions at Anzio. He was a fellow student in my syndicate at the Staff College and is my master now after six weeks' absence on sick-leave. I know both sides of the line well, and the enemy's were almost as good as ours weren't. Penney still maintains Anzio was a success. Magnificent English quality not to admit mistakes, misunderstandings and defeat, but misleading historically. The fact that, in quite a different strategical conception, Anzio paid a dividend should not obscure the initial strategical and tactical mistakes. He will see some of his own blunderings tomorrow, and I can be unbiased and impersonal. I wasn't there then.

22 June 1944 Temp. HQ, 3rd Infantry Brigade, Italy Disheartened. Foolishly I thought that if I accepted the fact that I had no future, I might be allowed to fight for the cause without my past being raked up to inhibit my future. Hoped for a square deal in a new environment. Am sure now that's not to be.

Penney was Auk's chief signal officer in '42. He's always regarded me as a dangerous thinker – critical, self-opinionated and uncompromising. As I showed him around he was agog for the least suggestion that anything on our side of the house wasn't perfect or that the enemy knew his stuff because that might reflect on the capacity of High Command. In effect he was laying for me. I was propriety itself. Walked in front wherever there might be enemy landmines. I 'Yes-Sir'ed and 'No-Sir'ed and 'As you say, Sir'ed and consumed my full ration of dirt.

On the way back we were alone in my car and the cat emerged from the bag. I had 'done myself no good' in the six weeks I was with Auk at Alamein I. I suppose Auk having been reinstated as C in C India[53] they've still got to have a whipping post for the rude things Auk said about the Cavalry and Guards when the scions of the ruling strata were showing up so poorly that there were even complaints to the King! So now I'm the dangerous, menacing disturber of the Anglican natural order. They're prepared to admit now that Auk saved the world – interesting that – and they're also prepared to focus their dislike of the Auk on poor me. So I 'did myself no good'.

This means in due course if they can't destroy me physically, they will give me the *coup de grâce* professionally. I felt it in my bones today that I'm the pre-ordained scapegoat. Well, there isn't anything I can do about that, so I'll just have to wait until I see how they do their dirty work. They are certainly not going to let me gain any military reputation, even in my present lowly job. All through that conversation ran a bogus morality. A &.B were 'no good' because they had affairs. I think they consider me immoral as well as intransigent. Give a dog a bad name – and they aim.

This same 'big boy' is assembling the religionists of my natal persuasion[54] under his command for a splendid audience with the holder of St Peter's See, and ostentatiously I have not been invited. This isn't my imagination. It's a statement of hopeless, incontrovertible fact and makes it desperately hard for me to go on effectively. Thumbs down against me everywhere, and no man can live at peace unless his neighbour wishes it. I can only expect dishonour or death, and there's no alternative. I'm the Middle East scapegoat – ME SG.

23 June 1944 Temp. HQ, 3rd Infantry Brigade, Italy Today there appeared a Corps commander 'old friend'[55] who was a fellow instructor at the Staff College. He went into our fighting fitness, and gather he was pleased with what he found. Afterwards he became friendly and personal. Said I'd had a damn raw deal, but it was a good show my working in this capacity and he hoped all would forgive and forget. When I replied that I knew of nothing for which people had to forgive me, he answered that he meant that I would forgive them. Told him I was damned if I would. So that was that. He was kinder than most and seemed pleased to see me, but I saw by his expression that he found me very changed. We haven't met since '37. He tried to talk of our then colleagues, most of whom have gone far. But I can't talk about a life I wish to forget, and as regular soldiers mainly converse about the past or their friends conversation wilted.

25 June 1944 Temp. HQ, 3rd Infantry Brigade, Italy Yesterday went to Ostia[56] and bathed. We've de-mined part of the Lido beach and the swimming is excellent. The Hun had made a sad mess of the place – he always expected us to land there. A group of British and Canadian nurses were bathing, swimsuits bought in Rome. Not beautiful but feminine, and to be looked at as women's bodies. Every male there was aware of them. It's grim to be womanless, but women suffer just as deeply – women who are physically awake. In spite of the disconcerting presence of female anatomies, enjoyed my swim.

The news gives no indication of our future tasks, but our business is clear. We must get out of Lombardy and Venezia, linking up with our French effort on one side and Tito[57] on the other. Hard fighting's ahead, and I don't yet see the war ending this year. Am going to try to write up our last little battle, a good example of a cheap, quick success. A nice little C of E padre to lunch today. He's been with one of my units for three years, and said many men are rapidly reaching communism, describing their restlessness, frustration and disquiet forcibly. They've a notable conception of the class war and social struggle. Against that he set the average young officer who has the political prejudices of the English middle class, and is in that respect totally out of sympathy with the men. Were I, or anyone like me, to speak to the men as a socialist commander should speak, I'd be sacked for being a politician. If I behave like the Tory I'm expected to be, I won't help them but won't damage myself. Firmly believe the fighting spirit of our armies would be 100% higher were the officers of the same political persuasion as the rank and file. An army of socialists led by conservatives is an army weakened by the cross-currents of war. Glad to hear the men aren't as torpid as they look, and can't find out for myself since I may not talk politics with them. Sensed he thinks me a queer fish – how right he is.

Am returning to the idea that the ex-soldier won't be given the political coordination on his own. Must think it out in terms of an organisation of ex-combatants, male and female, into something independent of Tories, Liberals and Labour that's capable of directing political action. It's something that has to be done, and not too difficult provided imaginations are caught. Have written a training directive to cover the next month's work, and got a nice note today (forwarded) from Dick O'Connor.[58]

27 June 1944 Temp. HQ, 3rd Infantry Brigade, Italy Most of the day I've been preparing my training programme for August. Last year I was hatching Exercise Paul for the South Downs, now for the *Campagna* I'm planning a similar affair. *Plus ca change.* Hope the next phase won't be delayed or we'll all get stale. In three weeks we'll be a pretty formidable party: experienced, coordinated, enduring, refreshed.

Suppose that in two weeks' time someone will have to report on me.[59] Don't know who can very well, for Ronnie Penney was away in England practically since I arrived. He is, however, well in with the powers. He's a narrow type who has been jealous of me since Staff College days, and is even now in my downfall jealous of my independence. Will they find some excuse for 'promoting' me into a dud job at home? Must see this business through.

Am interested to see how these chaps, whom very soon I'll have had for three months, perform when they next meet the enemy. Think they'll do well – better than most. All the armies, ours and theirs, are getting frayed at the edges. Manpower factors are telling increasingly, except perhaps with the Yanks. The enemy has to eke out his people with Poles, Austrians and Ost Deustch[60] that he'd not have looked at two years ago. Tonight learned that the Italian authorities have found the bodies of two Italian officers serving with the German para-division we beat up on 3 June. They commanded an attacked Italian battalion – the Folgore[61] parachutists – and the Hun had shot them both in the back. The day they were killed their battalion had fought bravely against mine – poor dupes, lying dead with bluebottles on their staring eyes. Anzio seems a long time past. How quickly one forgets.

29 June 1944 Temp. HQ, 3rd Infantry Brigade, Italy James, my predecessor who was sent to another brigade so I could be put under someone charged with my suppression, has been killed. If he'd stayed here he'd be alive. Another by-product of the Paget-Ritchie-Brooke scandal. When I'm fighting – what with a combination of funk and action – I can forget and take interest in the immediate moment. This life would drive one to drink, were there any to be had. This is a fierce implacable year of struggle, and it grieves me that I can give so little help. The enemy can't win but he can be lethal in failure, like a ringed wild boar.

30 June 1944 Temp. HQ, 3rd Infantry Brigade, Italy Frustration indescribable. For a grown mind, life in this limited field of action is equivalent to imprisonment mentally. Without possibility of professional development it's increasingly difficult to maintain enthusiasm. The only hope of sanity is renewed operational activity. Contact with the enemy is far more stimulating than contact with friends – also less effort. But it must be endured. Quietly amused at the sycophancy of seniors of the reformed church to the Supreme Pontiff. In reversed circumstances can't conceive of British RC generals seeking contact with archbishops of their church. Snobbery incredible.

2 July 1944 Temp. HQ, 3rd Infantry Brigade, Italy All day in the sun, and feel grilled. Now off to Rome for my 2-day break, so goodbye to my *Castello*. I've been happy enough here – or rather not too unhappy. The war is very remote and it's hard to feel any use to the cause. 'They also serve who only stand and wait' still applies. There's hard slogging in Normandy. Rommel seems to have effected his counter-concentration at the expense of some 35,000 casualties in the initial fighting; daresay he'd calculated on something of the sort. Now the real business begins, a lurid affair for France and England alike. Wish we could see into September, as by then the year's campaigning will begin to show its strategical harvest, and the German people will be thinking of their next winter. To be doing nothing in this crisis is complete anticlimax.

3 July 1944 Temp. HQ, 3rd Infantry Brigade, Italy Last night dined in Rome with four generations of the Vaselli's who own the *Castello*. A charming cultured family. There was the most adorable old scoundrel with a magic guitar, and after dinner they all sang *chansonettes* superbly. Dinner was perfect. *Hors d'oeuvres* (pre-war), soup, a good meat and vegetable dish, spaghetti, excellent sweet and savoury, and a new wine with each dish, ending with bubbly. Three efficient white-coated menservants, laughter, talk in three languages of politics, geography, travel. Pre-war liqueur whiskey and German beer afterwards. Unbelievable. Didn't get to bed till 1.30am. Slight hangover this morning and a big nostalgia for civilisation and women and table-linen and cut-glass and wine. I'm middle-aged at last, becoming 'all that ever went with evening dress'.

5 July 1944 Temp. HQ, 3rd Infantry Brigade, Italy Soon the farce of reporting on me will recommence, with all the advantage that procedure gives to jealousy and suspicion. I hunger for philosophy and hard thinking. Something in this life is narcotic to contemplation – the boredom, perhaps, and living with men twenty years my junior, having nothing mental in common.

6 July 1944 Temp. HQ, 3rd Infantry Brigade, Italy Near midnight, and start at 5am tomorrow. My new trailer caravan, a converted 15cwt[62] long chassis, has spring bed, writing table, washbasin, two cupboards, electric light. Tiny but perfect. Last time I inherited a caravan I got sacked almost at once, so any day now perhaps. Jeeped from Rome today through towns where in each the girls were prettier than the last, and that already a world high standard. They showed no signs of having been fought over by two lots – the girls, I mean, but then it was Sunday. Countryside delightful – too 'Old Master' for words.

So Monty's got Brest – nippled Rommel in the bud,[63] so to speak. I feel like imploring Winston, 'Leave a little war for me'. Have too little to do to keep my mind in trim. Professionally I've lost that active interest in novel methods and technique because something inside me realises that never again will my ideas be any use to the Service. Have passed into the limbo of those who have no ideas. Re-hatched the plot and scenario of a week's training exercise on the same lines as last year's on the South Downs. All that after being sent here in search of 'war experience'. Wish I was in France where there's fighting and might even be opportunity.

7 July 1944 Temp. HQ, 3rd Infantry Brigade, Italy A happy day spent at a medieval village on an Appenine hilltop, playing war with my chaps. Clean air, cool breeze, strong sun, scented hillsides. But this isn't 1943! Have to pinch myself. Am planning a terrific exercise next across all Macaulay's[64] place names and wish this tragic crusade was also a thing in history. Up at 5am tomorrow to visit more by-play.

11 July 1944 Temp. HQ, 3rd Infantry Brigade, Italy Outdoors all day working out a forthcoming exercise. The *Castello* is becoming as much part of my restricted life as once did Provender. We train away busily for future events and when it's over we'll be in fine fettle, ready for any Hun anywhere. Doubt my journey has been strictly necessary, but can't see how I could have done otherwise. How I dislike this life in a kindergarten.

15 July 1944 Temp. HQ, 3rd Infantry Brigade, Italy. Continue in my obscure vacuum. Occasionally more favoured contemporaries drift in dressed as generals to gloat and be called 'Sir' by me. I'm docility itself, like some poor old peasant whose been so beaten up that the only thing to do is to stand cap in hand when the great go past. Only, unlike the peasant, I have philosophy and Eve, so nothing can touch me behind that armour – except being refused all prospects of action.

19 July 1944 Temp. HQ, 3rd Infantry Brigade, Italy. The Russians seem to be doing remarkable things. Evidence points to the fact that the Hun has still to use some 180 divisions on that front out of his 300 available. By giving 70 to the Baltic and France and 50 to Italy and the Balkans, which can't be far wrong, there's little left for new tricks. No Luftwaffe and no reserves, so he's inevitably sunk unless he has a really effective secret weapon Mark II. The only question is whether we can complete his destruction before October, when he may be rescued by the winter. Meanwhile we'll go on having the Flying Bomb[65] till he retreats from the Pas de Calais. Gathered from between the lines of skimpy news that damage to London and elsewhere is considerable. As a trump card it's a pretty formidable ace and emphasises the need for speedy success in France.

It's a curious feeling to be surrounded by enemies, German and British, and to be totally friendless, knowing that my death in action would be received with greater satisfaction by Alex's HQs than by the Germans. Not conducive to good work. Were I to be badly wounded tomorrow, no-one in Italy would risk his career by even visiting me in hospital. Difficult to escape bitterness as in a brigade HQs I've no-one to talk to openly.

21 July 1944 Temp. HQ, 3rd Infantry Brigade, Italy Still the enemy withdraws in Italy, spreading misery and malice as he retreats. We hear that Italian partisans do good work against his flanks and stragglers.

Hitler seems to have some internal trouble.[66] Party v. *Wehrmacht* is an inevitable development, so hope the Party wins. I like the idea of Adolf Attila eliminating Junkers, but no Junkers means no army and the more officers he rubs out the nearer the Hun will be to packing up. If the Junkers won now and huffed Adolf, we'd have a Badoglio[67] capitulation in no time with the Hun army drawing rations in POW camps and disguised Nazis in control of AMGOT.[68] All the 'best people' in Germany would be palling up to the 'best people' in the army of occupation, with Big Business reinstating cartels in the background. Let's hope Adolf pulls off a sweeping revenge. They say Dick O.C. is commanding tanks in Normandy – a good smokescreen.

22 July 1944 Temp. HQ, 3rd Infantry Brigade, Italy I seem doomed to military make-believe. Really think I'm the most patriotically vain ass in the world. Why the hell didn't I join the civil service and have no pride at all? I'm not wanted here, merely keeping somebody's boyfriend, ten to one some young guardsman, out of a brigade command. One of these days they'll find an excuse to sack me and I'll be able to forget these leprous 31 years of soldiering, and begin to live all over again. But it will mean leaving the British Isles. Some complaisant military asses – could reel off a list – would call that sour grapes escapism and perhaps it is, but that still doesn't preclude its rightness. Liddell Hart stresses the absurdity of damaging one's life forces and he's right. A brain which tries to dismiss its content is only a stultified organ existing in a vacuum. Such fun Hitler bumping off ex-CIGs in Germany.[69] Who will now trust anyone in the Party or the army? Old Joe right again – he purged his generals before the war began and nobody's tried to shoot him. Adolf's in a jamb because the *façade* is cracking. Hope it's quick so I can plead senility and escape from a service as relieved to see the last of me as I of it.

23 July 1944 Temp. HQ, 3rd Infantry Brigade, Italy The *Sirocco* has gone, the *campagna*'s refreshed and I feel better, less taut. How can any German field commander ever again in this war make a genuinely honest military judgement? Lord knows, it's hard enough in our army, but in Germany now impossible. Their High Command is hamstrung, though like a dying boar it will fight on, there being no alternative. Met another rebellious padre today. Salvation may come from that unusual quarter. Rather than face the complexities of postwar civil ministry, he's applied for an army chaplaincy. He's one of the few men I've met who are prepared to speak their mind about WSC [Churchill].

Pottering about in the woods found a German burial place with a number of dead officers and men, all on 23rd May, the date of our first attack from the beachhead. Not distressed unduly. They were buried in the heart of a forest evidently to keep their losses from the Italians, because transportation and grave-digging must have been difficult. A beautiful day – bright, brisk and cool. Want to get up into the last high hills overlooking the Po valley where the enemy must inevitably fight for as long as he can. Hope to be really busy in the next few days – good practice.

24 July 1944 Temp. HQ, 3rd Infantry Brigade, Italy An air-letter from Basil LH, very pessimistic – not of victory but of the general atmosphere at home, and worried about Normandy. Says Archie Rowlands[70] writes that Auk talks of getting me to him 'when I can be spared from Italy'. Wry joke. What a hope of my going on soldiering. My outgoing boss [Penney] actually came to say goodbye today, and said that the Military Secretary was coming here soon and did I want to see him. Replied I, 'No. Why?' That's what they can't understand, these toadies and lickspittles. No, I've finished with the **** regular army as the right wing of a Conservative Party and the left-wing trades-union.

Tonight I'm playing at soldiers, camped in a park now pretty dilapidated. Huns left a whacking great ornate para-boy cemetery on the late king's front-doorstep. Nice reminder of a happy alliance – were I the Italians I'd plough it up. WSC[71] let himself go at an RAF mess in France. He's far more at home with RAF than with army – I think they do him better in alcohol. He seemed to suggest that Adolf might pack up before schedule, and if so a rush general election. Troops haven't responded to proxy registration, too politically vague and dispirited. No party is regarded as worth a damn to them.

25 July 1944 Temp. HQ, 3rd Infantry Brigade, Italy A long day stage-managing Exercise Rupture in perfect weather. People who've fought a battle or two tend to think they know everything, and their reaction to a new situation is to try to repeat what they did last time, so saving themselves the trouble of thinking things out. It confirms what I've always thought, that one ounce of imagination (balanced by knowledge) is worth a ton of experience (unfortified by stern principles). We're all busy and by July 28 some of us will be very tired – I for one, I expect. Good for me.

26 July 1944 Temp. HQ, 3rd Infantry Brigade, Italy Up to my neck in Exercise Rupture, which I think is going to prove some use. The troops refuse to be interested in something which hasn't got real death and real Jerries,[72] but make all the usual blunders which get them killed when we're fighting. We grind forward against an enemy still disorganised and desperately short of time and men. So we should be able to do him a lot of no good when next he tries to stand against us. My HQs for 'Rupture' are in woodlands which the Germans used while fighting us at Anzio in which they prepared that enormous cemetery. Lines of empty graves 6ft deep and 1ft apart wait to be filled hidden under the trees, and others have only a little earth over the recent body, no cross or indication of identity. Certainly a great many have been buried here, where no-one would suspect it.

Am going to be busy tonight and tomorrow keeping this show going, but it will be worth it. On July 29 I've an early drinks party in Rome and all my enemies will be there. Afterwards I'll take 2 days off, as haven't let up since I arrived. Want to rubberneck around the Vatican.

27 July 1944 Temp. HQ, 3rd Infantry Brigade, Italy A strenuous bout of training, quite like old times in England and far more exhausting than real war. In war proper, one doesn't have to run both sides or try to see how the junior leaders actually behave. Just throw off a one-sided operational order and leave the rest to the Valkeries. But on an exercise as director one operates in three dimensions – past, present, future scenario – both auditor and actor in one exhausting yet stimulating mental exercise.

28 July 1944 Temp. HQ, 3rd Infantry Brigade, Italy Back from 'Rupture' and as usual after four days of high activity – physical and mental – with very little sleep am suffering the reaction. News is consistently less credible. Brest Litovsk[73] fallen, Warsaw deserted. Where does the enemy mean to stand? Yet he fights on in Normandy and here. Are we seeing a new version of the 'stab in the back'? Alternatively the dynamic of events has taken charge and there isn't a thing he can do about it.

Last night Loewen – my new master, much junior – came when I was aching for sleep and talked for an hour about myself. Theme: 'I had lots of friends anxious to help, but why did I feel bitter? Others were in the same boat.' Really tried to be polite and hope I was. I did point out that anyone who'd been the recipient of special high-up malice without redress, and who consequently had forfeited his career, had a good reason for bitterness. That's the new tale, 'Chink's bitter.' Wish I had some bitter, pints of it, to wash away bitterness. But I did make him realise that the day we stop fighting I'm off, if it's humanly possible.

Penney had 'got away with murder', Loewen warned, and the phrase was unforgettable. Afterwards Chink's spirits plunged to the depth of considering suicide, until he felt histrionic and put his revolver away. By morning protective irony had reasserted itself.

29 July 1944 Temp. HQ, 3rd Infantry Brigade, Italy A little depressed by the rumour that these nasty yellow anti-malaria pills also suppress a very important physical reaction. Impotence – what a sacrifice to lay on the altar of my grateful country! This morning attended, unwillingly, the unveiling of a memorial brass to my crowd for their recent dead. An absolute Jehovah God of Battle invocation in the local C of E tabernacle – the only thing the Pale Galilean would have recognised was the Lord's Prayer. Looked in on an all-ranks dance tonight given by one of my units. Good party. Plenty of local girls turned up and so did plenty of Fascists to turn them away. Result:

I arrived in the middle of a free fight, soldiers versus Wops meaning mischief and prepared to cut girls' hair off. So much for co-belligerency! Wasn't armed or I'd have shot the blighters, but the chaps soon disposed of them. Shows how little our ex-enemies respect us. Damn Dagos!

30 July 1944 Temp. HQ, 3rd Infantry Brigade, Italy Today a VIP drove past and I remembered in 1914 a similar ceremony on a wet Flanders pavé road and being splashed with mud from the august car. Am more anxious than ever lest this show should end before we have action again. We're 100% ready.

Still think I needed to do this to clear my spirit of the dirt of the last three years. Couldn't go dingily into politics after a row with the Army Council, but now can say *Nunc Dimitis*[74] with a clean heart. Even beginning to regard my successful juniors passively, but can't feel quite real about it all. Bitterness attaches and detaches – one ignores the other victims.

31 July 1944 Temp. HQ, 3rd Infantry Brigade, Italy I see no hope for German arms, neither does the German General Staff. Germany is kaput, and I only regret that I had so little hand in the downfall.

Yesterday lunched with the Monarch [George VI]. What a business. Prewar – cold consommé, cold chicken, ham and lamb cutlets, a perfect salad, an excellent sweet. Wonderful show. Sherry, brandy, wine cup. Ate everything in sight and went off like a gorged tom cat. Never thought to have such a good lunch for many years to come. Gluttony! On Wednesday go into Rome for a day, so will escape from the military atmosphere and my soldiers will have a day or two in peace. Really, I've spent a criminally impotent summer.

1 August 1944 Temp. HQ, 3rd Infantry Brigade, Italy Dog eats dog in Germany, and how much easier for us all if the Junkers are liquidated. It's always more difficult for us to punish a plausible gentry class. Hope Himmler[75] and Goebbels[76] work quickly.

Was told this morning that Kenneth Bols[77] has been killed out here, not confirmed. Poor lad if it's true, and for Elaine to be a wife and a widow with two infants in four years of war is tragically fast-going in experience. Only hope the rumour is false.

2 August 1944 Temp. HQ, 3rd Infantry Brigade, Italy Mustn't hanker too much after Bellamont. It may not long be mine, and I find it hard to see myself flopping down into Eirish society, an obvious military failure. Must get out of all that life and can't go on always avoiding people who know I've bogged it. Every day here is an agony for that reason alone. 'Stupidly sensitive', perhaps.

'Lots of other chaps have had it too.' 'Everyone can't expect to run the show.' I know all the comforting slogans and they don't comfort me at all. Tonight I must leave my *castello* at 12.20am to see some pretty obvious experiment, and tomorrow do Cooks Tour agent to a party of Hussar officers who want to see Anzio and our front. In the evening I go into Rome.

4 August 1944 Rome, Italy Last day in Rome. Quiet parties, half political in character and not much to eat and drink but intelligent, charming episodes. Am not going to be pessimistic about the next phase, that's negative, but it won't unfold until the personnel who triumphed at the end of this war have been moved by events from centre stage.

8 August 1944 Temp. HQ, 3rd Infantry Brigade, Italy Tired last night. Our energetic new Divisional Commander[78] likes to do everything himself among an admiring, cooperative group of subordinates. Am constantly attending conferences, the results of which are cancelled almost before I can cover the slow ten miles back to my temporary HQs so return on crowded, skiddy roads to a fresh meeting. But who am I to jib at military restlessness? It's such a rare quality and the war is ending, so what's wrong with wasting energy? A perfect morning: rain-washed, clean, full of light and perfumes. Local population numerous and ubiquitous. Pretty girls in tantalisingly cool summer frocks watch the dusty, mud-splashed transport columns stream endlessly through shattered villages. The farmers are generous with pears and red wine. Everyone laughs. Gunfire sounds totally *de trop* – a land for lotus eaters. Lounging soldiers watch their enemy from villa windows.

Noon Been flying most of the morning. A glorious experience, over the house-tops under the enemies' noses and no-one shot at us. Great stuff, air supremacy! A beautiful sight, pink and white, and looked down into *palazzos* with green bathing pools shaded by tall Cypresses. Andomache waiting for Perseus, or some lovely Umbrian courtesan attending her next client. A bit of both, perhaps. But hadn't bought my Gorgon's head – no room for one in a Taylorcraft.

Returned to meet an even more senior boss[79] (who used to 'Sir' me in 1940). He commented on Fate's reversals and did not use my nickname. Replied that in the senior ranks of the army there had to be either knights or knights errant, and anyway I'd sooner finish the hunt on a bad horse than on none at all. Well, the hunt is almost over. Hope we have one spectacular finale, and then will shake the dust from my soul and come away. They can hardly prevent me, neither will they try. The collapse of this front might begin the end, much as in 1918 when the Austrians broke. Away to the north the guns are going –

how stupid war seems in all this beauty. Never understood before where the Old Masters get their models, but they're all here eternally, all Madonnas and Magdalenes in one. I'll send in my papers directly the fighting stops.

10 August 1944 Temp. HQ, 3rd Infantry Brigade, Italy Snug in my cabin on wheels slept unbrokenly, but this morning felt guilty, thinking of my poor troops in the open. Storms are frequent. Back in same atmosphere but totally different countryside – beautiful, but inconvenient for fighting. Roads few, bad and easily damaged. A queer war in villages full of Tuscans. High-breasted, gentle-eyed madonnas carry water down streets which any minute may become a lurid shambles. In the very near distance are the German-owned hills, outposts to the Gothic line. Motored up to various villas and farms this morning to peer cautiously at the enemy's hillsides, before being offered excellent wine by their owners in the cellars. Civil population is everywhere. Lord knows how we'll feed them. All going well, we must go on advancing everywhere before winter bogs us down again. Hard to believe there's an enemy opposite us at all. The real obstacle is a bridgeless river. Tomorrow move my own HQs a little nearer, more convenient for control.

11 August 1944 Temp. HQ, 3rd Infantry Brigade, Italy My HQs tonight in a villa high up which sees everything, including the enemy's present positions. My converted 3 ton operations lorry is a first class job, if irregular. Everything inside is always the same wherever we halt. Telephones and RT plug in, air photos etc all handy. It's money for jam. Flies are the trouble – everywhere. Spent today in the forward posts in front of vineyards full of peaches, apricots and pears. Full to the muzzle with fruit and vino which even the Yanks couldn't buy in England. Some promiscuous shelling, but nothing to worry about. Tea afterwards in a villa in No-Man's-Land attended by partisans in red shirts calling themselves Garibaldis! We await bigger events.

The enemy is firing lazy big stuff from far off at the waterfront. Feel a fraud because here I am playing at soldiers in a front line crowded with lovely bright-eyed laughing girls. Too indecorous! How consistently the heroic evades me, and I ache to be brave, if only to confound the Army Council. Out of touch with the outer world and hear nothing except the BBC, so Lord knows what's going on. Finland, Turkey, the Baltic States, Home Politics – all unknown quantities. A little patience, one more rousing good battle, and the show will be over. After which, 'Were you ever in the army, Mr Dorman-Smith?' 'No, never. Why do you ask?'

12 August 1944 Temp. HQ, 3rd Infantry Brigade, Italy How can one make war here? Today I sat in a drawing room in front of our advanced posts and spoke to a charming woman speaking really good English whose house was surrounded by Hun-made craters and liable at any moment to destruction by the outsize shells he was throwing about. Her brother had worked in the local British Consulate for many years, and her husband was doctor there. Nothing would make her move, so she conveyed my orders to the villages without a tremor. Have seldom seen anyone less scared or more mistress of herself. Hated leaving her in a Grade A target area but she wouldn't budge.

Basil L.H. wrote to say his heart and mine beat as one. Hear WSC is in Italy. Monarch's visit clearly a trial run. Thought he couldn't keep away – this man will be denied no possible satisfaction. Expect there'll be a Via WC in Rome, but then most 'vias' in Italy are WCs! The enemy's restless tonight. From my eyrie can hear his MGs and shells on my front. Big stuff. How could I stay on? Even if let, I could do no useful work under the present regime extended into the postwar world. A rapid exit is both patriotic and dignified. Anything else would be servily bread-and-butter.

The ultimate choice, however, was not to be left to Chink. The four-month report on him – either positive of negative – was due, and he would never learn the exact sequence of events. His Divisional Commander, Ronald Penney, later testified that long beforehand, 'I and many others of my and Chink's contemporaries had decided that brilliant as he undoubtedly was, Dorman-Smith should be kept a long way away from active command of troops in war.'

Penney had returned after six weeks' absence on sick-leave in June, already suspicious about Chink's 3rd Infantry Brigade role. After handing over command on 24 July to Loewen, newly promoted to Major General for the purpose, Penney summoned each of the three COs for a private interview before leaving for Rome. 'They unanimously expressed their anxiety about going back into battle under Dorman-Smith's command…. My own opinion is that he should not be in command of troops.' Direct interference in Loewen's division broke strict army code, and Penney had no authority to unsettle senior morale. That action was bound to result in a negative report, because the single-word answer had to be either Yes or No. In Rome Penney consulted the visiting Military Secretary, Lieutenant-General Wemyss. Wemyss advised him to talk to the Army Commander, the recently knighted Sir Oliver Leese. Penney was about to stay with Leese, an old friend dating back to their mutual time with Chink at the Staff College. 'Oliver and I knew all about Chink,' he countered Loewen's subsequent protest, 'and we would take whatever action we considered necessary.'

Undated August 1944 Rome, Italy This evening see Oliver Leese[80] informally and may be able to open his eyes to what's been happening. Tomorrow may know more about the future.

At the tense meeting with Leese in the Grand Hotel Chink handed him a typed emotional six-and-a-half page defence, covering twenty-three points. He concluded with an appeal to the Army Council for annulment of the adverse report and – a request that he would soon regret – permission to retire at once.

18 August 1944 Rome, Italy There's nothing more for me anywhere in the army. Don't care. Have done my best. But still hate failure for whatever reason and still feel ashamed. Last night in the Grand Hotel lounge (reserved for pur-sangs[81] only) I interviewed O. as Roland to Oliver. Amusing sequel to our last meeting at Delhi, when I was the senior. Now he held all the cards except the joker – that was my report on this incident in this batty and neurotic institution. He was a little shocked by it, and more by my request to quit. Unbelievable in wartime. 'But', I demurred, 'this isn't my war. You've not let me have any of it. Can't I even go and make a shell or put out a fire?' No. I mustn't even do that if it can be avoided. Leese declined any staff work in my present rank. Have declared myself quite disinterested, not only in my service future but also in nationalism *per se*. Both shocking statements.

The situation is now that I must wait here for seven days until elephants and conquerors and Uncle Tom Cobleys in general decide how best to bury Caesar, so I won't be a nuisance to the upper classes in the dangerous postwar years when the toilers are free of the army. What they'll do, I don't know. They think my bitterness unjust. They'll probably get a surgeon to remove my gallstones! The *denouement* reads like a Russian novel – all psychological and *Journey's End*.[82] The claustrophobia of Anzio had a psychic effect on chaps. The *New Statesman* would sell its soul for film rights. So that is that. I'm given special telephone formalities by the top boys, and I've kept my car, driver, servant. But I must not trust these guys at all. They'll doublecross me if they can.

To Basil Liddell Hart   Air-letter from Rome
                                                        22nd August 1944

My dear Basil,

The *Nombre de Dios*[83] expedition failed as feared, but partly I blame myself as being insufficiently adaptable to the part I was expected to play. It's over now, but the climax was the smartest (and least principled) quasi-instinctive frame-up I've yet seen in this pretty tricky service jungle. Even those who profited by the *denouement* to get rid of an awkward intruder showed their disquiet at the method chosen. We're a sweet lot of things, our class. Ancient Rome at its best had nothing on us.

I've created rather a difficult situation by requesting to retire, because there's a sneaking feeling that a Chink who 'retires hurt' with cast-iron evidence in his suitcase is in a pretty good position to make 'mischief' as a civilian. The army's first real gift to the left wing since Wintringham.[82] Equally there's a determination, worthy of better things, that Chink must not be given any opportunity of upsetting the war accident apple-cart from inside. So I'm coming 'home' by air, to a pretty future!

I'm relieved myself that the farce is over. [It] was a stupid piece of Celtic quixotry, and knights-errant stand little chance against knights-static. I've seen a little of everything to do with this war without having any responsibility for anything in it. The fighting war is nearly over, the Hun for all his ability and discipline can't compete for much longer against such *materiel* disadvantages. He is, as a soldier, finer in defeat than in victory. A wild boar at bay, ringed by not-too-brave hounds.

As soon as I'm a decent civilian, stripped of all the inhibitory allegiances of my ex-caste, we must meet and talk sense, as distinct from fence. I'm greatly looking forward to my coming freedom. I think that for me life is going to begin at 50. But perhaps that, too, is quixotry. Well, so long as there are sufficient windmills!!

Good luck Basil, see you soon, Chink

29 August 1944 *St Mawgans Airfield, Cornwall, England* Arrived too late to get to Town [London] last night. Now they're waiting to see whether I can fly there, or alternatively spend an endless day in the train. It's raining hard, so the shuttle service may not fly.

My trip to Italy has been brief and frustrated. Now it remains to be seen how they go about my case. I'm cured. Don't care one bit whether I go on in the service another day. I've become 'civilian' minded. Grown up, perhaps, at last. I've to report myself in writing to that dusty snake Wemyss [the Military Secretary] and wait, which I'll do, leaving any next move to them.

When his War Office appeal was heard in London on 11 October, his objective was to put his retirement impulse straight. 'I wish nobody to think,' his written statement began, 'that I would apply to leave the Service in wartime, except as a protest against gross injustice.'

The Army Council reply was dated 1 November 1944. No further suitable employment could be found, it was stated, so he would be placed on retired pay with effect from 14 December and granted the honorary rank of Brigadier.

To Basil Liddell Hart   15 Half Moon Street
                                                                    London W1
                                                             1 November 1944

My dear Basil,

The Army Council have come to the conclusion that I must either run or leave the service, and naturally enough decline to abdicate. So I am finally retired on 14th December and that is that. Soit, I've had it. I'll go quietly for it's a little undignified to appeal to our fragile monarch against an army council that runs him. At 49 I begin where I was at 19, only this time without a calling. They were kinder to Socrates. So I become a vagabond without any particular means of support, and I'm not sure that I bring to this new adventure as much go-getting enthusiasm as seems to be demanded. Meanwhile the orthodox aristos triumph. Leese to Burma [as C-in-C of Allied Land Forces, South-East Asia] and Dreary McCreery to the Eighth Army [taking over command from Leese]. No wonder it's me for the hemlock.

I have not decided where or how to live yet. For a month I'm going to [my parents], then will come to London to look for work. The PM has successfully prolonged the war for another six months. I notice too that Alex is talking about 1945. Poor Alex, his apologies for the campaign in Italy won't wash, at least there won't be much colour left in it when we're through with him. The question arises whether the extraction from Italy of the force which included Southern France wasn't almost as bad a strategic blunder as APW's removal of 4th Indian Division to the Sudan after Sidi Barrani, or the break up of the ME forces in Cyrenaica to 'assist' Greece. PM again I daresay – <u>divergent</u> strategy.

Good Luck to you,

Yours ever, Civilian Chink Esq!

Civvy Street:
1945–1958

The drop in emotional tension of peacetime meant no more diary entries, such ventilation – as we would say today – no longer being needed. Chink soon had a project in mind. Conversations with discontented army padres and servicemen had predicted a strong postwar reaction, and by 1945 he was actively seeking a congenial political party to puncture Conservative complacency. Basing his plans on promoting a League of Angry Men, he met journalists from the rightwing *Morning Post*[1] and socialist *Tribune*, and canvassed every newspaperman he came across, like the editor of the *Bournemouth Echo* who asked why the army didn't use his services. 'Something fishy was the suggestion – suspect I'll get a lot of that.'

Labour looked good at first, but Beveridge's social welfare proposals duly won his support for the Liberals. By the time Germany had surrendered in May he had been accepted to contest the Tory stronghold of The Wirral in Cheshire for the liberals against John Selwyn Lloyd for the Conservatives and Lois Bulley for Labour.

The wet weather was grim and his northern hotel and bedroom grey – 'the Liverpool monsoon is as depressing as Simla's' – and he longed for it all to be over. Interest from the *News Chronicle* and Bill Grieg of *The Mirror* in his League of Angry Men came to the rescue – 'He says politics is as full of tactics as soldiering, and both require knowledge and experience. He's right' – but otherwise the National Press saw him, he liked to say, as the new Mosley come to judgement.

Energetically he got stuck in, enthused by the stream of letters and telegrams from 'Angry Men' and soldiers' wives. On Election Day he polled a creditable 14,302 votes, better than average when across the country the Liberal Party was almost wiped out. Selwyn Lloyd won The Wirral for the Conservatives, Labour took power nationally, and Chink relished every newspaper headline about Churchill's public humiliation.

There was the consolation of a new start. Eve was prepared to live in Ireland with him, he longed for children, and had hopes of a divorce from Estelle. He moved across the Irish Sea with Eve and her young daughter Elizabeth, and they rented a house in central Dublin; run-down Bellamont offered weekend escapism but belonged to his father during his lifetime. There was ample time to study the past, and Chink attended lectures on archeology within walking distance at University College. Becoming a full-time student was ruled out, ironically, because UCD refused to accept his Uppingham and military credentials.

Fulfilling the Haifa palmist's reading, a son, Christopher, was born in 1946, followed nineteen months later by a daughter, Rionagh. From a distance, Estelle reluctantly consented to a divorce after the birth of both children, and when Eve's divorce also came through, their wedding took place at Westminster register office in London. Letter-writing once more became his favoured means of communication. Physically, he might live in Ireland but his mind stayed absorbed by his old world elsewhere.

To Basil Liddell Hart   2 Mount Street Crescent, Dublin
                                                           December 1945

A Basil, A cara,

Which is the nice way the Celt has of beginning with Oh Basil, Oh friend. This already is a cry *De Profundus*;[2] the green mould of Erin begins to settle on me. I have cast off all urgency of living; drowning, I no longer clutch at straws.

All of which only means that I'm leading a wonderfully idle life and steadily accustoming myself to the idea that nothing I ever do henceforward will be of the least importance to anyone; least of all to myself. That is the way to live in Dublin, it just takes a little getting used to. Occasionally, like the Dead in Hades, one reaches out to where there is still warm blood running – and then I trouble you with a letter.

Cheers, Chink

To Basil Liddell Hart   2 Mount Street Crescent, Dublin
                                                           1 June 1946

My dear Basil,

You ask about my future plans. I have none. After the election I went to Dublin and wallowed in its *Dolce far Niente*. We inspected Bellamont and decided to renovate it for our eventual headquarters. Life there is cheaper, and at least we have a roof!

There is no work for me in England, that I can find anyway, and without work at a reasonable income and a home I can't stay in that country. I'm sorry in some ways. I'm still over-busy in heart and brain to sink into the bogs of Cavan, and can't bear doing nothing, but it's Hobson's Choice.

Far more important is the future of Liberalism. The party went into the Election totally unprepared. The electorate quickly saw this, and although it approved both of our policies and candidates, it rightly assessed our powerlessness. This was an anti-Tory election, and the people made a good job of it.

Anyway, I've had my experience of politics. It was an interesting struggle, though fore-doomed as are most of my undertakings.

Good luck, Chink

To Basil Liddell Hart   2 Mount Street Crescent, Dublin
                                                                2 June 1946

My dear Basil,

I've just seen an article by you on Monty in the latest *Strand Magazine*. He can't yet be compared with any independent commander so can't be assessed on performance for he has always been under the command of a higher operational headquarters, as distinct from political headquarters. Wellington, of course, was an independent commander by that yardstick. So was Auk, Alex, Eisenhower, even Jumbo Wilson. (But Alex and Jumbo had rarely, if ever, any scope for manoeuvres.) One more point in assessing Monty. He always fought with 100% air ascendancy and about the same artillery supremacy. Even so, he never achieved a 100% decisive victory. He is perhaps incapable of [that].

I think there's a certain danger to the public if someone of your unique reputation for objective evaluations of the military animal ever blurs the picture!

Yours ever, Chink

To Basil Liddell Hart   7 Mespil Road, Dublin
                                                            10 December 1946

My dear Basil,

I've suddenly concluded that much of what's being said by Monty and his group is founded on the most natural of all facts in war: the tendency to see a military situation in dark colours to the direct degree to which one is remote from it. To Churchill, Brooke, Monty and Alex, certainly to Smuts, the situation in early August 1942 appeared far more difficult and dangerous than it did to me. Next to Auk, perhaps – even more than Auk for I was more detached – I was best placed to know the real position. Two good brains both now dead, Strafer Gott and Kisch, confirmed my own considered opinion that having rescued ourselves from Ritchie's deplorable dispositions at Mersa Matruh, the crisis of the Axis follow-up had ended by 3rd July.

Churchill and Brooke had brought no staff with them capable of a proper examination of the detailed position. Winston, in Auk's caravan, tried to urge

Auk to begin a premature counter offensive immediately (Auk refused), but there was little or no consultation between Brooke and Auk. Wavell saw the thing whole and accurately but said nothing, probably because it's his nature to say nothing, and anyhow he trusted neither Winston nor Brooke.

Monty's people – Alex, Monty, Horrocks, Leese, all new to the desert – thought they were called on to retrieve disaster – not, as in fact, to shoot a sitting rabbit already corralled. I'm sure that all, in their way, felt that the situation was almost lost when they took a hand in it. The general opinion in Khartoum was that we'd lost Egypt and the war. The further away from a losing battle, the worse it looks for the losing side. The nearer one is to a winning battle, the harder it is to realise the extent of the enemy's defeat. But the Alamein misunderstandings are in the end psychological, fostered hugely by Churchill's gauche and dramatic emotionalism, and Brooke's determination to give Alex the ME command at Auk's expense.

I genuinely think that Monty thought he knew what he was doing and why, even when he was in fact only confirming his predecessor's arrangements, which says something for the British military 'Common Doctrine'. In certain situations it's useful! But largely I believe the whole Alamein controversy is one of misunderstandings. When are Auk's Despatches to be published?

Good hunting. Ever, Chink DS

To Basil Liddell Hart   7 Mespil Road, Dublin
                                                  26 February 1948

My dear Basil,

Auk's Despatch does awaken old, sour memories. I try to avoid all contact with anyone who was in my life before 1945 except you (who have a professional interest in my memories), and my ancient Mother. My Father alas has died, before this Despatch was published. It might to some degree have reinstated me. He was deeply grieved at the collapse of my military career. Ridiculous, of course.

The Appreciation of 27th July 1942 was written by me on my own initiative, largely because I wanted to impress on the Auk that the first phase of the battle for Egypt was over. The situation had stabilised [and] the initiative had passed to us, provided we were mentally and physically ready to avail ourselves of the inherent advantages of the strategic position. I told you all this in October 1942; odd to see it confirmed in 1948.

Of course, it gives the lie to Monty, why wouldn't it? I don't think he's so much a wilful intelligent liar, as a wishful self-deceiver who, like Jack Horner, thinks he's a good boy for pulling a plum out of a ready-made pie. Monty goes further than Jack in trying to pretend he was also a good cook. I'll write again in a few days to return the Despatch.

The change of government here is timely and proper. Queer how the British papers lament Dev's[3] passing and throw cold water on the new Coalition. I came out in favour of Clann na Poblachta,[4] the reunion of this Ireland, and complete separation from Britain. I couldn't think of anything more likely to annoy Alanbrooke and Alexander. But it will come one of these days.

Good Luck, Eric

The death of his father meant the end of Chink's regular academic routine. Isolated Bellamont Forest and its estate now passed to him, and he took up residence with his small family straightaway. Death duties stretched ahead for seven years, Dublin's wider social life was far away along bad roads, Bellamont's restoration and upkeep needed capital, and his lack of income, compared to Eve's, stood out. This was not to be a happy-ever-after ending. The reality of life under pressure with her was replacing fantasy with rapidly growing intolerance, and mutual discord behind the scenes intensified. Bestselling military memoirs continued to appear in the UK from members of Montgomery's team, keeping Chink on the boil, and he longed to write his own account. Meanwhile his increasing IRA support in letters to correspondence columns of *The Irish Times* and *The Manchester Guardian* was making readers uneasy.

The answer for him, as always, was to keep busy. He notified *The Times* of his change of surname from Dorman-Smith to O'Gowan, stood in the local elections for Clann na Poblachta on a United Ireland ticket (and lost again), and more successfully planned an all-expenses-paid anti-Partition trip – essentially public relations – across the United States with official Irish government backing.

The day after his arrival in New York, he was browsing in Scribner's bookshop when he learned that a new Hemingway novel *Across the River and Into the Trees* was being serialised in *Cosmopolitan*. On impulse he called to the adjacent publishing house, to learn that his old *aide-de-camp* was staying at the Sherry Netherland Hotel nearby, but about to return to his home in Cuba. Speed was essential, so he phoned straightaway. 'Come in, Chink,' he heard with relief when he arrived. 'I've some poetry to read to you. I think it's good. I can't talk poetry to anyone but you.'

A couple of hours together were enough to restore their camaraderie, despite a procession of strangers: lawyers acting for previous wives, sons by a wife Chink had never met, and a new wife, Mary. And those two hours were all they had, because Hemingway had to leave for his flight at noon. 'Hem was gross, lavish, champagne like water, grey bearded', he jotted afterwards.

'Very interested in the Rommel story [and] recognised necessity of telling it on principle. Very well informed on British layout, including my past!'

To Ernest Hemingway   Hotel Commodore
Finca Vigia   42nd Street at Lexington Avenue
San Francisco de Paula, Cuba   New York 17, NY
                                                              6 April 1950

Hem,

If it hadn't been inevitable it would have been extraordinary – destinies aren't extraordinary if one just lets them work. So this one has worked, and what is 25 years or so? Nothing at all – or, at best, only something to bleach us and bloat us, silly anatomical tricks which don't count, any more than the fact that we were once in binders and will be, in due course, in a coffin. All of which means that in my opinion it's high time we united to take charge of the planet again. We've let it rip for too long, wherefore all sorts of indifferent exploiters have taken over at the point we left off, and I don't think they've made a very good job of it all.

We won't lose touch now anyway. That was a damn good poem, Hem. I'd like to hear it again. Somehow it has the spirit of all and every crusade. About time we reexamined the 'crusade' idea versus the 'conqueror' idea. I won't go on with that one now.

Thanks for the Bubbly, it was lovely. One day we'll drink beer again in buckets.

All love, Chink

To Basil Liddell Hart   Hotel Commodore
                                               42nd Street at Lexington Avenue
                                                               New York
                                                            7 April 1950

Basil,

I got here on Wednesday, after a round-about fly which at one time reached the Azores. I am, of course, on Irish business, as a private individual from Ulster. It's no coincidence that Northern Ireland premier Basil Brooke arrived relatively simultaneously. This is in fact 'Operation Badger-Hunt'. It's immensely amusing, particularly after my retired life, to become a stormy petrel again. I'll be speaking in Boston, Pittsburgh, Washington, California, as well as NY.

Yours ever, Chink

To Basil Liddell Hart   Hotel Commodore
                                                  42nd Street at Lexington Avenue
                                                                        New York
                                                                    16 April 1950

Basil,

I hope by now you've got the Rommel Papers safely [returned]. With so fine a border of airmail stamps they could almost have taken off solo. I'm off to Washington today as a sort of unpaid, unofficial, temporary private ambassador! A curious but interesting assignment.

Good Luck, Eric

To Ernest Hemingway   Washington
                                                                     13 May 1950

Hem,

I'm still laughing at your letter. In Ireland it will be framed and placed among the Armorial Bearings. Meanwhile an assurance: I'm not walking out on you; there is too much to talk about and quite a lot yet to be done.

I don't think I'm gunning against the English out of pique. I hope not. Anyway, I'm not going to be psychoanalysed to find out, it is of no significance, for the justification is too obvious. So if you feel like a fight again, it is all ready for you, including that rare experience, a good cause.... I think you ought to come in on this play as the O'Hem, a mythical figure from the American underworld; we will provide sub-machine guns, road mines, depth charges, cyanide and unlimited champaigne [*sic*]. Belfast and environs pay in taxes a revenue of some £36 millions and as much in rates. You needn't feel the game is too small to shoot. Bellamont, 11 miles south of the damn Anglo-Oirish border makes a perfect GHQ. The overall intention is to get the British out of our last six counties and, having taken Belfast, to march on Dublin and purge that city of neo-Georgian fakers. It's a damn good double event and if, during the campaign, you die of alcoholic poisoning, we'll deify you as the O'Lafayette. You'll have to drink the hell of a lot of Guinnesses' stout: we need the bottles for molotof [sic] cocktails!

I will give your book[5] to Christopher as soon as he can read. One must start a chap somewhere. Do you wish it to be banned in Ireland? It improves the circulation a lot, and we can always appeal later. In which case you must clearly conduct your own appeal in Dublin. Naturally you will stop in Phoenix Park with the American Ambassador [to] defend your title as the world's only unintentional pornographer. We'll get every barrister in Dublin on the job. [Irish] literary censorship [is that] we keep a young priest, healthy and celibate, as a guinea pig. Any literature which makes him wish he weren't a priest is immediately withdrawn. The criterion is purely physiological.

All good luck to you, Hem, and remember that beneath a flippant exterior beats a broken heart, or damn nearly broken anyway.

Ever, Chink.

PS I no longer permit anyone else in the world to call me 'Chink', but you're too damn big to be stopped, so I'll let it be.

San Francisco PPS. This ought to have gone days ago. But I found an instalment of yours in May's *Cosmopolitan* and began to read. Damn you, it is devilish good. Too good for an aging gent like me. How do you know things I only know? That is great writing. You understand sorrow. Why didn't you tell me?

My love, Hem. C

Chink saw little connection between Cantwell, the bitter hero of *Across the River and Into the Trees*, and himself, despite the disconcerting glimpse during their New York meeting of his old friend's awareness of his lost career. But the book was to a great extent drawn from Hemingway's sympathy after coming across a description of Chink's military fall in de Guingand's *Operation Victory* in 1947, published in the US, too, by Scribners. Knowing Chink as intimately as he did, Hemingway had indeed 'understood sorrow', as the novel's dialogue reveals. 'I thought Monty was a great general.' 'He was not,' the Colonel said. 'But he beat General Rommel.' 'Yes, and don't you think anyone else had softened him up?' Chink might have left the army with the rank of Brigadier but for two months in early 1944 he had dropped down to Colonel; even Cantwell's rank was correct.

Chink, however, was soon absorbed elsewhere on his US tour. The guide assigned for the hectic schedule, Elise Morrow, was thirty years younger, mad about books, and married to the assistant editor of the *Saturday Evening Post*. Chink's link with Hemingway drew them together and in Washington they became lovers. Living in the present was all that mattered, and Chink's weeks off the leash began to speed.

To Ernest Hemingway   Bellamont Forest
                                                        Cootehill
                                                        Co. Cavan
                                                        Ireland
                                                        3 June 1950

Hem, the plane is nearing Shannon and bucking like a salmon – it would happen just as I begin to write to you. My American tour is over. I've done a lot and learned much too....

The ending of *Over the River* [*sic*] is interesting. I've not yet seen the earlier instalments and don't know what was wrong with the Colonel. Angina, it reads like, but I suppose he died mainly because there was nothing particular to live about. Conceivable – yet something always seems to turn up to start a new interest in life. At least it does so with me.

I'm feeling sleepy-stale after the night's flying.

Good hunting, Chink

To Ernest Hemingway   Bellamont Forest
                                                                                    21 June 1950

Hem, you old rogue, I take it you really intended to send me that proof of *Over the River* and weren't just pulling the leg of your unfortunate publisher. In whichever case, thank you; between the endless small chores that have to be done when one is over-housed and under-staffed, I managed to read the proof during the day of its arrival.

As a failed two-star-general myself, I know quite a lot about Cantwell's appropriate emotions. The curse of the next-oldest of the professions is that they take away our tools. Harlots, however superannuated, don't suffer that supreme indignity. My sole complaint is that you don't hammer Monty hard enough; I hope you are reserving that for a later effort. But behind Monty was Winston, a pretty combination. All that we will discuss in due course....

Ever, Chink (MCS)

The touches of bravado in his letters to Hemingway were as much to reasssure himself as his old *A de C*. At home, his already difficult marriage to Eve was seriously threatened by her discovery that he was writing to Elise, and being independently wealthy she was spending long periods away in England, out of contact. His bitterness contributed as his anger built at Montgomery's popularity and the widespread obliteration of Auchinleck's achievement. Retreating into his shell, it was a relief to be in touch again with Hemingway, and to compare his own military experience with increasing press criticism of the Korean War.

To Ernest Hemingway   PASSED BY CENSOR
                                               Brigadier E. Dorman-~~Smith~~ O'Gowan
                                                                                Bellamont Forest
                                                                                    3 August 1950

Hem, the fact is the maintenance of military forces in high readiness is almost impossible in a Parliamentary Democracy. I don't mind if only regular troops are beaten up, we are paid to be beaten up anyway, so what? I object to conscript kids and half-trained occupation-forces and poor sods of that calibre getting shovelled into the front without warning. Besides it is damn bad that these North Koreans should be able to make us do what they want, both in combat and strategically. I don't like seeing blokes getting beaten up piecemeal.

I am naturally an arm-chair strategist [and] Cantwells such as we are could, if any listeners there be, ask some pretty pertinent questions. My contribution to date has been to write [to] your Ambassador here a week ago to suggest that we raise an Irish contingent to be incorporated as volunteers in the *cadres* of the Army of the USA. There would be no difficulty in raising a damn good regimental Group, and the kick the Irish would get out of fighting under the Stars and Stripes instead of the Union Jack, particularly if paid in dollars not pounds, would be terrific....

Do I gather correctly, Generalissimo, that you are already at work on a further book on the recent war – to include a blowing down Monty's neck? It is easy to over-rate him. He came aboard when the rough weather was ended. That was not his fault. I used to know him well. If you are really handling him in any profundity, as distinct from a purely natural American reaction, I am at your service for consultation. There is the hell of a lot of background to consider. I am not fond of the said Monty, but I want to see justice done him. It hurts more.

Love to you both,
Chink

To Ernest Hemingway   Bellamont Forest
                                                          7 October 1950

O'Hem,

The hell of you being so far apart in space and history is that I can't explain my motives. Did Cantwell have motives? ... Professionally I'd rather surround and starve my enemy than blast him off the globe. I'd rather beat him at a game of which he didn't even know the rules than have and win a slogging match. The fewer people one kills the better. That is why I'm glad my brain had quite a lot to do with the destruction of Graziani's army.[6] That was an intelligent thing to do, only we had quite a lot of work to persuade our own side to scrap their stereotype. You'll like that story one day.

It also amused me to stymie Rommel, appropriately at the last hole, and after Afrika Korps had signalled *en clair* to the ladies of Alexandria to prepare for the victorious army. It would have been amusing to have seen Rommel dragging Africa Korps out of Alexandria to go on with the war.

I regard unnecessary bloodshed as the hallmark of the amateur; the job is to know what is necessary. The emphasis is on skilled prediction of the technique most likely to save lives, at least on one's own side and if possible on theirs too. For a man to have to become a killer is a loss of ground. Killing is necessary, not essential.

My complaint in the late war is firstly that Churchill broke up the Graziani campaign to go off to Greece, when we could have carried the war into Tripoli (ever studied Belisarius?[7] You'd like him) and that he could not see that we'd already in early July 1942 put paid to Rommel. So Churchill went off to visit Uncle Joe trailing clouds of defeat when he should have told Joe that we'd 'Moscow'd' Rommel before Alexandria, and it was up to him to do the same with Adolf. I dislike Churchill [and] my failure was to let that dislike become too obvious. Professionals should not obviously dislike their clients, and politicians are our clients.

[By] August 1942 any competent professional Staff General on either side could have proved conclusively that Germany had lost the war, because she could no longer win it. The technique then was to help her lose it neatly. That is political judo. So when I gird against the fighting you describe[8] it's for the good reason that it should not have taken place. We fought that phase as if we were Ulysses Grant ending a civil war in the only way one can end civil wars: outright. That was clumsy, and led to a lot of men we could use today being killed unnecessarily....

So, my old and necrophilistic friend, pray receive the salutations of one who designs means of execution for the killers to employ, the salutation of the professional for the enthusiastic exponent. Get fit quickly, and the autumn perfumes of a wet Irish woodland to reinforce the scents of Cuba.

All luck to you, Chink

To Ernest Hemingway   Bellamont Forest
                                                         3 November 1950

O'Hem,

That gross brute Winston Spencer Churchill is serialising his third volume and getting around to the part I know, about Tobruk and Auchinleck. He is being so damn unjust and so bloody inaccurate that I ache to weigh in to the discussion. I will when it begins. But what is clear is how little WSC knew about the truth of things, and how much was kept from him.

He moans about the disgraceful loss of Tobruk, which was quite indefensible in any case, and which, as early as February '42, Auchinleck had determined would not again be permitted to be invested. If the Eighth Army could not keep Rommel west of Tobruk, it was to be evacuated. I cannot believe the War Cabinet was unaware of that intention. Of course at the last moment, when Ritchie had already withdrawn the bulk of his forces almost 100 miles to the

East, WSC wires Auchinleck to hold Tobruk, and then blames the soldier for the chaos and disaster which ensued.

The curse of being in my sad calling is that one is at the mercy of politicians, and I do believe that British politicians are more unjust to their subordinates than yours. Had Hitler treated Rommel in Libya as Churchill did Auk, we would have had nothing more from R. after his first defeat by Auk. Hitler, until he cracked, was a bolder man than Churchill. But what a history it all is, and the curse of it is that I who was in no way responsible for the disasters, and warned against them in advance, got the sticky end of the business after it was all over. I never forget what you once told me in Paris about the crookedness of the Britian [*sic*] ruling class. You were damn right.

There are a lot of duck in, and a good showing of wild pheasants. I wish you were here to shoot them with and for me. Good luck old hunter, get well quick and we'll shoot Winston as vermin.

Love to you both, Chink.

To Ernest Hemingway   Bellamont Forest
                                                        19 November 1950

Hem,

I had lunch with Auk in London last Friday, first time I'd seen him since we dined at Mena House by the Pyramids just after Churchill removed him from command. Tried to persuade him that to be double-crossed by politicians is an occupational mishap of senior soldiers in democratic states, [so] it's the duty of us who have undergone that *triste* experience to put up signposts for the next lot along, and not just to be little gents and let the crooks get away with it. This business of turning the other cheek can be overdone. My other cheek has been practically smitten off [and] I'm getting tired of it. He has a lot of good dope in his private files and he remembers items I don't and vice versa. I believe he'll play ball with that stuff [as] I want him to do.[9] Then we have very good documentation indeed, what with Rommel's Diary and papers almost ready for publication and Churchill's very misleading and selective books. Auk, ever the gent, fears that I will be too bitter; he needn't worry, I'll be good.

The point is, what is the target? You and I have this in common, we belong to a defrauded generation; we have seen the crooks and the slim boys get away with the spoil. We saw this coming as we talked in 1924; we still see it. You had far fewer illusions than I and I admire the way you insist that anyone who fought cleanly, no matter on what side of the line, deserves more respect from us than those who did not fight but grabbed whatever was going. In the last 35 years everyone has had the chance of fighting; there is no excuse for not having that experience. Yet broadly speaking the no-fighters have scooped the pool, that is what Cantwell was mad about, rightly. If only we could marshal the fighters of the world, but the fighters of the world don't unite and by and large they are not good politicians....

Our love to you both, Chink

To Ernest Hemingway   Bellamont Forest
                                                                          3 April 1951

Hem,

This is the hell of a gap between letters, for which I apologise. The fact is that I began to write my angle on the Auchinleck, Rommel, Churchill and all, story; impelled equally by your advice and by a letter from Hugh Morrow[10] of the *Saturday Evening Post* to the effect that Rommel was news and I, he believed, had a slant on him. When I began, of course, I saw the story of 1941/42 could not be told in one 5,000 word installment. So I let the mill rip, much as I would tell it to you but fortified with quotes from Auchinleck's official despatches and Churchill's new war volumes. I've already fired across the water no less than ten 2,000 word instalments and am by no means through. Even if nothing happens to them I'll have got something off my chest that has been too long there.

It is not a nice story, because it's about a not nice politician as much as about fighting. Auchinleck, brought in as a stopgap and in due course ejected quite inevitably as a scapegoat, is a fine, dramatic figure. Rommel becomes a fairly minor character whose entrances and exits largely derive from mistakes made as much in London as in Berlin.... I saw a lot of what was going to happen without being able to help until the disaster happened; then I was permitted, in the way a Chief of Staff is permitted, to take hold and influence events. But Churchill, to avoid criticism in England, had already decided on poor Auchinleck as a scapegoat, and I got the backlash of that decision.... Morrow wrote encouragingly of the first instalments, but by now is probably looking for dramatic meat.

At the suggestion of finding a publisher, I'd like to publish my first and only book in America; that links us up too. If anyone takes it, I will indent on you for a foreword and appreciate the jest of one from you on the nylon-smooth level.

Ever, Chink

Concentration on long articles in which the *Saturday Evening Post* was losing interest, let alone a potential book, was interrupted, however. *The Hinge of Fate*, fourth volume of Churchill's popular *History of the Second World War*, went on sale in 1951 with the charge in print that Auchinleck had lost the confidence of Eighth Army. 'I am sure we were headed for disaster ... the

Army was reduced to bits and pieces and oppressed by a sense of bafflement and uncertainty. Apparently it was intended in the face of attack to retire eastwards to the Delta....'

The appearance of Chink's name in cabinet dispatches, proving that he had been sacked on Churchill's recommendation, was a shock, but it was a minimal one in comparison to the slur on Auchinleck's honour. An action, for libel was ruled out when access to Auchinleck's papers was denied. Undeterred, Chink retained solicitor and senior counsel for a case in his own name on behalf of the achievement of First Alamein; as a bonus it would have to be heard in Ireland.

In May 1953 ammunition was fired off against Churchill and the publishers, Cassells, attacking 'the very serious wrong' to his name that had been done. Retreat to the delta was a calumny, and any settlement would have to produce a public apology and withdrawal of the book. Churchill's team fought back under the leadership of the top QC, Sir Hartley Shawcross,[11] acting privately. Chink shut down his personal correspondence and legal letters went back and forth. Shawcross flew to Dublin twice, appalled by the prospect of his elderly and physically frail client having to appear in a Dublin dock, and on his second visit on 2 May the following year met Chink alone. Face to face, he astutely appealed to Chink's sense of chivalry which wide-ranging interviews had revealed as his weak point. The ploy worked. A settlement[12] was reached over inserting a footnote with the wording agreed.

To Basil Liddell Hart   Bellamont Forest
                                                              4 July 1954

My dear Basil,

Twelve years ago to this day Rommel found the game was up. A good omen. Churchill and Cassells, who have quarrelled so I'm told, are legally bound by their agreement to put that footnote in all future editions. Had Churchill ever promised me voluntarily that he would do so, without any legal agreement, I would naturally have taken his word. I have not seen him for 12 years. Shawcross was doing his best for a powerful client and his own reputation, so whatever he said about Winston I naturally took with a grain of salt. Had Churchill intended goodwill he could have written to me to say he was sorry, and I could not have abused that approach.

When Shawcross appealed with regard to the author's ill-health and high responsibilities, I responded at once. Instead of seeking apologies, damages etc, I merely sought for a correction. I dislike the idea that in a case of this sort I would be guided by mercenary considerations. I may have been a mercenary soldier, this is another matter. Shawcross then concentrated on the minimum form of footnote, which after 12 months of twisting and turning, Churchill was forced to accept. I'm not seeking [his] approval of me as a soldier, he has no competence in that regard. We have Rommel's implicit approval.

Yours, Eric

But that apparent solution would lead to a sharper grievance when the corrected edition failed to appear. 'I fell for that,' Chink would tell Hemingway. 'I was a damn fool.' Chivalry counted in that aspect, too, and his concern all along had been for Auchinleck's reputation rather than his own. His last illusions about British power were extinguished by the case.

Within seven weeks of the final meeting with Shawcross, a successful IRA raid on the armoury of Gough Barracks, the British Army base in Armagh, Northern Ireland, took place. Publicly, Chink's acclaim was carried to the Letters pages of the *Irish Times* and voiced loudly elsewhere, courting deeper unpopularity. Smuts had worn Boer and British medals side by side, he would point out if challenged, and the border was a strategical flaw in the Atlantic Pact. Rumours spread, and it is now known that he took his IRA support further once the campaign began. He gave advice on tactics, fitted out one of his cellars as an Ops room complete with maps and sand table, and allowed his estate within eleven miles of the border to be used for training.

Old loyalties, though, remained in place. Use of Ultra in the war was still concealed, and Chink kept that secret. No erstwhile colleagues in high military positions were ever compromised, and there he could have done great damage but was silent. It was a high wire act doomed to lead to disenchantment, and his few close friends – like Liddell Hart, Hemingway, the historian Hubert Butler and the pacifist Irish politician Owen Sheehy Skeffington – were patient.

To Hubert Butler   Bellamont Forest
                                                                    [undated]

My dear Hubert,

I've been trying to persuade my brother Reg, the Privy Counsellor,[13] to get the top-level British to realise that the Irish Problem is scotched, not killed, and if nothing is done other people will be killed too. So far only one extra battalion has gone to Northern Ireland, but that is a converted artillery unit. They have to scrape the platter. Are they going on indefinitely doing this, or will they look the whole question over once more? I believe the 'southern' politicos are almost as scared of a final settlement as Stormont, too many vested interests are involved in the *status quo*. In terms of Machiavelli this would be the time for a stepping up of 'outrage' in the Six Counties; I don't believe the men concerned have the resources for a really sustained effort. This political stuff is distracting, for I would far rather get on with the work in which we are both interested but to do so now is ivory-castle stuff.

Very sincerely, Eric

PS I was delighted at the Armagh job – so neat, a really good job and will put an end to this unreal and totally unprogressive ticking and tieing [sic] with the quite impossible fossil Orange mind. It is useless to parley with the six county Unionist – he is not interested in Ireland, but in Britain. No man can have two countries.

To Senator Owen Sheehy Skeffington Bellamont Forest
Dáil Eireann Autumn 1954

My dear Sheehy Skeffington,

Liberals are too few, too valuable, in Ireland for them to fall out in public. So I am not, repeat not, going to tangle with you in public. I respect your outlook, though I do not always accept it as infallible, and I'm sure you feel the same about me, only more so.

You know perfectly well that the Orange Oligarchy, who serve the purposes of Whitehall for reward, would never agree to a redrafting of the border since that for them would be fatal. To shift the Orangeman one has first to shift Britain. Since Gandhi-ism can only work once, and is scarcely adjusted to this climate, it's difficult to see how the pacific appeal can be effective. Partition is not going to be kissed away. By all means organise passive resistance in the Six Counties; certainly extend the hand of friendship which no Orangeman will ever take; assuredly see your problems in a world perspective – but realise that having done these things *ad infinitum*, you won't have got one whit nearer re-unification. If we can separate our opponents, divide Whitehall from the Orangemen, we may find a better response. Until that can be achieved, the Orange Order just sits pat – and why wouldn't it, since it hold all the cards?

Do not think that I've changed my views as to the desirability of having men like you at the head of affairs. I have not changed at all in that respect.

With all good wishes, very sincerely yours,
Eric O'Gowan

To Basil Liddell Hart Bellamont Forest
17 September 1954

My dear Basil,

My new and too faithful girlfriend, Malaria, caught me a swinging blow, sending me to bed with her for a week, some of which I don't now remember. Am rather floatily about again – another present from Anzio.[14]

Along with your papers came a PC from Heidelberg from Afrika Korps in reunion there, signed by Kesselring, Frau and Manfred Rommel. '*Dear Sir, from the meeting of the old Afrika Corps veterans we send, in memory of the hard days near el Alamein during July 1942, best wishes to the former Chief-of-Staff to F.M. Sir Claude Auchinleck.*' I sent an appropriate reply by cable. That was nice of them.

How consistently Monty wrote for the record[15] and how cynically he shaped the narrative of events to suit the record he wishes to establish. Wonderful to be so obtuse.

All best, Eric

To Ernest Hemingway   Bellamont Forest
                                                                    15 November 1954

O'Hem,

The Churchill case is a long story; let's keep it until you can see the file. The old scoundrel had the insolence to publish the Cabinet Despatch he sent from Cairo in which I, among other and better men, was stigmatised as having lost his confidence. The tale of those days has yet to be told. One cannot tell it in Britain, there Churchill is fenced with phony Press Barons and an adoring public. *Cha sempre ragione*[16]....

Needless to say, I remain British Army Scapegoat number 1 plus plus. I do not mind anymore, though it's hard for a professional soldier to have no more profession just because he tells the truth to a politician who has damn nearly thrown away the whole war. So far as Britain and I are concerned, I have had it. Which was what you once predicted over that damn sawmill in the Quarter.

Very well then – let us mutually reach for the decanter prior to driving the British out of the top left-hand corner of Ireland, where they have over-stayed the welcome they never got from us.

All love to you both from me, Eve, Christoir, Rionagh[17] and one 11 year old black spaniel. Ever, Chink.

Hemingway's regular affectionate letters postmarked Cuba – sometimes typed, sometimes handwritten, always with the odd spelling he remembered – showed that their correspondence was mutually beneficial, if increasingly nostalgic. 'We always had a lot of fun, Chink,' Hemingway reached out one day. 'Do you remember beer drinking at Aigle and the horse-chestnuts 'like waxen candelbras?'[18]

To Ernest Hemingway   Bellamont Forest
                                                      6 January1955

Hem

Of course candelabras aren't waxen, and yet that is how horse-chestnuts
in flower still look to me – but then the horse-chestnut lives with one foot
in faeryland, what with its little horse-shoe markings, the spring flowers and
the autumn nuts; all quite useless and infinitely charming. No childhood is
complete without a horse-chestnut memory somewhere.

Indeed I do remember Aigle, and that day's fishing is like yesterday – all
the rest too. Perhaps we ought to give them a going over once more before
we forget them altogether and finally. This business of growing away from all
these good things, which is what age does to us all, is not funny at all; least
funny when one has done a lot and has not got a routine behind one to make
the days pass, or the backing of a family to dominate and bore so that one can
still feel superior.

It would do neither of us much harm could we get together again; we may
both be missing whatever quality our early friendship put into living which
living seems now to lack. Eve and I would come out to you like a shot, but you
know how it is, Hem, the dollar world is strictly *fermé* to non dollar-holders;
not just for having no dollars but because whatever we could get our hands on
would just melt away at American standards. That closes up one's world a lot.

Alternatively you come here. With that book brewing inside you, you are
not likely to move, and perhaps while you're not all that fit you ought not to do
too much yet awhile. So we could postpone that until we have gone over the
possibilities of combining our menopauses in Cuba.

I thought your Stockholm piece[19] a first class contribution.

Love, Chink

To Ernest Hemingway   Bellamont Forest
                                                      18 February 1955

Hem

Your letter came this morning. It snowed a little last night but today has
a blue sky and a crisp clean air, with patches of ice forming on the lakes, a
little like Les Avants[20] before the snow came properly. So the letter seemed
appropriate for all that you were a long way off and not having the sort of fun
we used to have when the world was younger.

You seem to be running some species of marathon over this next book. No
good asking whether you have to go that pace, what is in has to come out, like
too much mixed alcohol, only less unpleasant. But 45,000 written words in one
week is plenty and does not leave much of one for other employment. I have a
hunch that the sooner we make contact the better for both of us.

'... you and I are old;
   Old age hath yet his honour and his toil;
   Death closes all: but something ere the end,
   Some work of noble note, may yet be done,
      Not unbecoming men who [*sic*] strove with Gods.'

And they say, nowadays, that Alfred Tennyson was not a poet. I suppose we did 'strive with Gods'!

Am reading Aldington's latest book on [T. E] Lawrence. I was never greatly enthused by Lawrence myself, never met him, but Wavell thought a lot of him and he was no bad judge of a man's inwardness. Aldington plies a pretty muck-rake, hopping nimbly from mendacity to paederasty, with a frequent side kick at anyone who ever held different views....

More than anything, I hope we shall be able to make an encounter this year. Years are becoming precious.

   Ever, Chink

Hemingway replied by return, scribbling 'Answered Chink March 14 1955' on Chink's envelope with its heavily-franked Irish stamp. 'Years go by', he made a point of agreeing at once, 'like snapping your fingers at nothing.'

This time, however, Chink did not write back promptly. Other matters intervened, including a proposed visit that year, during Eve's customary absence, of Elise, his Washington love. He kept postponing his overdue letter to Hemingway until, without any conscious decision, it seemed too late: apologies, after all, were not what their relationship was about. The award to his old friend of the Nobel Prize for Literature in 1954 had emphasised their huge imbalance, both financially and in worldly success, and a trip to Cuba now stood out as the fantasy it had been all along. Elise's stay failed to materialise despite their elated plans, but local happenings on a smaller scale proved equally compelling. In the end he never did reply to Hemingway, and for that omission he would castigate himself.

To Eoin MacNeill   Bellamont Forest
The Irish Military College   26 March 1955
The Curragh
Co Kildare
Dear Eoin MacNeill

I don't mind how many junior officers write to me in the tone of your very kind letter which I greatly appreciated. The important thing is to keep the pot boiling by attacking everywhere all the time, one never knows when the other fellow's nerve will crack, and our Orange friends are none too happy. You would laugh to see the indignant and insulting letters I get from that quarter. A very good sign!

I wish I could look in on your training. If ever I am down your way, I shall, if I may, pay a call on the Mess. Good luck to you all.

Sincerely, E. Dorman-O Gowan

To Hubert Butler   Bellamont Forest
Kilkenny   31 March 1955

My dear Hubert,

Partition – based on the ability of the Belfast Protestants to out-vote the Catholic areas of Tyrone, Fermanagh, Derry and the backing for strategical reasons of Whitehall – is the ultimate expression [of privilege]. To sustain that it's necessary to hold as hostages the Irish Nationalist areas in the Six Counties [so] liberalising elements can have no expression while this situation continues. Apparently no Minister [in the Commons] is ultimately responsible for law and order in Northern Ireland unless military force is used. By substituting B Specials for military force, the Minister of War is kept out of the picture. The Northern Ireland Unionist claims to uphold the British way of life, but nowhere could be less British. Can anyone imagine B Specials being permitted to function in Britain, let alone the regular police being allowed to carry arms openly?

We're not going to get out of the asylum in a hurry. So long as Britain is behind him, the Northern Unionist won't ever budge. Wooing is no use, return to the sacred cow of a Commonwealth would be no good. We have been there before. This can't go on indefinitely. The North is damn near bankrupt and the rest of us are castrated. I can't see anything left but that the IRA in the North should prepare to make one God awful stink until people take notice. That is what Egypt did, and it worked. But what the hell?

All the best, Eric

The IRA campaign would not be called off until 1962, but the repressive Special Powers Act was reintroduced in 1956, resulting in the arrest and re-arrest on a string of six-monthly sentences of IRA men whom Chink knew. By then contact with him had been formally broken off, and he had been aware for some time of disillusion. 'I am not in the secrets of the IRA, thank goodness', he corrected his MP brother Reg in 1958. 'Some of their transactions defeat me, as do their politics.'

His proposals for all-Ireland political frameworks that recognised two separate states, or for joint consultations between young delegates from all

Irish parties, north and south, got nowhere. He continued his opposition to partition and the Northern Ireland *status quo* through the letters pages of newspapers and university debating clubs, but by the campaign's end he could find humour in it. 'I now favour its retention as a tourist attraction,' he mocked, 'like the Beefeaters and the Horse Guards. One could sell it as a genuine Victorian Toryism. Nothing could be funnier than the Twelfth of July and the wee men in bowler hats. I think the whole thing ought now to be consciously preserved as an imperial monument, and paid a subsidy as such.'

Gradually, with lingering distrust on both sides, he and Eve were entering a marital truce. Elise Morrow's marriage did not survive and her love for him would be partly responsible, but there is no indication among his prolific correspondence that he ever knew. When, out of curiosity, her Harvard student son Lance visited him during a visit to Europe he was made welcome; Lance's lasting impression of his mother's erstwhile lover was of 'a knight'.[21]

Chink himself was looking inwards, and in 1958 there was a valid reason for his introspection when a freelance journalist, Larry Solon, suggested writing his biography. Chink's response at first was cautious.

To Larry Solon   Bellamont Forest
c/o Picture Post   3 April 1958

Larry,

Nobody can be entirely objective about himself. That's the cogent argument for oral confession in the Catholic Church. It's a great psychological advantage to me that I should have to address an apologia for my active life to you, because you necessarily look in on my world from outside. Part of my make-up was conditioned by it, and if I've always been in some sense a rebel, a rebel must also belong. I had a deep sense of injustice, and behind that a fear that some weakness in myself which I would not admit might well have brought it upon me. I do not like admitting failure, and hate to think that some fault in my makeup led me to let the Service, for which I lived, down.

For whom are we telling this story? Most retired generals write for each other and it is generally damn dull. Nobody has written anything on the level of my own story and made it come to life. There should be something in the saga from Los Angeles to Calcutta if we strike the right note. It is for you to decide the approach. I will never tell the story on my own.

I became very conscious when I went to the Staff College that I had moved out of the idealistic world of regimental life, where unselfish duty to a very limited *cadre* was the rule and competition was impossible and unproductive in any case, into a very competitive field. Not just competitive in terms of pure merit, but where anything went: influence, social position and downright climbing on one's neighbour's back by hook or by crook. I was ill-equipped for that atmosphere. I think I differed from my contemporaries who, in retrospect, seem definitely to have had more pronounced ideas as to their futures.

I was, all the time, interested in the interplay of politics and war. I had studied the Peninsula War, for Wellington with his cold detached efficiency has always intrigued me, and went on to the war between the States with its political overtones. World War I seemed too mechanical to be interesting, except in the technical aspect which did interest me a lot. I looked on politicians as a low form of life parallel to lawyers, stockbrokers, businessmen – how little I knew.

My treatment from August 1942 until August 1944, during which I was almost completely boycotted by men who formerly had been friends, suggests that once Churchill had turned his thumbs down I became an object of mass dislike. There is no mechanism for redress when such things happen; the victim drops out of sight. You will remember the case of the Archer Shea boy, immortalised in the play *The Winslow Boy*.

Larry, I have no idea of the framework you wish to build on, how much you want to write; no idea of anything, in fact. Gradually we will build up a mosaic from which you can devise a pattern and a plan. I think that I'm writing the truth, but one can never be sure about oneself – which is where I came in.

All the best.

Very sincerely, Eric

To Larry Solon   Bellamont Forest
c/o Picture Post   6 April 1958

My dear Larry,

You will realise that though all generals look alike my evolution differed from, say, Eisenhower's, Omar Bradley's[22], or Rommel's. It also differed from the Establishment-type contemporary in the British army, as well as from the young entry from the more obvious middle class – Auchinleck or Montgomery, for example. My moral outlook was distinctly my own [and] I evolved an ethic based largely on my earlier reading about chivalry. In effect I relapsed into a 17th century paganism, not altogether inappropriate to a military career. I also preserved a certain detachment of outlook, a determined if unconscious noninvolvement.

As I go on, I think you will see how these letters are important to my mosaic, if the story is to present a real person and not a uniformed dummy. But if you prefer the dummy, say so, and I will give you an historical *facade* instead.

Sincerely, Eric

The book would fail to materialise, but the same month he received an unexpected approach from a military historian who was clearly impatient with that very same historical *façade*. Eyebrows raised as he quickly scanned the letter, Chink saw that an interview was being sought for a prospective book on British Generals in the Western Desert.

This was more like it! He invited the author, Corelli Barnett, to stay straightaway. An old enemy was at last in sight and he could give it a name: Bad History.

Fighting Back:
1958–1969

To Basil Liddell Hart   Bellamont Forest
                                                              [Undated]

My dear Basil,

Last week I had an amusing young visitor, one Correlli Barnett[1] who is doing a book on the British generals in the Western Desert. Someone told him that I might know things about that time, the first occasion on which anyone but yourself has ever asked me about the last war. He's been well received by Auk, and is in touch with Gatehouse and Galloway.[2] Barnett rates me as 'angry', read history at Oxford, is very intelligent with something of the 18th century in his thinking. Aged 30, he has a pretty low opinion of his elders, though he is fair-minded enough. Independently he has formed the opinion that the events of July 1942 in the Desert represent a turning point in the war, and the turnaround at El Alamein is a very dramatic story.

I liked the lad, wishing I was his age again. He left me to see Auk once more and then Dick O'Connor. I think it is all to the good that the next generation should interest itself in writing about these things, but it made me feel very old. Lord knows what he will make of it all, he writes nicely and his questionnaires are penetrating enough. He hopes to interview Ritchie also.

I think you would like Correlli Barnett. He has the youth we've had to surrender in part, and he has a sort of feeling for truth. We will not be here all that much longer, and I would like to see the rather smoky torch we carried handed on to someone young and modern – be kind to him if you meet. He probably laughs at us all up his fashionable sleeve, but at least you have a pleasant taste in waistcoats in common. I did like the boy. What would you give to be thirty years old once again?

All good wishes.

Very sincerely,

Eric

To Correlli Barnett   Bellamont Forest
[Known to his friends as Bill]   27 May 1958

Dear Bill,

I agree with you that the First Battle of Alamein, the battle for Egypt, was indeed the turning point of the war for Britain. Had we lost Egypt I cannot see how a successful recovery could be expected, certainly there could not have been a landing in NW Africa.

The Appreciation envisaged the Alam Halfa battle, and set out the general principles of the operations for August/November 1942 period afterwards, acted upon by Alexander. Plans for Alam Halfa were set in motion in discussion with Gott before the end of July. Let that fact whet your appetite.

Yours, Eric

To Correlli [Bill] Barnett   Bellamont Forest
                                                                 14 June 1958

Dear Bill,

The subject is so distasteful to me that I rarely go back to it dispassionately. But 14 years have passed and one should be able to consider it without too much internal bleeding.

Yours, Eric

To Correlli [Bill] Barnett   Bellamont Forest
                                                                 18 June 1958

My dear Bill,

Had I known I was entertaining an angel/devil unawares I would certainly have watched my step most carefully. There you sat with butter not melting in your damn mouth, coldly observing a superannuated brontosaurus making military noises (obsolete) as it rotted away in the ancestral swamp. I am sure you are quite pitiless. My only consolation is that I was once, in my military way, also an 'Angry Young Man'. The hell of it is that the British Army either disposes of Personality minus Brains or Brains minus Personality. I feel that I am one of the few with the latter qualifications, but personality largely depends on having the adequate badges of rank and sufficient social pull. The army also has its tycoons.

Ever, Eric

To Correlli [Bill] Barnett   Bellamont Forest
                                                                  2 July 1958

My dear Bill

This day sixteen years ago, listening to the very agreeable noise of a large amount of our own guns in full blast as I watched the situation map, and knowing that the Australian Division was coming up from Alex, I began to feel that we had Rommel where we wanted him.

Ever, Eric

To Correlli [Bill] Barnett   Bellamont Forest
                                                                3 July 1958

My dear Bill,

The first battle of Alamein – July 1–26 1942 – was a major strategical tactical victory for the Commonwealth. At its conclusion the Germans were so sewn up that they could neither hope for success in a renewed offensive nor take the necessary decision to pull out. They had lost the initiative and their fate was sealed. This is really Auk's immense contribution to World War Two, for which he has never had credit.

Ever, Eric

To Correlli [Bill] Barnett   Bellamont Forest
                                                                14 July 1958

My dear Bill,

You must realise my sense of injustice done to Auk. Churchill's grudging, cattish description rankles – 'breakfast in a wire-netted cage' etc. What did he expect? That was where we had eaten, in the open, for almost a month without worrying too much about our comforts. What was good enough for a C in C ME should have been good enough for a visiting PM. But a slap up lunch in relative luxury (food ordered from Shepheard's Hotel) listening to RAF denigration of Auk suited him well.

Ever, Eric

To Correlli [Bill] Barnett   Bellamont Forest
                                                                19 July 1958

My dear Bill,

I do not want you to think that I am subjectively pro-Auk. Auk never brought me any luck; indeed it seems to me as if he did his best not to employ me in any active capacity during his period of command. Moreover when in 1953 I tackled Winston regarding the Cairo Purge, Auk would have been an important witness, but he certainly would have run out on me, and did decline to become involved. Since mid-August 1942 I have only seen him once, when he lunched with me in London and I talked with him for a while afterwards in his flat. A bleak encounter.

Anyhow, if I speak well about Auk it is not out of gratitude, for I owe him none. When he became, once more, C in C India, he made no apparent effort to rescue me from my purgatory of an infantry brigade in Britain. I can be objective about him. I have never been given to hero worship and feel no such thing for Auk or Wavell. Of the two, Wavell was incomparably the more intelligent and able, with a far superior brain when he chose to use it. Auk remains a great soldier. I think he felt me to be cool, remote, revolutionary in my thinking, and dangerous to orthodoxy, [whereas] Wavell and I got on well. It was quite a different relationship.

Ever, Eric

To Correlli [Bill] Barnett   Bellamont Forest
                                                  5–20 August 1958

My dear Bill,

Here is another of my 'Dear Bill' letters. At this rate you will be able to publish a book called 'Dear Bill. Letters from an Angry Old General to an Angry Young Writer.'

(*20th*) [Auk] was the spider at the centre of a web so big that he could not properly test the strands. It is impossible to over-emphasise the burden he carried throughout July while his enemies sniped behind his back in Cairo and London. I know of no other man who could have carried it and remained cool, calm and cheerful and completely determined to succeed. No other commander did carry a comparable burden, so far as I know. Having me there helped a little, but his was the final responsibility. Remember always that nobody anywhere could help Auk; only he could have lost the war in an afternoon. Instead, his determination made all the subsequent successes of the war possible and sewed up Rommel into the bargain.

Ever, Eric

To Correlli [Bill] Barnett   Bellamont Forest
                                                  6 October 1958

My dear Bill,

I'm reading Monty's memoirs[3] with fascinated horror. Such humourless egoism suggest the reptile rather than anything human – this strange, unlikeable man.

If I am 'unsound' and yet produce a tactical solution before, but almost identical, to Monty, which of us is being unsound? His luck is that while I took up my post after a total defeat and in a period of great demoralisation, he and his team, with good British Divisions and oodles of tanks, artillery and well-rested troops coming to bite the Axis tail, took on his after we had restored the situation.

Ever, Eric

To Field Marshal Sir Claude Auchinleck   Bellamont Forest
                                                          16 October 1958

My dear Claude,

Even at our lunch at the Mohammid [*sic*] Ali Club on June 25th 1942 I felt that if only we could draw back from Matruh without too much damage to what was left of Eighth Army, we would be able to stop Rommel decisively at Alamein. I agree that the possibility of failure had to be provided for on the lines we know of. But I did not think that we would fail. And I well remember your natural exasperation at the tendency which you sensed at Eighth Army HQs at the end of June to make a virtue of retreat. You said to me, 'These damn English have been taught too long to be good losers. I have never been a good loser. I am going to win now.' Not quite as succinct as Nelson's 'England Expects', but good enough.

The Appreciation was my last contribution to the Army as a thinking soldier. But I have kept my panache. I have not yielded to fate. I have not kissed the rod, not even the Black Rod.[4] I hope I still have a sense of humour. Whatever attributes I still possess are as much at your service as they were on June 25 1942. To fight Monty will also be to fight for the honour of this whole officer corps of the British Army.

What Freddie is saying[5] is that he sat up all night over my Appreciation. Even though he denigrates it, it's a dollar to a divot that he passed the contents on to Monty when Monty reached HQ Eighth Army. My assumptions can't be too far out [because] they accord with normal Service practices.

Ever, Chink

To Basil Liddell Hart   Bellamont Forest
                                                          16 October 1958

My dear Basil,

I have spoken to [Auk] on the telephone to say that he could count on my full support.[6] He suggested that he might well have told Monty of the earlier phases of the struggle before sending him up to Eighth Army for a look around. The idea was, so it seems to me, that Monty should then return to Cairo until 15 August when he would formally take over Eighth Army at the same time as Alex took over the reduced Middle East. In that case, presumably, Auk would have briefed Alex on the lines of the Appreciation, and Alex would have briefed Monty similarly. This Monty did not want to happen, so he pirated Eighth Army.

Fortunately Alex's own Despatch gives the lie to Monty. Anyhow, it's hard to see how Auk can refuse to proceed against Monty for libel. Not to do so would destroy his reputation.

Everything will depend on the interpretation of my Appreciation and if that is accepted then Monty is sunk. He well deserves to sink, because nothing less chivalrous than his scathing strictures on Auk can be imagined. He clearly doesn't want anyone else to claim the glamour of Alam Halfa. Interesting, but also very, very nasty.

All good wishes, Eric

To Correlli [Bill] Barnett   Bellamont Forest
                                                                                                 18 October 1958

My dear Bill,

Auk is hopping mad at Monty's book: 'A tissue of malicious libels' ... 'Astonished and disgusted.' In the eighteenth century there would have been a duel. In the event of a libel act, which I personally hope happens, Auk should be sitting pretty so long as he sticks to my Appreciation.

Ever, Eric

To Field Marshal Sir Claude Auchinleck   Bellamont Forest
                                                                                                 20 October 1958

My dear Claude,

When Wavell asked, 'You are very strongly posted here, have you considered making a fake withdrawal to lure Rommel into the trap?' I replied that I had thought of it, but that anything tricky of that sort might be too much for Eighth Army, and anyhow there was only one direction a new Rommel offensive could take and we were catering for that. Archie agreed.

Ever, Chink

The Editor   Bellamont Forest
The Daily Telegraph   5 November 1958
Sir,

My attention has been drawn [to] certain strictures made by General Montgomery in his *Memoirs*. Allegedly it was General Auchinleck's intention that, in the event of a fresh attack by Rommel delivered before Eighth Army was fit to take the offensive, [he] would 'fall back on the Delta; if Cairo and the Delta could not be held, the army would retreat southwards up the Nile'.

The *Memoirs* do not make clear the fact that Auchinleck's operational responsibilities included not only Egypt, but the Middle East up to the Caucasus. The loss of Cairo could have been borne, the loss of the Persian Gulf oil would have been fatal [had] the Germans in Southern Russia done the right thing and pressed forward to the Caucasus instead of dividing their thrust. Throughout his period in command, Auchinleck had always to be ready to face both west to meet Rommel and north to meet invasion from the Caucasus. His forces were never sufficient to meet more than one. He could not assume that Hitler's strategy would be quite so disastrously foolish.

The whole controversy seems to [stem] from General Montgomery's pardonable ignorance of the complicated tactical and strategical situation. The unfortunate discussion has arisen from that somewhat acrid, and perhaps inaccurately remembered, passage on page 94 of Lord Montgomery's interesting and revealing, *Memoirs*.

I am, Sir, your obedient servant,

E. Dorman O Gowan

To Field Marshal Sir Claude Auchinleck   Bellamont Forest
                                                             16–23 November 1958

My dear Claude,

Monty only had to glance at the map on the wall of your caravan to get all the clues. His memoirs make no reference to that map; his readers are invited to believe that he found Eighth Army in paralysed chaos. By rights the BBC should give you time to state your own case – don't suppose the BBC under Jacob[7] would permit <u>me</u> to speak.

(23rd) I don't so much mind about Monty, but I do want to see proper acknowledgement for the decisions and tactical execution which wrecked Rommel during the July fighting at El Alamein.

Ever, Chink

To Correlli [Bill] Barnett   Bellamont Forest
                                                            8 December 1958

My dear Bill,

I don't know about writing my own story. I don't want to write as most generals write. My life does not hinge on the last war entirely. It was an episode, as was the first world war, but not a completion. All in between was equally, if not more, valid. Possibly one could take the attitude of one who in WWI was Western Front cannon fodder [and] thereafter set out consciously to do away with that idea of warfare, and who failed.

Ever, Eric

To Field Marshal Sir Claude Auchinleck   Bellamont Forest
                                                           2 January 1959

My dear Claude,

I envy you the sunlight and friendships of [your trip to] India. It would be great happiness to walk with you once again in the early morning of a hot day. If you come to Bellamont this Spring we can do just that, cold, fine or wet.

Others have also said I should write. My difficulty has been to find a publisher because I'm generally recognised as having 'a chip on the shoulder' at coming unstuck at the PM's hands. General opinion, both in and out of the Service, is that he was justified.

It is highly amusing to consider what might have become of us both if we had been dishonest enough to our profession and the Monarch to have accepted the command under Churchill's very odd conditions, so militarily unsound.

Until this Monty fracas I could not even claim authorship of the Appreciation unless you admitted it, because either you or somebody in the War Office had charmingly removed my signature. You alone are in the position to get the reversing done. If you really feel as you say you do, you might be able to help me a lot.

Ever, Chink

To Correlli [Bill] Barnett   Bellamont Forest
                                                                                             22 April 1959

My dear Bill,

While I was a soldier I never thought of, or about, myself. If I had good ideas (and I was a diligent student because I did not have a lot of money so made soldiering a hobby as well as a duty) I put those ideas into the kitty for anyone to use.

Professionally I was quite single-minded. I could not bear the thought of anything in which I served not being top-notch, or that it should fail to be excellent through any fault of mine. So I struggled to shape first my regiment, then the Army at home and in India, to the highest standard of fitness for use that I could envisage. A very hard worker, I would never have compromised with inefficiency high or low. So I did, I fear, make some enemies. Still, I never looked on myself as 'clever', and I was surprised to see some very private notes which Archie John Wavell has sent me written by his father, that I was 'one of the army's most brilliant brains', or word to that effect.

Ever, Eric

The pattern of self-analysis and harking back over the past for Correlli Barnett's book, to be entitled *The Desert Generals* and now near completion, was emotionally draining. It was also disconcerting to come across curt mentions of 'Dorman-Smith' in many postwar memoirs, but that gave him an idea. Referring to himself distantly in the third person might help whenever he felt in need of self-protection.

To Correlli [Bill] Barnett   Bellamont Forest
                                                28 May 1959

My dear Bill,

Until you appeared and began to ask questions, I had pretty well lost touch with Dorman-Smith who had become well submerged under O'Gowan [and] was quite different in many respects. You have largely built up a picture of D-S from external sources, none of whom knew very much about him. I was interested that you should deem him 'an intellectual', for I have always distrusted intellectualism. I would rather have thought him merely intelligent, a reasonably good craftsman.

Dorman-Smith was primarily an infantry officer and believed in the importance of that arm as the cement of battle, around which the auxiliary armament – artillery, armour – orbited. For that reason he wished to mechanize the infantry so that in open country they could move and manoeuvre with similar mobility [because] the proper cooperation of all arms won battles. Experience of the 1914–18 war had set his face against bloodbaths. Instead he looked to manoeuvre, stratagem, flexibility and the rear attack, applied intelligently, to offset the 'frontalism' [of] military orthodoxy. Any resistance [came] from those conservative types who believed in the mystique of the horse in war. Dorman-Smith believed in an overall effective army, not in special forces and commando subtractions to an elite.

He emerged from the Cairo Purge as the only 'victim' over whom Churchill exercised complete constitutional control. To be justifiable, victims had to be made to appear inefficient people worthy of punishment, described by Lord Boothby[8] in a recent TV interview as normal Churchill technique.

D-S spurned doctrine or precedent. [If] his tendency was towards the caustic, he was never sarcastic to juniors. He was always ready to exercise independent judgement and responsibility, and was impatient of 'staff ritual'. The important thing was that a good idea should not be lost. There was so little time for what had to be done, for from 1934 onwards war seemed to him inevitable ….

I write this just to complete your own concept of a profile, [because] few of the men who have told you about me knew me at all well. If anyone challenges you, you have this additional data to support your personal assessment.

Ever, Eric

To Correlli [Bill] Barnett   Bellamont Forest
                                                [undated] 1959

My dear Bill

If there was to be an Auchinleck dinner, convened by his friends and Generals to recognize his rehabilitation and staged with the proper ceremony

– white tie, full decorations and loyal toasts – I could not attend. I am still in precisely the same disgrace as I was officially on return to Britain in late August 1942. If I did attend, I would be placed among the junior and undecorated brigadiers and colonels. This scapegoat-status would be unacceptable. I am still in the position that I cannot attend a regimental dinner, even my own, or enter a Service club (except by the 'service' door to wash dishes). Moreover [it] cost me not only my career's fruits, but also a slice off my pension. Being prematurely retired, I did not make the full pension-period. But consider what has happened to [my contemporaries]:

Auchinleck: F-M
Ritchie: General, OBE
De Guingand: Major-General Sir
Norrie: a Baron
Arthur Smith: General, Sir
Galloway: Lt. General Sir,
Harding: FM, Lord
Whiteley: General, Sir
Even Tom Corbett remains Lt. General Corbett, CB, MC.

Dorman-Smith is not only the Prime Minister's scapegoat, he is everyone else's scapegoat – including the Auk. It was, and is, my dear Cassius, a very dirty world.

Ever, Brutus

But the years in the Western Desert continued to exert an irresistible pull. 'Your contribution is uniquely valuable', pleaded the military writer R.W. 'Tommy' Thompson by post, who was embarking on a study of Churchill at war to be called *The Yankee Marlborough*. Thompson's letter reached Chink on a cold, frosty day the following spring, and at once he rose to the bait.

To R.W. 'Tommy' Thompson   Bellamont Forest
                                                              16 February 1960
Dear Thompson,

I am delighted to have your letter this morning. All that I have and know, which is a great deal, is at your disposal. The official historians have now done their best and worst, so we haven't to fear any new revelations from that quarter – what is more required is a complete reinterpretation of intentions and events which will break down the official mythology. Young Corelli Barnett's *Desert Generals* is to be published in June, but will [not] complete the study since it purposely omits the upper strategical and political stratas [*sic*]. I feel that this top-level crest line, as yet unexplored, will interest you most.

I will be in London for a week from February 29 and Bill Barnett has asked me to be his guest at the Military Commentators Circle. My re-emergence from ghostly scapegoatery should be amusing.

It is a serious undertaking to challenge so well-established a myth. Still, the dead will not sleep easy until that is done. They did not die for myths.

With all good wishes,

Very sincerely, Eric O'Gowan

To R.W. 'Tommy' Thompson   Bellamont Forest
[undated] 1960

Dear Thompson,

The Circle episode was too hectic. I'd hoped to be let sit back with my host, young Barnett, and look on, [but] instead got swept into a vortex of people whom I'd known in the past, [since] separated by time and space. It was annoying to find myself in the centre rather than on the circumference. Still, it was probably proper that I should meet Auchinleck again after so many years in just that sort of unembarrassing setting, if only because Auk has rather a guilt complex about [my] fate. He need not, if only because at least I've been able to extract myself from the rat race in which, with the 'big lie' around my neck, I'd have had small chance of survival.

I got the impression while in England that the British do not seem to know where they are in the world.

Yours very sincerely,

Eric O'Gowan

To R.W. 'Tommy' Thompson   Bellamont Forest
24 March 1960

Dear Thompson,

Winston says himself, 'Before Alamein we never had a victory, after Alamein we never had a defeat.' In his mind this is undoubtedly how [he] saw it. But no blame attaches to his direction! We had not 'sustained an almost unbroken series of military defeats'. Whenever the soldiers had been given a fair chance, they had done surprisingly well.

The Cyrenaican campaign was a striking victory which Winston jettisoned. Auchinleck's 'second Cyrenaican campaign' was no defeat, whatever happened afterwards, and much of that was due to bad direction from Whitehall. July '42 at Alamein was not a defeat but a major strategical victory. Churchill ignores these because they were not his doing. But Alamein II, that superfluous battle, was his contriving; so that is the 'first victory'. It came in time to save the Prime Minister, but not to save Britain as a great power.

But the myth had been established. It is still with us. We do owe it to the men of my own generation who were boys in 1914/18 and commanders in 1939/45 (those who experienced two wars directly in the field) that the truth should not be buried under the Cenotaph in the interests of the 'System' and the 'Establishment'. There is much to say.

Very sincerely, Eric O'Gowan

To R.W. 'Tommy' Thompson   Bellamont Forest
                                             8 June 1960

My dear Tommy,

I am glad that you're in personal touch with Claude Auchinleck. He has refreshingly noble qualities in this mean and naughty world; one of which is an almost complete lack of suspicion and an almost pathetic trust in his fellow men. I've really no idea about how he regards myself – he never knew me very well personally. He probably thinks I'm a bit of a bore for harping on a subject which he had done his best to forget. I intend to continue the harp music, if only in the interest of common justice to Auk himself – anyhow, somebody has got to do it. John Connell[9] took the matter part of the way, but he did not fit it into the bigger picture.

Yours sincerely, Eric

To Correlli 'Bill' Barnett   Bellamont Forest
                                             23 June 1960

My dear Bill,

Major General R.A. Allen (now 74 and pretty rickety) was at Chatham Garrison when I was Instructor of Tactics to the School of Military Engineering. He knew Brooke, Wavell, Auk and Monty pretty well. [Recently] he came over to Bellamont to find out what had happened to me, [and] now he's temporarily convinced that 1st Alamein was 'a strategic crowning mercy' and 'my battle'. This he volunteered at exit door of airport.

He has a residue of nice bawdy wit, a lot of worldly experience and writes a good letter. He would insist on taking early tea with me in the hall at 6.30am when I like to be alone! I forgave him. Ever, Eric

To Correlli 'Bill' Barnett   Bellamont Forest
                                             8 July 1960

My dear Bill,

General Allen suggests *The Desert Generals* [should be] serialised in a serious newspaper. Must have made a deep impression. The point is that the chaps now running the three Services were all pretty small fry when the last war ended. They have all been fed on the Big Lie and the odds are that this has distorted all their thinking. By wiping the slate of [it] on various levels, we may make it possible to look at the current problems with new objectivity. We aren't just rattling the bones of the dead for the sake of necrophilia. We're signposting for the living.

Ever, Eric

To R.W. 'Tommy' Thompson   Bellamont Forest
22 July 1960

My dear Tommy,

I'm glad you're looking into the psychological factor. During the first week of August in Cairo in 1942 Smut's representative in Cairo, General Theron,[10] told me that he had talked to Winston in the Embassy garden that morning. I asked Theron what his impression was. Theron replied: 'Had he not been the Prime Minister of Great Britain I'd have said that he was drunk.' The story told by Jacob[11] of finding Churchill stamping around his room saying 'Rommel, Rommel, Rommel, what else matters but besting him?' indicates serious imbalance. I have myself wondered whether Churchill was ever strictly sober by the late evenings. Did that produce his perpetual irritability?

Ever, Eric

To R.W. 'Tommy' Thompson   Bellamont Forest
30 July 1960

My dear Tommy,

Horrock's book *A Full Life*[12] is due out in September. I used to know 'Jorrocks' well as a younger officer. Age! Well, there is the calendar. But this morning I did 4½ miles across rough, hilly tracks inside 45 minutes, which is not bad for 65.

All the best. Ever, Eric

To Correlli 'Bill' Barnett   Bellamont Forest
29 August 1960

My dear Bill,

A signed copy of *Desert Generals* arrived in the post, opened and re-sealed by Irish Customs. An old Irish custom in respect of books sent to me through the post as a result of their once discovering that a friend had sent me from London a copy of the Kinsey Report[13] (a masterpiece). I think the whole customs branch read my Kinsey before returning it to Britain. When I eventually recovered it, it was thoroughly fingered (as must have been their wives!). An amusing and enlightening experience.

Ever, Eric

To R. W. 'Tommy' Thompson   Bellamont Forest
                                               1 October 1960

My dear Tommy

I dined with Auk [in London] last Wednesday. He was in great form, and is still basically numb. I sometimes wonder what he really thought at Alamein 1. He told me that at the Ismay *Memoirs*[14] launching dinner, very top-level but no Montgomery, Churchill made a point of coming across the room to Auk and saying nice things. Auk said that he was 'deeply touched'. How very English to be sure – the two have not spoken since Cairo in 1942. But it may be a preliminary to a subtle de-scapegoating process which seems to be gathering momentum.

I don't know how I'd have received a Churchill approach, since he has nowhere, except under compulsion from myself, repudiated his dispatches to the Cabinet of August 6th and 20th 1942 when it was quite possible to disprove them. Until this is done I'll remain a hostile scapegoat on principle. I don't see how senior commanders and staffs can ever function in warfare under that nasty precedent. If there is any de-scapegoating it will have to be on a PM's level and by the Crown. So there.

All good wishes. Ever, Eric

The hint in that last sentence was more than wishful thinking. Spurred on by public interest in Barnett's *The Desert Generals* when it was published that autumn, and increasingly indignant on Chink's behalf, his brother Reggie was raising the matter of a pardon with the Secretary of State for War in Harold Macmillan's government, John Profumo.[15] Among those about to give references was John Connell, who had been in touch with Chink while researching his biography of Auchinleck the previous year. A biography of Wavell was Connell's next project, so 'Jack', as he was known, knew the background better than most.

To John 'Jack' Connell   Bellamont Forest
                                               1 October 1960

My dear Jack,

Brothers rush in where angels might reasonably fear to tread. I've just had a letter from Reg enclosing [one] from John Profumo, and Reg says that he's in touch with you. I think you realise that I will not be party to any action to 'vindicate' my military reputation unilaterally.

My concern has always been with Claude [and] the men of all ranks who fought under him at the decisive battle of First Alamein. What happened to me afterwards is not important, except as an example of how the authorities can break and re-break a man until there's precious little left of him. There was a day in August 1944 when I seriously contemplated suicide[16] – but not for long.

Moreover, I don't consider that the cancellation of the Cairo Purge is a matter for the Secretary of State for War, anyway. What was done by a Prime Minister can only be undone on that level and by the Crown. It's necessary that justice be done, and be seen to have been done. This is way over Profumo's head, as I see it.

Beyond challenging Winston over his statements in *The Hinge of Fate*, I've never defended myself personally, although I have defended Claude. What has been important in all this business is that three writers of history, Basil, yourself and Bill Barnett, all working independently with different material, have formed the same conclusions, [and that] the official history supports [them]. No statutory Court of Inquiry would be half so objective or decisive.

The person who really matters is Claude. For that reason I was glad to hear that Winston had gone out of his way to be polite to him at the Ismay *Memoirs* launch party, and that Claude had responded. Such transactions probably come into the theological category of 'special graces'.

Ever, Chink

To John 'Jack' Connell   Bellamont Forest
                                                            October 1960

My dear Jack,

The best way out is to recognize Auk's immense contribution to the winning of the war by some signal honour. For myself, I have only this to add. My father died still thinking that I was a disgraced failure. How could he think otherwise? My mother is alive, only just, and she thinks in her heart of hearts as my father did. If anything is to be done about me, it would be merciful to do it quickly. I would like her to see the truth before she goes. Still, I would accept no recognition until I was fully satisfied that Claude's services had been publicly recognised.

What I would really like, and Claude will understand, would be to be able to take the General Salute of my old battalion of the Fifth Fusiliers on the St George's Day Parade, [and] that would mean that I should be restored to the rank I held at First Alamein. After all, I was Claude's acting CGS. He told Brooke that himself.

I suppose Dreyfus wrote this sort of thing to Emile Zola.[17] And probably bored Zola stiff.

Ever, Chink

To John 'Jack' Connell   Bellamont Forest
                                                            4 October 1960

My dear Jack,

Your charming and undeserved telegram [of encouragement] was read to me from the Cootehill Post Office last night – my face was very red, too! Firstly, <u>Auk is not wearing a white sheet</u>. We've heard much about his 'mistakes' but little about mistakes of those above him. Primarily Churchill in May/June 1942 placed him in an impossible strategical position.

I just do not want to fall into the War Office's hands once again. I think you'll agree that I've had enough. In any case, the issue is, as I've already said, much bigger. Fortunately the System has its precedent in Gough,[18] if it cares to use it.

All the very best. Ever, Chink

PS I think Profumo would agree with my attitude.

To John 'Jack' Connell   Bellamont Forest
                                                            7 October 1960

Jack,

Reg phoned last night to say that he was shortly to have lunch with Profumo. So far as he is concerned he is rooting for Chink alone. He thinks, I feel, that I'm stupidly chivalrous about Auk. I disagree, for chivalry is all-important. When the time comes, if ever it does, I can maintain my principles about loyalty to my Chief and my profession.

Have you seen Crossman's[19] review of Ismay, Horrocks, Barnett in the *New Statesman*? His theme is 'Historians Take Over' and I entirely agree. 'What *The Desert Generals* proves', he concludes, 'is that the generals of World War II should stop recollecting their emotions in complacency and leave the job of writing its history to the historians.'

What I would welcome now, as the proper role of impartial historians, is a joint demand, signed by all of you [Connell, Liddell Hart, Desmond Young,[20] Barnett].

1) That First Alamein be recognised as the important and decisive victory it was.

2) That in the full light of history the Cairo Purge should be liquidated.

3) That the 'retreat to the Delta' *canard*, now recorded in a Cabinet document, should be expunged.

A joint memorandum addressed to the press and circulated in the Commons would have considerable impact.

Good hunting, Chink

To Sir Reginald Dorman-Smith   Bellamont Forest
                                                         8 October 1960

Brother,

You told me yesterday that you and Jack were all set to beard Profumo in his War Office den. I would prefer neutral ground. In the WO, he will be buttressed by those committed to the preservation of the myth, which is the Churchill/Montgomery line, (i) that Alamein 1 does not officially exist, since there is only one official Battle of Alamein (Oct 23/Nov 4'). (ii) The business of the '8' on the Africa Star[21] has been been finally settled. (iii) The 'Cairo Purge' cannot be reopened.

The last is, I understand, a principle, because a great many people consider themselves to have been unjustly treated, and to re-consider one would be to open a floodgate. It will be urged that by the end of the war the best officers had come to the fore, backed by Montgomery, Alexander, Wilson, Alanbrooke. Those outside had necessarily failed.

And please, dear Reg, remember this. I'm not asking anybody for anything, not even for justice; I'm used to the wilderness and rather like it. But I still do care for the Service to which I gave the best part of my life. It is the practice of making scapegoats of soldiers to save the bacon of politicians, however eminent, which is finally at issue. It is the system which is on trial. Tell the buggers that and you'll have my gratitude. But do not let them stall on all this; it's their funeral.

Ever, Eric

To John 'Jack' Connell   Bellamont Forest
                                                         8 October 1960

My dear Jack,

Reg is still worrying about my post-purge career. Whatever the gossip may be, he seems to think that they have something against me in Italy. Well, they haven't. The boot is on the other foot. He's inclined to conform to higher Authority – that's his nice nature and why he was second top boy at Harrow and an Under Officer at Sandhurst. But he can fight if he understands the issue and feels it just. He needs backing up. What would matter a lot to me would be for Claude to insist, on his own volition, that his former staff officer should get justice. Nothing else would matter compared to that. So poo to Profumo! Indeed, if Auk and the historians can't pull this off, I'm content to go on as I am.

What we have to protect is the integrity of the Service. What we have to defeat for all time is the bad precedent of making scapegoats of soldiers by politicos. Who are these chaps anyhow?

Ever, Chink

To Sir Reginald Dorman-Smith   Bellamont Forest
                                                                  28 October 1960

Brother,

What has to be done is to recognize the importance and value of First Alamein; to recognize the Old Desert Army and the Old (pre-Montgomery) Eighth Army; to recognize Auchinleck's predominant part in the wrecking of Rommel and in turning the tide of war; to cancel out the Cairo Purge and the unjust stigma under which the scapegoats have remained ever since. If we all get that far we will have at least achieved historical justice. The personal injustices done to me in England and Italy during 1943/4 are irrelevant. Do not let us be deflected into trivialities.

Anyhow, Brother, good hunting on Monday.

Ever, Eric

PS One point which may be useful. It is not generally known that during the July fighting Eighth Army under Auk took over 10,000 Axis prisoners. Eighth Army casualties were just over 14,000. Now these Axis prisoners were taken not in a round-up after a breakthrough, but in straight fighting. The 1914/18 ratio was prisoners 1, wounded 2, killed 1. On that basis, Rommel's July casualties may have approximated 30,000, they can hardly have been less than 20,000. So I think Rommel is hiding up when he states that Eighth Army losses were larger than those in PAA.[22] He nowhere states what reinforcements [he] received during July. They must have been considerable.

To John 'Jack' Connell   Bellamont Forest
                                                                  28 October 1960

My dear Jack,

Given that the cart went before the horse because of Reg's perfectly proper intervention, it seems to me that the Profumo meeting went well. His penultimate words to Reg, which you may not have heard, were 'I appreciate your attitude. If my elder brother had been treated as on the face of it yours has been and I had the power to do so, I would threaten to pull down Buckingham Palace.'

I agree that patronage triumvirate – Brooke, Monty, Alex – did reject a number of able men, men far more able than myself. But we were summarily rejected by a Prime Minister, our cases unheard and without any opportunity to defend our actions, and on hearsay evidence. When the next interview comes, your experience of this last will make it all the easier.

Don't over-do. Ever, Chink

The reply addressed to Reggie two months later was not signed by John Profumo, but by the Parliamentary Under-Secretary for War. 'I do not deny that [he] may well have been unlucky, but it is another thing to accept that the changes of staff ... were necessarily wrong.... No doubt historians will continue to express different views....' Chink had expected that answer, and was not too discouraged. He took it on the chin.

To R. W. 'Tommy' Thompson   Bellamont Forest
                                                                11 January 1961

My dear Tommy,

My gossip is that Bill Barnett has just sent me a French translation of the *Desert Generals* [and] the Germans, too, have bought the book. He's sold 10,000 English copies already. Not bad for a young 'un. See how the whole counter-offensive goes forward among you writers, naturally and inexorably.

Ever, Eric

To John 'Jack' Connell   Bellamont Forest
                                                                20 June 1961

My dear Jack,

It is hard to believe that Archie Wavell was only 67 when he died, and I will be sixty-six next month. He was only twelve years or so older than myself. In the army twelve years means a great gap in seniority and experience; in civil life there isn't that extreme difference. When I first knew him, in 1931, I was 36 and he 48. 48 doesn't seem all that old now, but to me in 1931 he was highly mature. Had he lived he would today be 78, which is not old these days, and he'd certainly have written the story of his campaigns; perhaps the story of his life, perhaps a historical study of his time. I can't think that he'd have let Winston, for example, get away with his own 'Jack Horner' version of the Second World War.

But it has fallen to you to tell his story. I hope he'll stand at your elbow as you write it – I think he will. Only, if so, the words will come very slowly!

Ever, Chink

Mortality was impossible to ignore. On 2 July 1961 Hemingway's death was reported on the BBC news, and Chink wept openly in front of his teenage son, Christopher. 'Sad about Hem's death', he admitted to Thompson. He never had any doubt that it had been suicide. 'I feel guilty,' he would blame himself, 'because on re-reading his letters it seems just possible that I could have helped Hem over that difficult stile into the sixties of a man's life.' He refused all media requests for an interview, which redoubled when *A Moveable*

Feast was published posthumously with its many elegiac references to him, such as 'I hope Chink will come. He takes care of us.'

To Professor Carlos Baker[23]   Bellamont Forest
Princeton, New Jersey   24 February 1962
My dear Professor,

Your most interesting letter and the Hemingway correspondence have just arrived. I'm glad they proved useful. Hem's total post-WWI letters are impressive, all of it stuff which the recipients were glad to keep. They should shed a bright light on his mature thinking, while correcting some impressions that he only thought with his viscera!

I'm not clear just what the names are of Hem's other two 'heros', Gustavo?[24] of the Spanish War and General?[25] of WWII. I'm equally unclear as to how I could possibly have got into that illustrious bracket. As to my own broken-backed saga which has never been formulated for the good reason that nobody would be the slightest bit interested in it, bits can be reconstructed from books which I'm sure are to be found in Princeton's immense library – flattering and the reverse. I'll probably be found under Dorman-Smith. I did not resume my Irish family name until after my father's death. After your appetite has been either whetted or dulled by these, should you want anything more I'll be happy to provide it. What you'll find is a fair profile of that part of the iceberg which shows above the surface.

When next you write to Hadley, do please give her my love.

With all good wishes, Eric O'Gowan

To John 'Jack' Connell   Bellamont Forest
                                                                5 June 1962
My dear Jack,

I've been reading Barbara Tuchman's *August 1914* – clever, clever. I was in that episode at Mariette, where my platoon was more or less astride the bridge. The Huns had got behind us and we only just got through them. For a while the half-battalion formed up around a slagheap and watched German columns moving south half a mile away in the same direction. We marched through them in the dusk. It was very much like the parallel retreating and advancing by Eighth Army and Afrika Korps on June 30 1942. I think I'm about the only Fifth Fusilier officer of that battalion living now – most were killed off on the Aisne or at Ypres.

I'm well. I rise at five, breakfast lightly and then run across country from 6 to 7am, meeting deer, badgers, hares and various birds as I go. I admit to a siesta – I'm apt to doze over television. I read *The Irish Times*, *Guardian*,

Telegraph daily; the *Spectator, Sunday Times, Sunday Telegraph* and *Observer* weekly, and find myself getting daily squarer. No theatre, cinema or race-going. Antiquarianism and history, plus warfare and philosophy, form my reading. But I still dislike the thought of being 67 in July. One seems to have wasted so much life by selecting the wrong profession.

So there, Jack. Now I've surfaced again – us atomic submarines, Ma'am.

Chink

To Basil Liddell Hart   Bellamont Forest
                                                          25 June 1962

My dear Basil

This year and these coming weeks see the 20th anniversary of the desert summer of 1942. There is a sense in which Alamein 1 was almost as great a 'miracle' as the Marne. Had we failed, the war might well have been 'lost'; though only you seem to have observed that fact. It is not popular yet to suggest that July 1st 1942 was the real turning point. Mussolini realised that right enough as July wore on – so did Rommel; the fact really broke both men. Alamein 1 is twenty years in the past, Mons almost 50 years away, and so much in between – I begin to feel old.

I need not ask whether you are active, for you never seem to be without some fresh ploy.

Ever, Eric

PS By the way, so far as I can make out, modern western military organisation seems just about to have got around to that which I proposed to Auk in a paper of July 20 1942. Quick work.

To R. W. 'Tommy' Thompson   Bellamont Forest
                                                          27 July 1962

My dear Tommy,

Alas one can't go back. Even if one could take away the accumulated responsibilities, one would take away the original savour. I remember one spring morning in 1915 when the battalion was out of the horrible trenches around St Eloi being allowed off for the day, and riding through relatively clean country to Baileul just to have lunch at a decent pub.

I was alone, I had a decent horse, the day was lovely and I was still alive. That was enough – more than enough; being alive after all that death was perfect. I've had other 'perfect' moments since then (less and less as life proceeds) but I've never forgotten that day, for Lt Dorman-Smith, Fifth Fusiliers, would not be 20 years old until July 1915, for all that he'd already been wounded twice. Just to be alive – that's it!

We must needs make the best of what remains – controversy mainly and looking on at other people's dogfights. No matter, we've had lots, and success bloats a man. I don't think I'd have enjoyed high military rank, had my career survived.

Ever, Eric

To Correlli Barnett   Bellamont Forest
                                                                              19 March 1963

My dear Bill,

Gough of the Curragh Incident and 5th Army at Passchendaele has just died. Fundamentally a very decent sort. His scapegoating was unjust – at least some amends were made a lot later. He did live to enjoy them.

Ever, Chink

To R. W. 'Tommy' Thompson   Bellamont Forest
                                                                              3 December 1963

My dear Tommy,

Monty [Thompson's next biographical subject] – he'll be a problem. It's easy enough to write about a man having the quality of nobility, but Monty lacks that. Wavell had it conspicuously in a homespun Norman way which in the 12th century would have made him one of the better barons or even an intelligent abbot. Auk too has the sort of nobility worn by a crusader's effigy.

Moreover, Monty never held supreme command in the field in a war theatre, as did Wavell, Auchinleck and Eisenhower. Monty was always a subordinate – in the ME and Italy an Army Commander, in NW Europe one of several Army Group Commanders. He was always buttressed against Winston by someone higher, and adversity had tamed Winston by then. Monty's luck was that he never had to be great, just adequate to the time – the time for greatness was over. But he was efficient, safe tactics, safe public relations, no risks, no recriminations. He came in on the turn of the tide.

Remember, too, that he did not end with the end of the War. He went on in Germany, then CIGS, and later in the forerunner of NATO etc. Indeed, this last 20 years of Monty reveals him in balance; it would be well to seek out his admirers as well as his detractors. Of course the foolish vain man has written about himself!

Ever, Eric

To Correlli Barnett   Bellamont Forest
                                                         9 December 1963

My dear Bill,

'It was a frightfully hot day' and 'We were in a garden in Mons' in *In Our Time* represented Hem's attempt to reproduce myself telling him about the fighting at Mons on 23 August 1914. Hem was developing his style in Paris around 1923 in that darn flat over the sawmill and used to sit up typing all night just to produce these snippets of conversation by morning! If the BBC like I'll come on and read them for the 50th anniversary[26] – after all, they're a bit of me and I was there.

Ever, Eric

He did accept the BBC invitation to take part in their commemoration of the 50th anniversary of the outbreak of the First World War, and read the two Mons cameos aloud on air. The inscribed first edition in his pocket, dedicated to him in ink, was now extremely valuable.

Meanwhile a fresh legal battle held his attention, this time over Alexander's *Memoirs*. The point of contention was that the book denied an existing plan for Alam Halfa, implying that Auchinleck's intention had been to retreat.

To Correlli Barnett   Bellamont Forest
                                                         31 December 1963

My dear Bill,

During 1963 my lawyers have been tangling with legal representatives of Cassells and Alex. Ernest Wood[27] is about the best man on libel practice in Dublin, Bill Finlay[28]: a youthful and attractive rising star, is more orthodox, and junior counsel Ulick O'Connor[29] has literary leanings. Wood is the one who understands I'm a realist, unconcerned with my own reputation but concerned with keeping history straight. I'm approaching my 69th year, and if I can twist the Alex/Monty tail it will be sufficient *nunc dimittis*.

Ever, Eric

After much preparation he was advised that the matter was too specialised for a jury to comprehend, but emotional reward would result. Contact with Auchinleck over it enabled them to become close again, and this time his invitation was accepted.

To Correlli Barnett   Bellamont Forest
                                                         19 May 1964

My dear Bill,

I had Auk on my hands till Monday last. He's a remarkably good visitor, no trouble and most appreciative. He has seen little of Ireland so I motored him about a bit, to Bundoran in the West and the Boyne tombs in the East. He fairly eats topography, loving to ride with the map on his knees. Luckily the weather was fine.

Auk has changed but little: a slow tenacious old badger of a man; shy but a fighter-back. Intelligent to a point, and then numb. The youngsters loved him. I think he enjoyed himself; if anyone felt strain it was I. He talked a bit about the past, still wondering about Ritchie! He certainly does not show his years. He was amused by your discussion of your encounter with Monty.

Ever, Eric

To R. W. 'Tommy' Thompson   Bellamont Forest
                                                         29 May 1964

My dear Tommy,

A book mailed from Hodder & Stoughton, *Generals at War*, had a [handwritten] inscription by Freddie de G.[30] on the title page:

'To Chink, in recognition of the part you played.

With happy memories, Freddie'

Blimey! First reaction was to re-pack it and send it back to H & S. Second reaction was to write Freddie a nice note of thanks in the hope of stringing him along and perhaps getting more data from him about the August 15/21 1942 period. Third reaction, very faint, was to be Christian and forgive my enemy. But the gnawing problem remains; why did he do so odd a thing? This is on a par with the tone of what he says about myself in the book – conciliatory and all that. Possible reasons are:

1) Change of bandwagon. But why now?
2) Campaign somewhere to get all Cairo Purge boys buddies again.
3) Fear of what I may eventually write!
4) Freddie getting older and more lonely.

Really no good reason suggests itself.

Our love, E

The reply suggesting that he attend a forthcoming Military Commentators Circle meeting, however, got a dusty response.

To R. W. 'Tommy' Thompson   Bellamont Forest
                                                              10 June 1964

My dear Tommy,

When in 1945 I closed the door on my past life entirely, I turned to other thinking [until] I was painfully dragged back by the demands of writers such as yourself, John Connell, Bill Barnett. Now I'm in a halfway house, for I don't want to go back to what I've left behind, for all that I miss it. My reaction is against getting involved in a Service atmosphere again, as a sort of Marlay's ghost dragging the chains of my degradation behind me, prickly with discomfort and the aggression that engenders.

This not just self-pity, it is that I want to be able to walk into my old club the Senior without the feeling that I'm walking into groups of my aged contemporaries who have taken part in that 'vicious vendetta' of which de Guingand belatedly writes.[31] My instinct is to hit them a hearty wallop, and then I'd be removed by the club porter.

It's like being run-out by a fraudulent umpire when you felt that you had a useful innings still in you, and then of being chucked out of the team by spiteful competitors who questioned your competence as a cricketer. The answer is to put the bat away. Besides, if one accepts the umpire's verdict it makes the whole game crooked. This is the only service I can give the Service for which I was reared and conditioned.

All our love, Eric

To R. W. 'Tommy' Thompson   Bellamont Forest
                                                              23 June 1964

My dear Tommy,

I do not wish to denigrate or underrate Montgomery. He was a tonic to Eighth Army at that time. He had the right seniority for his position, he really commanded without any nonsense or argument, and his grasp of the situation was excellent. But his grasp was much facilitated by his predecessor's actions during July/August 1942. There is no small doubt that he took credit for Auk's work and plans. Both he and Alexander owe the springboards to their important military careers, as do many others, to Auk. Indeed, without him victory in WWII might well not have been achieved.

Ever, Eric

To John 'Jack' Connell   Bellamont Forest
                                                            13 September 1964

My dear Jack,

I've read your biography *Wavell, Scholar and Soldier* with great care, enormous interest and some nostalgia; that last because that lost world was my real home. APW is there, eternally alive in history with his strength, spiritual beauty, his integrity, loyalty, humility, love and weakness. This is a portrait of a great Englishman, and a worthy companion to your *Auchinleck, A Critical Study* on any bookshelf.

The genesis militarily of both Auk and Archie is interesting and important. They were NOT the products of the 1916/18 western front of WWI. Mespot,[32] Palestine, Russia – both your men had wide-ranging experiences. So they both brought to their high command a horizon of thought and feeling totally different from that of Monty and Alexander, and also of Churchill.

Anyhow, Jack, you've rung all the bells, for which I congratulate you. Now give yourself a chance. We've all been too solemn too long – time we had a binge.

Ever, Chink

Apparently out of the blue, but in fact thanks to Carlos Baker as go-between, a letter came from Hadley Hemingway asking if she and her husband Paul Mowrer,[33] might stay at Bellamont in September. In his role as her son Bumby's godfather, Chink and Hadley had remained loosely in touch but not met for almost forty years. He invited her at once. It might have been an awkward experience, but to their mutual delight they found themselves as compatible as before, despite new partners.

To Hadley Hemingway   Bellamont Forest
Chicago, USA   [Undated] Autumn 1964
My dear,

So damn easy to be sententious and to miss the lad we knew. Paris, Les Avants, St Bernard, Milan (Milan before you!), fishing at the end of the lake. How could I tell a professor[34] about Milan 1918? We were all so young.

What happened to us all? I don't want to open old sores; you need not respond if this is what happens. But we shared such riches in Hem's innocence. On looking back, what full years they were. Surely the times when we were together were the best of them – for me, anyway.

You see, Hadley, the pundits who write about Hem seem to skip these important years, thinking that since Hem wasn't 'important' then, they weren't important either. Hem was even more important then, to ourselves and to himself too, perhaps, than he ever was afterwards.

With love, Chink

Churchill died the following January, but Chink felt no similar fond impulse towards the past: certainly no trace of re-evaluation.

To R.W. 'Tommy' Thompson   Bellamont Forest
                                                             30 January 1965

My dear Tommy,

As I write, the old man's remains are undergoing the last charade, thereby adding to heraldic necrological [*sic*] lore. RIP. That carrion crow, Richard Dimbleby, is qualifying for his 'K'. This strange necrophilia will have its reaction. People will want to look behind the oceans of turgid eulogy recently inflicted upon them. Ever, Eric

To R.W. 'Tommy' Thompson   Bellamont Forest
                                                             [undated] 1965

My dear Tommy,

Long time no hear! So this is 'to enquire', as was the Victorian formula. Apropos of that, here is a bit of gossip for your files. As you know, I've pretty well worked out the background of Churchill's actions during August 1942, and in correspondence with Auchinleck I gave him the gist of my deductions, just to see whether from his angle I might have dropped a stitch somewhere. This is his reply:

> 'I see your point now very clearly, and do not think I know of any facts of which you are not aware.... As to the Iraq/Persia Command, I honestly believed it was militarily unsound in the conditions then obtaining and I refused it for that reason. Moreover, I felt that I could not do myself justice or exercise command properly if I appeared as a commander discredited in the field. It would not have been fair on the troops. I felt this very strongly at the time. It was not pique. Also I think I was pretty tired, though I would have thrown that off in a month or so. Churchill's visit tired me much more than Rommel ever did. His obvious inability to understand why we at Tac HQ lived as we did and not in Cairo affected me greatly ....'

Now I think that this is the only comment that Auchinleck has ever made on the whole sorry episode, so it does have considerable historic value.

Ever, Eric

To R. W. 'Tommy' Thompson   Bellamont Forest
                                                    14 April 1965

My dear Tommy,

By the way, you'll be amused to know that The Auk has invited himself here for mid-May and for a week. I'm delighted to have him, if somewhat surprised. It will be an odd psychological event.

Ever, Eric

To R. W. 'Tommy' Thompson   Bellamont Forest
                                                    26 July 1965

So, my dear Tommy, I have successfully negotiated that most disagreeable of turning-points, my 70th birthday. That day seventy years ago, the birth of an heir to the local 'squire' was signalled by the ringing of the estate bell and a bonfire on the Hoop Hill, plus, doubtlessly, the issue of free porter to the numerous employees. Last Saturday no bells rang, there were no bonfires, I collected three tiny bottles of assorted liqueurs, three pipes, plus a tin of pleasant tobacco. Enough. One more door closed behind me with a sour chuckle from the immortal porter. Damn it, I do not like being 70, but then I disliked being 60; fifty was disagreeable enough, forty seemed a watershed; so perhaps I'll get over it.

Ever, Eric

With time speeding up, his relationship with Auchinleck was developing the warmth of congenial friendship, if not exactly familiarity.

To Field Marshal Sir Claude Auchinleck   Bellamont Forest
                                                    1 April 1967

My dear Claude,

I never quite get used to addressing my late Chief by his christian name; no matter, I bravely struggle on! I do apologise for not answering your letter due to a determined attack from the authorities of Manchester University, who want my papers. Legal advice from Dublin says, 'Do not part with anything which may be used before we have combed through them for dangerous material'. Since 1954 a voluminous correspondence has grown up: Basil LH, Jack, Thompson and Bill Barnett being the main contributors. Additionally there's the slightly funny file of my challenge to Winston in 1953–54 (had I known then what I know now I'd have been harder than I was), and I gave Alex a rub for altering a quote from your Despatch to suit his memoir as to your alleged intention to retreat from Alamein if Rommel attacked in August or September.

Anyway, Manchester's top librarian[35] was here yesterday. I fended him off by saying that my lawyers had forbidden me to release anything without their approval, but that I had no objection to permitting Manchester's History Department to carry out research here under my supervision. This will possibly develop into a symposium to be held either at Manchester, or perhaps on the ground itself, into the whole crisis period of 1942, say May-September. The idea (prompted by myself) is that Manchester in association with a German and an Italian University should conduct an academic study, aided but not controlled by such ex-military folk as myself, Bayerlein[36], and assorted Italians.[37] Soldiers are too service-conditioned to see an episode of war from its many sides: the geographical, psychological, historical, industrial, political facets. The Yanks are already well forward, with universities working in with the Pentagon.

I would love to see you if you thought of coming to Ireland again. I've bought all the Dartry lakes and remaining 50 acres, [and though] the big house has been pulled down the steward's house remains. So if you take it into your head to move [into it] we'll set about its modernisation. However, you will probably end by living in a castle in the Atlas mountains. Before you do so, please come here once more.

Yours ever, Eric

To R. W. 'Tommy' Thompson   Bellamont Forest
                                                       30 May 1967

My dear Tommy,

Your letter came when I was about to be wheeled into the operation theatre of a Dublin hospital. Cause: the discovery of a large and menacing hostile growth situated on a tangle of adhesions, the result of a 1934 appendix operation. Though nobody will admit that it was cancerous, I suspect that it was.

After a month in hospital I'm back home a stone lighter, slightly groggy but otherwise fit enough, although my brain does not seem to work quite as quickly as I'd like. Consequently I've been completely out of action, living in a small pink-walled box, and at the mercy of assorted females prepared to outrage one's privacy without compensatory concessions.

Ever. Eric

To Basil Liddell Hart   Bellamont Forest
                                                       3 June 1967

My dear Basil,

You and I have penetrated the 'zone of mortality' when the facts of life seem to be our own and other people's deaths. Still and all I suspect that we will be about for a while yet, hell bent to bore the rising generation with our obsolete aphorisms on warfare to which that generation will pay not the slightest attention – probably to their ruination. Following this warning that some bell or other is poised to toll for me, I really will have to sort out my accumulation of papers. I rather dread working through it, though it's surprising what interesting things appear – things one had forgotten completely.

Ever, Eric

To Field Marshal Sir Claude Auchinleck   Bellamont Forest
                                                          1 July 1967

My dear Claude,

This morning 25 years ago I wrote in my pocket book 'Battle of Alamin'. I got the spelling wrong, but the fact right. Yesterday, also 25 years ago, I had noted down '2 great squadrons, Bde, 21st Australian Canal'. What a slender reinforcement, and how few decent tanks! But not even the armour was in hand till the late afternoon of that day. All you had ready for battle were some 3000 dicky South Africans, a green if doughty Indian infantry brigade, and the remains of 13 Corps to the south, barely in touch. Not an encouraging assembly on which to base a stand. Thanks to your indomitable guts and your tactical redisposing of the South Africans <u>outside</u> the El Alamein box, it was enough. Do you remember how hot it was, a *Khamsin* hiding everything? I wonder whether you know that <u>on that day</u> Rommel announced he had broken through at El Alamein and was on his way to Alexandria?

Anyhow, if nobody else does, I salute you – you saved literally everything on that day. On the 25th Anniversary of your decisive battle of First Alamein we in Bellamont Forest remember how much the country owes to you from that time.

Ever, Chink

To R. W. 'Tommy' Thompson   Bellamont Forest
                                                          14 July 1967

My dear Tommy,

The Auk in his old age is coming out with some very interesting statements in correspondence. It's rather touching the way his mind goes back now. I wrote on 1 July to salute the victor of 'First Alamein', and his reply was typical.

'Your letter brought back those days when you and I were very much together and very much <u>alone</u>. It's nonsense to give me the credit you do. I know what I owe to you and realise very clearly, as I always have done, that without your wise and indomitable thinking always at my side and in my head, we could never have saved Egypt – and India and all the rest.

I think what sticks most clearly and vividly in my head is that drive back along the dark and deserted road after we had taken over from Ritchie, and the next morning when, I seem to remember, we were very much alone in the Desert. Also I remember often our tent at our HQ and our bedding rolls on the sand. I do not think I could have stuck it out anywhere else, or without you to advise and plan ahead. Again I thank you.'

Leaving out the bits about myself, those few words do describe the loneliness of supreme command in a crisis as stark as was that. Not all the heavy staff paraphernalia lets a top-level commander off the hook of loneliness.

I swim daily and am very fit again. But my surgeon has now discovered via X-ray that I have something in my bladder which may call for a return to the knife. I do NOT appreciate that idea.

This is between you and me. Ever, Eric

To Field Marshal Sir Claude Auchinleck   Bellamont Forest
7 August 1967

My dear Claude,

Escaped from hospital, and back home. Quite a party. Two major abdominals between May and July are about sufficient for someone my age. Now got to put back some weight lost in two weeks. You, as a seasoned Oriental, would have laughed because five days after they cut me open I developed a beautiful attack of malaria. It shook my surgeon who discovered that it could happen after an 'op', confirmed by the registrar, an old IMS[38] man. Have to be back in their hands in a few months. Carcinoma is a nasty mistress.

I do hope you are well. After all these goings-on one begins to look around at the dwindling ranks of contemporaries and note the gaps. It really is depressing how many have gone already. I'm glad that I've lived long enough to get a glimmer of truth into what [Compton] Mackenzie[39] calls 'Bad History'.

Ever, Chink

To Basil Liddell Hart   Bellamont Forest
19 November 1967

My dear Basil,

I haven't heard from Claude Auchinleck lately. [It had been reported in the press that Auchinleck was moving to Morocco.] I think he feels that England is too narrow, too full of people, too fussy. He is very much an oriental – arid open spaces and all that.

Your Israeli mail [about the Six-Day War in June] continues to be enormously interesting. I like the point about the mental level of indirect approach, because to my mind it's really a method of thinking strategically and tactically: the hunter setting a trap for his quarry, the enemy. Right thinking translated into right action dislocates the enemy and keeps him dislocated, mentally and physically. Freedom of action results.

Ever, Eric

To Correlli Barnett   Bellamont Forest
                                                    5 December 1967

My dear Bill,

After malaria, a blood clot. Now everything appears ok. Well, my contemporaries seem to be falling like leaves in Vallombrosa, so I can allow myself a chuckle.

I'm getting tired of shaving. I've grown a beautiful set of early nineteenth century side-whiskers which hide an ear with a hearing aid and absurdly contrast with white hair – what's left of that. They are warm to wear in a cold wind, and Cootehill has got used to them. Dublin stares a bit!

Ever, Eric

To R. W. 'Tommy' Thompson   Bellamont Forest
                                                    25 February 1968

My dear Tommy,

Just now I've been watching from the hall windows five stags on the lawn, two of them fighting until one gave up to be chased by the victor over the grass while the others watched. A lovely sight on a morning of white frost. Wild animals are so much nicer than stupid humans (present company excepted, of course).

On Thursday last I had my second inspection: into the bladder with gun and camera. I appear to be doomed to [these] three-monthly for the rest of my natural. But, Tommy, I do not get used to being wheeled into the Op. Room knowing that I'm going when unconscious to be up-ended before assorted nurses whom I see dimly masked like oriental females. I recognise the theatre nurse, she has nice eyes, and she winks as I subside into oblivion. It is not fair; I ought to be allowed a return match.

All the best, old chap!

Ever, Eric

To R. W. 'Tommy' Thompson   Bellamont Forest
                                                             6 November 1968

My dear Tommy,

I had Professor Michael Foot, Department of History, University of
Manchester, here for a week. He sat in a re-arranged Ball Room and ferreted
through my paper accumulation of years. Then he bought a suitcase and
departed with it bulging. I got a nice letter from the Vice Chancellor saying
that my tendentious bumph was now with Auchinleck's papers and history.

It was a lovely autumn; still the oaks keep their leaves, and the last few days
have even been sunny and calm. Look after yourself a bit, Tommy.

Ever, Eric

To Basil Liddell Hart   Bellamont Forest
                                                               3 March 1969

My dear Basil,

Years since I've communicated with you. How are you, and did you shake
off your autumnal malaise? I saw a crocus yesterday among our snowdrops; a
signal that one should cast off winter's lethargy. Still here, winter. The big lake
below the library window is well iced up this morning – sign of a hard frost
in the night.

Recently the Irish Military College, which combines the Cadet School,
the Infantry course and the Staff Course, and which seems to specialise for
its military history on the Desert War, asked me to speak to the Staff and
Infantry courses, not so much for instruction as for 'local colour'. I found
them a nice lot, very <u>professional</u>, with a well-run show on a shoe-string. Odd
speaking to soldiers again after all these years. They are a queer mixture: one
instructor trained at Leavensworth, another from Camberley, a third just off
for a year at the German Staff College. Thus they look in on our 2½ years
Desert junketings with refreshingly open minds. What they want to know is
about the personalities, and how these reacted to their problems on the several
levels of action. I dare say I'll see more of them later in the year.

Do let me know sometime how things go with you.

Ever, Eric

To R. W. 'Tommy' Thompson   Bellamont Forest
                                                          St Patrick's Day 17 March 1969

My dear Tommy,

This winter past I've been refreshing my WW2 thinking, having concluded
that one cannot understand the vagaries of our warfare unless one returned
to the study of Churchill himself. Wasn't he a bit like a headmaster of a

public school in the organisation of his 1940–45 administration? He took the 6th form, so to speak, leaving admin to the bursar, matron in chief etc. War Cabinet as housemasters, chiefs of staff and field commanders as prefects – bad boys could be sacked! Did he unconsciously model himself on a Harrow School system? It's a thought. Also he had favourites – ultimately Alexander and Mountbatten. And Monty as Head Boy!

In a separate letter, when you are ready, I want to tell you about the major decisions of the War Office in 1934 which resulted in mechanisation. Changed it in theory from a manpower plus horsed-power walking army to a group-weapon plus machine-moved fighting machine. When in 1933 we mobilised a division at Aldershot and marched it over the Hog's Back – horse, foot and guns, miles and miles of walking infantry – there was almost no difference between them and a division of 1918. The whole tactical technique changed almost overnight, far too quickly for a great many senior officers. I suggest that right up to 1942 Churchill continued to think in terms of the fighting he'd experienced. I doubt whether he understood the real implications of the Desert War, still in WWI blinkers.

And so the tapestry weaves: Winston as a purple thread among the golds and reds and blues of other men, near and far. What he could never understand was the calibre of the top soldiers and sailors of 1940–45, how different they were from the 1914–18 vintage. Good men among those too, but the WWII lot were better trained, more experienced, more educated, less socially affected; mostly they were poor men, rich in experience and humanity but not in goods. The age of amateurs was past.

Very well. I'm going to stop here for the present.

Ever, Eric

To Professor Carlos Baker   Bellamont Forest
Princeton, New Jersey   2 April 1969

My dear Professor,

Your 'Ernest Hemingway' is the most alarming, fascinating, and saddening document I have ever read, besides being the quintessence of biographical scholarship. Sad as is the known ending, a strain of sadness pervades the narrative; the sort of sadness one feels for the bull blinking for a moment before hurling himself into the arena to do the only thing he knows how to do – to kill.

To [live] in Hem's way one has to stay young; it was his tragedy that his art necessitated an endless *da capo*[40] until he couldn't anymore return to his beginnings: the circle became a spiral. When he got to the limit of the spiral he wrote *The Old Man and the Sea*. His 'beanstalk' was spiral, I think, but when he got to the top there was no longer a castle, a giant to be killed, and a lovely to be consoled. Instead, nothingness, a condition he could never take. Logically there was only one thing left to do, and he did it. *Da capo* is not enough. But what a story, and how excellently you have winnowed out the grain from the chaff.

For myself, I had quite forgotten the Hemingways' visit to Cologne, as also the incident of the *laissez passer*, which shows how unreliable memory can be. My moustache was red, but my hair black (a characteristic of the Irish of the north). As well, I was the eldest son of my father, and never, alas, a Lt-General. These are small matters for your amusement.

En passant, your handling of Hem's 1944 exploits rings true to an alien military mind; more *da capo* stuff in effect. The doom of Peter Pan in life is to become the Peter 'bed pan' before petering out. I think that Hem saw that and had no way of escape from his self-created arena – if it was self-created, and not genetic destiny.

Yours very sincerely,

Eric O Gowan

PS I'd also quite forgotten the Henry IV passage. How you do dig things out to be sure!

Afterword

Four weeks later, after a sudden haemorrhage at home, Chink was taken by ambulance late at night to Lisdarne Hospital in Cavan. The haemorrhage, though, could not be stopped. He listened to radio news bulletins and asked for the papers to be read aloud to him each day, alert but growing markedly weaker, and on 10 May he requested the Last Rites. He died early the following morning.

The ecumenical funeral in the local Protestant church[41] near Cootehill had been planned by him in advance as a final tilt against sectarianism. The service was attended by Eve, his son Christopher, his daughter Rionagh, and his stepdaughter Elizabeth, as well as wider family members, neighbouring mourners, and two Irish Army officers. His quiet grave there overlooks a lake, part of the long chain of water linking the lands of Bellamont Forest and Dartry.

The Times obituary by Basil Liddell Hart appeared on 13 May 1969, the number which had dogged Chink throughout his life. 'He was not what the Army calls a 'commander', Liddell Hart began judiciously, 'in that he had not those qualities of patience, understanding and tolerance demanded of those who aspire to train men.' 'A great shock,' he wrote privately to Eve. 'So many memories come back of him and his keen mind, charm and debonair manner.' Correlli Barnett dedicated *Britain and Her Army*, the book he was working upon, to his memory, and 'Tommy' Thompson was grief-stricken. 'Jack' Connell had died in 1965.

In Marrakesh, too far away to attend funerals back home, Auchinleck was desolate. 'Two days before the cable,' he wrote to the family at once, 'I had said to myself that I must write to Chink. It is a long time since I had news of him. He was tragically mistreated and betrayed in the end. Envy and malice pursued him, but he never gave in. I am glad his end was peaceful, but I am very sad.'

Acknowlegements

I am indebted to Chink's son, Christopher Dorman O'Gowan, who has not only given me full access to the vast number of private letters written by his father during the Second World War, but made no attempt to influence my opinion in advance. Nor did he take issue with my conclusions, giving full copyright permission for their inclusion. The mountain of letters to his mother Eve, which form so much of the War Diary, are in his possession, as are those to his grandparents and his Uncle Reginald. Without that free hand from the start, the sheer profusion would have proved too strong a disincentive for this book to take shape.

My deepest thanks are also due to the late Correlli Barnett, author of *The Desert Generals*. To him go the laurels for transforming Eric Dorman-Smith's military reputation, and drawing public attention to Auchinleck's lasting achievement and the importance of First Alamein. 'Bill', as friends knew him, not only wrote the Foreword to my original biography *Chink*, but he was equally enthusiastic about this book from the start, and provided invaluable support.

Military historian John Lee has cast his eye over diary entries in the lead up to that crucial battle, and I am indebted to him not only for those contributions, but for his professional assessment of its lasting historical value.

Frederic Raphael's illuminating Foreword brings Chink vividly to life on the first page. Professor Sandra Spanier, General Editor of the vast Hemingway Letters Project, has been generous in her assistance throughout, as has the writer and author Lance Morrow in New York. I would also like to thank barrister and biographer Charles Lysaght for sharing his study of the 1953 legal case taken by Chink over *The Hinge of Fate*, fourth volume of Churchill's *History of the Second World War*.

Lastly, I pay tribute to Lester Crook and Jonathan Williams, whose mutual care for the complexity of history has enabled this collection to reach the wider public.

As the search for Chink's letters widened, the archivists at the following libraries and museums were immediately helpful when I contacted them, and I am grateful for their interest and expertise.

John Rylands Library, Courtesy of the University of Manchester, where Chink's papers are held: (FM Sir Claude Auchinleck (16 October 58,

20 October 58, 16–23 October 58, 2 January 59, 1 April 67, 1 July 67, 7 August 67), Larry Solon (3 April 58)

The Liddell Hart Centre for Military Archives at King's College, London: Basil L.H. (18 December 34, 26 January 35, 1 February 36, 15 February 36, 18 February 36, 20 April 38, 6 November 38, 29 August 41, 14 October 42, 24 November 42, 2 April 44, 22 August 44, 1 November 44, 1 June 46, 2 June 46, 10 December 46, 26 February 48, 7 April 50, 16 April 50, 4 July 54, 17 September 54, 16 October 58, 25 June 62, 3 June 67, 3 March 69). R.W. Thompson (16 February 60, undated 60, 24 March 60, 8 June 60, 22 July 60, 30 July 60, 1 October 60, 11 January 61, 27 July 62, 3 December 63, 29 May 64, 10 June 64, 23 June 64, 13 January 65, Undated 65, 14 April 65, 26 July 65, 30 May 67, 14 July 67, 25 February 68, 6 November 68 17 March 69.); General Sir Richard O'Connor (15 December 40).

Churchill College, Cambridge University: Corelli Barnett (27 May 58, 14 June 58, 18 June 58, 2 July 58, 3 July 58, 14 July 58, 19 July 58, 5–20 August 58, 6 October 58, 18 October 58, 8 December 58, 22 April 59, 28 May 59, Undated 59, 23 June 60, 8 July 60, 29 August 60, 19 March 63, 9 December 63, 31 December 63. 19 May 64, 5 December 66).

The Becky and Lecky Libraries of Trinity College Dublin (Hubert Butler (undated 1954, 31 March 55). Owen Sheehy Skeffington (Autumn (undated) 54).

McMaster University Library, Hamilton, Ontario, Canada: John Connell (1 October 60, Undated October 60, 4 October 60, 7 October 60, 8 October 60, 28 October 60, 11 January 61, 20 June 61, 5 June 62, 13 September 64).

John Fitzgerald Kennedy Library, Columbia Point, Boston, USA: Ernest Hemingway (9 December 23, 15 December 23, 3 June 26, 13 May 50, 3 June 50, 21 June 50, 6 July 50, 3 August 50, 7 October 50, 3 November 50, 19 November 50, 3 April 51, 15 November 54, 6 January 55, 18 February 55,) 15 letters.

The Fusiliers Museum of Northumberland, Abbots Tower, Alnwick Castle (fusnorthld@aol.com) holds the backnumbers of St George's Gazette with its 'First Battalion Notes' (1921 – 1924).

And to hear, rather than read, a variety of interviews and opinions, 'The Brigadier' (9 September 2017) is a podcast in the RTE Documentary on One series, and available online.

Lavinia Greacen

Maps

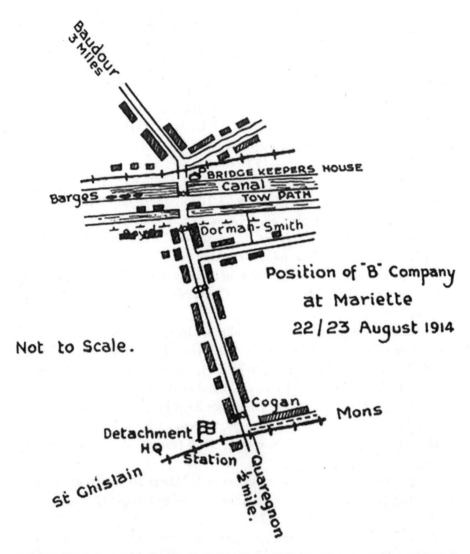

Baudour
3 Miles

BRIDGE KEEPERS HOUSE

Canal

Barges

TOW PATH

Dorman-Smith

Position of "B" Company
at Mariette
22/23 August 1914

Not to Scale.

Cogan

Mons

Detachment
HQ

Station

Quaregnon
½ mile.

St Ghislain

Aged 18, Chink disembarked in France with his fellow officers and men at 3am on 14 August 1914. The next train took them to the front. On 22 August, at Mariette in Mons, came his first opportunity to put theory into practice when authorised to hold the bridge against rapidly advancing German infantry. He ordered his men to shoot, only to be told that the first shot was his, and he counted three bodies before handing back the rifle. [Handwritten map by Eric 'Chink' Dorman-Smith.]

The following maps first appeared in *Crisis in the Desert, May – July 1942* by John Augustus Ion Agar-Hamilton and Leonard Charles Frederick Turner, published by Oxford University Press, Capetown, 1952. (With thanks to Christopher Dorman O'Gowan, who possesses his father's annotated copy.)

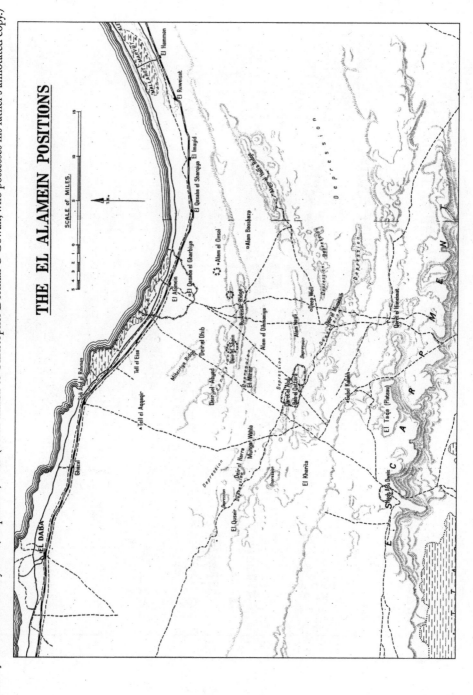

THE EL ALAMEIN POSITIONS

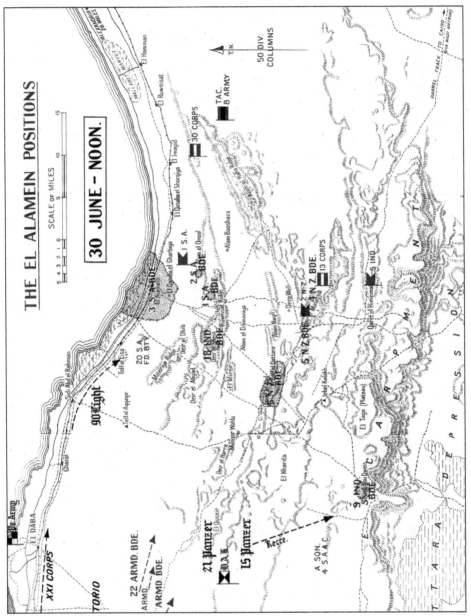

THE EL ALAMEIN POSITIONS

30 JUNE – NOON.

SCALE OF MILES

8th Army dashing back to the Alamein lines with an exhausted Afrika Korps in pursuit. (John Lee)

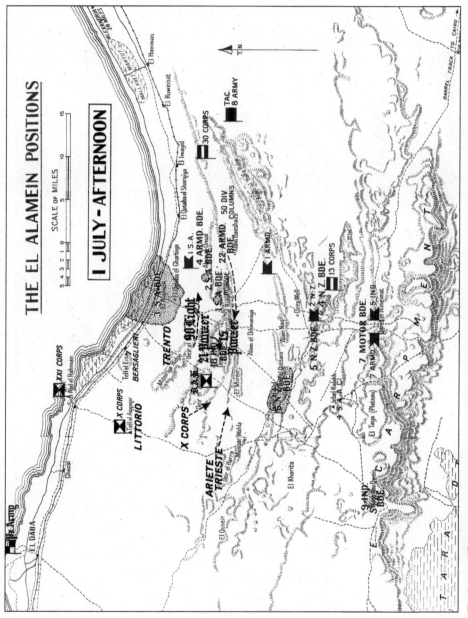

Evidence of Rommel's 'big push' being stopped cold in its tracks. (John Lee)

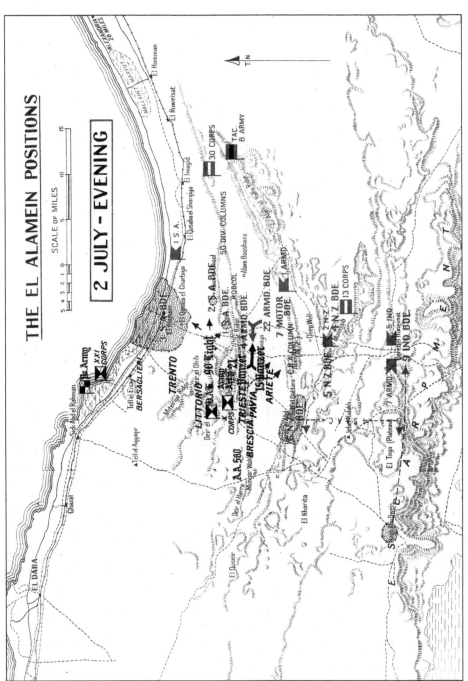

Rommel's attacks firmly contained by 8th Army artillery, acting as Chink had long wanted. (John Lee)

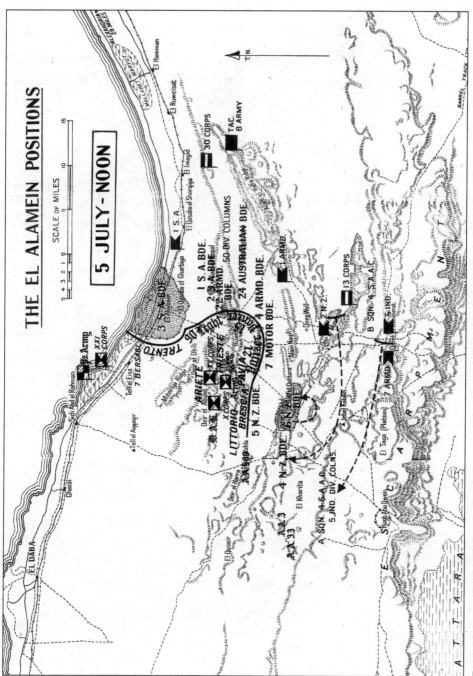

Rommel put under intolerable pressure by 8th Army counter-attacks targetting his army's Italian formations. (John Lee)

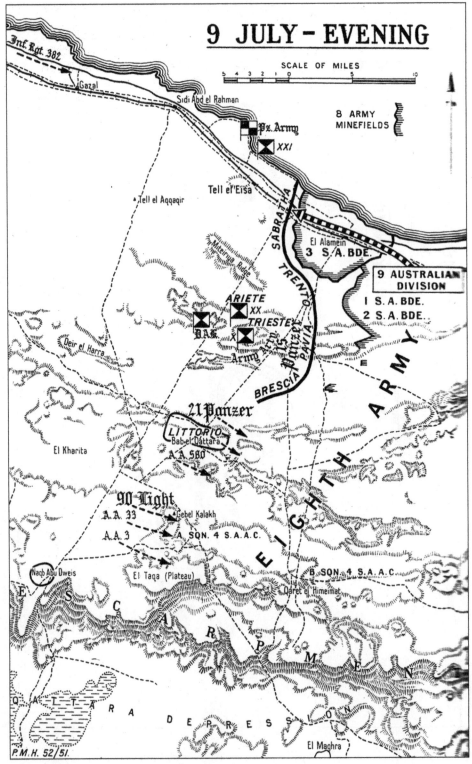

8th Army masses in the north to strike the weak Italian formations in Rommel's army, a product of Chink's reading of Ultra decrypts. (John Lee)

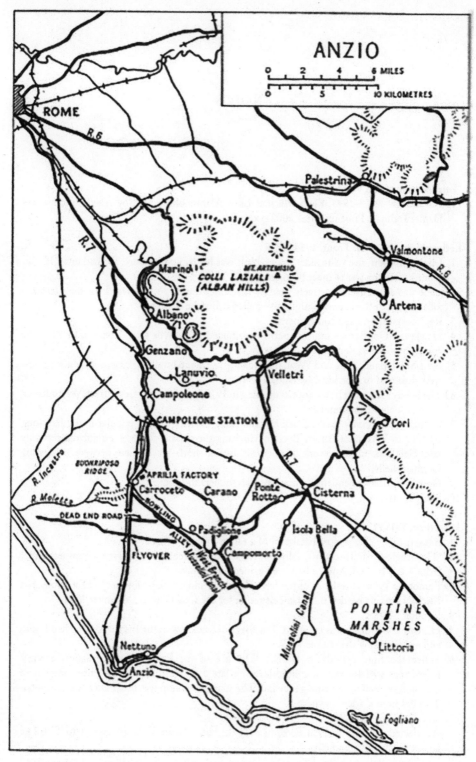

This map featured in The Italian Campaign 1943-45 by the late Col. Gilbert Alan Sheppard, Arthur Barker Ltd, 1968.

Notes

Introduction to the War Diaries
1. *War Diaries 1939–1945*, Field Marshal Lord Alanbrooke, edited by Alex Danchev and Daniel Todman. Phoenix, Press 2002 p xii.

Letters Home from the Front: 1914–1918
1. Lindsay Barrett, five years older than Chink, was his regimental hero. He had brought him home before the war to meet his parents.
2. Boche was a derisive term freely used by the Allies during World War I for the Germans (collectively), or the enemy. Chink mis-spelled it Bosche at first.
3. Regimental friend and contemporary.
4. Hawkes & Co, famous military outfitters situated at 1 Savile Row, London. In 1974 the company was renamed Gieves & Hawkes.
5. Near Sunderland, Tyne and Wear, this was the regimental training base, before newcomers were deployed to individual battalions.
6. The family's favourite hotel since its opening in 1910, situated in Berners Street, Westminster.
7.   A long prison sentence.
8. German anti-aircraft shells. Black humour slang deriving from a popular music hall song credited to RAF pilot Amyas Borton, who sang the defiant chorus – 'Archibald, certainly not! Get back to your work at once, sir, like a shot!'– as he flew between exploding German shells.
9. Gastroenteritis was endemic among British soldiers in Italy.
10. Slang for Italians.

Peacetime: 1919–1939
1. *Aide-de-camp*: The personal assistant of a senior military officer.
2. 'The Anglo-Irish Treaty, officially the Article of Agreement for a Treaty between Great Britain and Ireland. A truce had been agreed on 11 July.
3. Eamon de Valera, President of the Irish Republic, undertook preliminary negotiations, but Michael Collins headed the Irish delegation for the final talks in December 1921.
4. London.
5. His date comparison was incorrect. The Siege of Lucknow in the Indian Mutiny had begun and ended during the year of 1857.
6. When the time period for DORA (Defence of the Realm Act) legislation expired, internment without trial was introduced in late 1920, resulting in the imprisonment of men and women under emergency law. The Curragh Camp was in County Kildare, about 30 miles from Carlow town.
7. The Truce, 11 July 1921.
8. An elected member of Dáil Éireann held the title Teacha Dála, shortened to TD. The largest party was Sinn Féin.
9. The Treaty between Great Britain and Ireland, known as the Anglo-Irish Treaty, was signed in London on 6 December 1921.

10. One of four such barracks at Bordon Camp in Hampshire, where a complete infantry battalion could be housed.
11. Europe.
12. One consequence of the signing of the Anglo-Irish Treaty was that Crown forces would withdraw from Ireland.
13. The regiment did not become the Royal Northumberland Fusiliers until 1935.
14. A derisive term freely used by the Allies for Germans (collectively), or the enemy. For more about retention of this word, which readers may find offensive, turn to 'Apologia' on p.xii.
15. A small rocky hill usually on the veldt, deriving from the Afrikaans *koppie*. A familiar Army term from the Boer War.
16. A vulnerable cornerstone of British India, near the northern frontier with Afghanistan.
17. Railway Transport Officer.
18. The first German government had fallen, a state of emergency called, and a second cabinet formed. Three days before this comment, a separatist government was declared in the Rhineland Palatinate, recognised by the French government.
19. A town in North Rhine-Westphalia, between Aachen and Cologne on the river Ruhr.
20. 'My Old Man', 'Up in Michigan', and 'Out of Season', were privately published in *Three Stories and Ten Poems* by Contact Publishing, Paris, 1923.
21. Herr Gangwisch, owner of Chalet Chamby near Montreux where Chink had spent Christmas 1923 with the Hemingways.
22. Director of Staff Duties.
23. Military Transport.
24. By a critical article by Liddell Hart in *The Times*.
25. Giffard Le Quesne Martel (1889–1958) was Assistant Director of Mechanisation at the War Office at that time.
26. The Carthaginian general Hannibal Barca, whose exploits in the Second Punic War resulted in lasting fame as one of the history's greatest military strategists, was the son of Hamilcar Barca, a leading commander during the First Punic War. Until the rise to power of his son, Hamilcar had been considered the finest general Carthage had produced.
27. General Sir Robert Archibald Cassells (1876–1959) was Commander-in-Chief, India, and on the Executive Council of the Governor General.
28. Comings and goings.
29. Archibald Wavell (1883–1950), whom – unusually – Chink greatly respected, was GOC-in-C, Southern Command in England in 1938. Chink had admired Wavell's military thinking after serving as his Brigade Major at 6th Experimental Brigade, Blackdown, Aldershot, in 1931–32.
30. Where Chink had worked closely and harmoniously with Wavell at 6th Experimental Brigade.
31. The Chatfield Committee in London (1938–39) examined the Defence of India under three headings: modernisation, re-equipment and financial implications.
32. The Marquess of Linlithgow.
33. Brigadier-General Sir Henry Montgomery Lawrence (1806–1857), whose foresight and decisiveness to protect British residents under his care when the Indian Mutiny broke out in May 1857 led to the Siege of Lucknow. He died there from shell-shot wounds before relief could finally be achieved, leaving a poignant epitaph in the Residency graveyard: 'Here lies Henry Lawrence who tried to do his duty.'

War Diary 1940–1944

1. John Lee has an honours degree in Modern History from Birkbeck College, London, an MA in Social and Economic History from Birkbeck, and an MA in War Studies from King's College, London. He is a member of the British Commission for Military History.

The Middle East: 1940–1941

1. A predecessor of the famous Baháí Garden terraces on Mount Carmel.
2. A tactically important sea-port in eastern Libya, about to be (3–5 January 1941) the focus of Operation Compass, the first operation in the Western Desert Campaign.
3. Marshal Rudolfo Graziani (1862–1955), Commander of troops in Italian North Africa.
4. Wavell.
5. Lord Linlithgow.
6. Sir Harold MacMichael (1882–1969), High Commissioner of the British Mandate of Palestine.
7. Brigadier General Staff – but for temporary duty, only.
8. A major port city in eastern Libya, captured on 30 January 1941 by Australian troops.
9. His family home in Cootehill, County Cavan.
10. Brigadier John 'Jock' Whiteley (1896–1970), who was then BGS under Wavell at GHQ Mid East Command in Cairo.
11. Lieutenant General Sir Arthur Smith, Coldstream Guards (1890–1977), Chief of Staff at GHQ Mid East Command, Cairo, whom Chink had often clashed with in the past,
12. Brigadier Francis 'Freddie' de Guingand (1900–79) had been Military Assistant to the Secretary of State for War before being posted to Haifa during Chink's tenure there. On Chink's recommendation, he was now on the planning staff at GHQ ME.
13. German troops invaded Yugoslavia and Greece on 6 April 1941 and quickly subdued the Balkans. By the end of the month Yugoslavia had signed an armistice and mainland Greece was under Axis control. British forces, sent to aid Greece, were being forced to withdraw.
14. A biblical king.
15. Philip Neame VC (1888–1978) had been appointed General Officer Commanding and Military Governor of Cyrenaica only two months earlier. Wavell had sent O'Connor to advise him.
16. German soldiers, collectively.
17. Wavell.
18. Dick O'Connor.
19. In retreat, Allied forces had fought to delay the German advance and allow ships to be prepared for evacuation, using a variety of small ports. On 21 April Wavell took the final decision, and by 30 April the evacuation of 50,000 soldiers was complete, despite the Luftwaffe sinking an estimated 26 troop-laden ships. Some 7,000 British, Australian and New Zealand troops, however, were captured.
20. Bernard Freyberg VC (1889–1963), Commander of the 2nd New Zealand Division. In the hasty retreat from Greece, Churchill gave him command of the Allied forces in Crete. Freyberg would later face public criticism for not basing his tactics on a new German sea approach, instead of defending Maleme airfields as Ultra (the term used by British Intelligence for information gained at England's Bletchley Park from encrypted enemy signals) had warned. But had he done so, that would have revealed that the Allies had broken the code. Freyberg continued to command 2nd New Zealand Division with the Eighth Army.
21. Crete was attacked on 20 May 1941 in a major airborne assault on the three main airfields and, after tough resistance, Allied evacuation was ordered, leaving Crete under German occupation.
22. A broadcast Fireside Chat on 27 May 1941, in which Roosevelt predicted Nazi intentions to gain control of the Western Hemisphere. Citing Hitler's broken promises of ceased aggression and subsequent invasions, he said that 'freedom of the seas' was essential, and support of Britain played a pivotal role in preserving that.

23. Allied troops under Lt. General Alan Cunningham had entered Addis Ababa, the Italian-held capital of Abyssinia, on 6 April, and British terms of surrender were accepted by the Duke of Aosta, Commander-in-Chief, on 16 May 1941.
24. Tanks.
25. When DMT in India, Chink had worked well with Auchinleck, who was then DCGS. A regular pattern of dawn walks together, seeking solutions to military shortcomings that both detected, had soon evolved.
26. Auchinleck had met the high-spirited Scottish beauty Jessie Stewart on holiday on the French riviera when she was 21 and he was 37, and married her within five months. The couple had no children. He would soon confide to Chink that she wanted a divorce, and that he was trying to save the marriage at long distance.
27. Eugenie (née Quirk), whom Wavell had married in 1915. They had four children, and she would outlive him, dying at the age of a hundred in 1987.
28. Mutual term for their previous marriage partners.
29. The Purana Qila in New Delhi dates back to 1538, and its peaceful gardens shelter red-stone ancient monuments.
30. Mena Camp, 10 miles from the centre of Cairo.
31. Liddell Hart had served briefly as personal adviser to Leslie Hore-Belisha, the Secretary of State for War (1937–40) but fell out of favour after Hore-Belisha's removal in January 1940. With his ideas largely ignored, he was living in relative obscurity and working mainly as a journalist for the *Daily Mail*.
32. So be it! [French].
33. Will Hay (1888–1949), an English comedian and actor whose theatrical sketch as a jocular schoolmaster, Dr. Muffin, was known as 'The Fourth Form at St. Michael's.
34. Chink's regimental hero and role model, killed in action in 1916.
35. Elaine was his affectionate stepdaughter, and Kenneth Bols her fiancé. Bols, a suave Cassells' *aide-de-camp* with whom Chink had already had several 'sharpish sessions', was due to be posted away soon after the wedding.
36. The adventurous treks of Freya Stark (1893–1993) into western Iran and southern Arabia in the nineteen thirties had resulted in popular books, from *The Southern Gates of Arabia* (1936) to *A Winter in Arabia* (1940). Her war-work with the British Ministry of Information was in a propaganda network aimed at persuading Arabs to support the Allies, or remain neutral.
37. The Staff College at Quetta, set up originally by Kitchener in 1905 to be like Camberley, was on the western border of India (now Pakistan). It was still recovering from the major Quetta earthquake of 1935.
38. Having impulsively sent in his name to Wavell's request for a successor of brigadier's rank for the post of Commandant at Haifa Staff College, despite the loss of pay, Chink had thought Cassells might refuse to let him go. Cassells made no such attempt. 'I handed the Old Guard the dagger,' Chink fumed, 'and they did not hesitate to use it.'
39. Freddie de Guingand had not only been a favourite pupil at Camberley in 1936, but on the Directing Staff at Haifa the previous year, until moving on. Chink found him so charming – a diplomatic European, rather than a conventional Establishment figure – that with Freddie he relaxed his usual defences.
40. Chink's code for whoever held the appointment of Commander-in-Chief, Middle East. Pharaoh 1 was Wavell, Auchinleck became Pharoah 2, and as yet there was no indication that Alexander would become Pharoah 3 in August 1942.
41. Pross was a Greek restaurant on the corner of al-Karmil Street (today Ben-Gurion Boulevard). 'If you fancy dishes that cannot be found in other restaurants', newcomers to Haifa were advised, 'you must go to Pross.'
42. Chink longed for an influential role at GHQ ME.

43. General Georges Catroux (1877–1969) had been Governor General of French Indochina until transferring his allegiance to de Gaulle's Free French movement. Appointed High Commissioner to the Levant in 1941, Catroux had taken control of Syria for the Free French after that summer's defeat of Vichy forces, and had recognised Syrian independence.

44. *Kitchener* by Colin Ballard (Faber and Faber, London, 1930). Ballard, (b.1868) was the second son of a general, and a decorated senior soldier himself, rising to the rank of Brigadier-General before retirement and prolific military writing. News of Kitchener's death in 1941 may have drawn Chink to the biography.

45. An *embusqué* was someone who avoided military conscription by taking a government job, a shirker or dodger. Limoges, capital of the Limousin region in France, was the traditional provincial bolt-hole.

46. Arthur Smith.

47. Wop: a derisive term freely used by the Allies in North Africa for the Italian army (collectively), or the enemy. For more about retention of this word, which readers may find offensive, turn to 'Apologia' on p.xii.

48. Historically, a heel-tap was a shoemaker's term, and over time any drink remaining in the bottom of a glass took on the same name. If a toast was offered with the instruction 'and no heel-taps', all glasses had to be drained to the dregs.

49. A park within 18 miles of Haifa, with the ruins of a 4th millennium city. For Auchinleck and Chink, interest was heightened by its proximity to the battle won in September 1918 by General Edmund Allenby, commander of the British forces in Palestine, which had led to the final defeat of the Turks in the Middle East.

50. A popular but expensive Haifa restaurant, known as much for its menu as for its daring cabaret.

51. He had met Amery (1873–1955) in 1925 on a Swiss mountaineering holiday; both had been staying in Maloga, and they climbed together. Amery was Colonial Secretary and Secretary of State for Dominion Affairs at the time. His indiscreet anecdotes about Winston Churchill, whom he considered disastrous at the Treasury, had illuminated Chink's already scornful opinion about politicians in general, and Churchill in particular.

52. Archie Nye (1895–1967) had been born in Dublin in the same year as Chink, during his father's regimental posting, but rapid promotion since meant they were no longer on a par. As Vice Chief of the Imperial General Staff, his job was to represent the Chief of the Imperial General Staff, Alan Brooke, and in Brooke's absence to attend War Office and top military committees. By now Brooke, whom Chink had long antagonised, relied heavily on Nye.

53. Deputy Chief of the Imperial General Staff.

54. Commandant of the proposed Higher Command Staff College in Pretoria.

55. Major General Douglas McConnel (1893–1961) had been appointed acting Major General on 16 October 1941.

56. The Duke of Wellington.

57. Gustavus Adolphus (1594–1632), King of Sweden, who led his country to military supremacy during the Thirty Years War. He was killed in battle.

58. Strictly forbidden.

59. Air Chief Marshal Sir Henry Brooke-Popham (1878–1953) had been appointed Commander-in-Chief, British Far East Command in November 1940 at the age of sixty two. In a more demanding role than previously, his responsibility in the joint command was the defence of Singapore, Malaya, Burma and Hong Kong. London's decision to replace him was being postponed by the critical Malayan situation, but within three weeks of Chink's comment, Brooke-Popham was to hand over command himself, to Lieutenant General Sir Henry Pownall.

60. The *Prince of Wales* and the *Repulse*.
61. The Royal Scots Greys, a cavalry regiment.
62. Major-General Angus Collier, Queen's Own Cameron Highlanders had been in the same syndicate of four as Chink at the Staff College fifteen years earlier.
63. Snakes.
64. Major General J.F.C. 'Boney' Fuller (1878–1966) had awarded Chink 1000 out of 1000 in his entrance paper for the Staff College. 'Fuller taught me to see in the first place,' he would always acknowledge.
65. General Piers (Pat) Mackesy (1883–1956), whose army career was destined to match Chink's for disillusionment and enemies in high places. As head of the sole infantry brigade in the expeditionary force to relieve Narvik after the German invasion of Norway, superior command lay with the naval commander and Mackesy was faced with chaotic organisation. Appalled by the potential Arctic Gallipoli of an immediate frontal attack, he chose to relieve Narvik from the rear, using special forces, which took longer to plan. Although successful, Mackesy was recalled home and reprimanded by Churchill. He was retired in November 1940. Chink's admiration for Mackesy's military brain sharpened his reaction.

GHQ Cairo: December 1941–June 1942

1. Field Marshal Sir Henry Wilson (1864–1922), born in County Longford, Ireland. One of the most senior officers in World War I, appointed CIGS in 1918, who argued in 1920 for martial law in Ireland, Wilson was assassinated by the IRA outside his house in Eaton Square, London on 22 June 1922.
2. An island in the Nile, within easy walking distance of the city centre, and location of the popular sporting Gezira Club.
3. A village in Cyrenaica, Libya, almost 274 km east of Benghazi and 80 km west of Timimi.
4. A Libyan village near the coast in the north-east of the country, 60 km west of Tobruk.
5. Lines of Communication.
6. Benghazi, the largest city in Cyrenaica, was a vital Libyan port on the Mediterranean. It had changed hands several times since the outbreak of war, most recently a week beforehand on 29 January, when recaptured by Rommel.
7. Tanks.
8. el Agheila, a coastal city at the southern end of the Gulf of Sidra in western Cyrenaica. Captured by O'Connor in Operation Compass and since retaken by Rommel, it was the base for the new Afrika Korps offensive.
9. Sir Reginald Dorman-Smith (1899–1977), Chink's youngest brother, had been appointed Governor of Burma on 6 May 1941. An MP since 1935, 'Reggie' had risen to Minister of Agriculture under Chamberlain, but been excluded from Churchill's government.
10. Rangoon (now Yangon) was the administrative and commercial capital of Burma (Myanmar).
11. Lt General Sir Thomas Hutton (1890–1981) C-in-C Burma, was in Brooke's view, too, unsatisfactory, so about to be superseded (by Alexander). Chink had known Hutton in Delhi and Simla after Hutton's appointment as DCGS India in 1940.
12. Lt General Sir Philip Neame VC, (1888–1978) had been appointed GOC Cyrenaica in February 1941, and been captured, with Dick O'Connor, within 2 months on 6 April 1941. Both generals were currently prisoners of war In Italy.
13. Lt General Sir Noel Beresford-Pierse (1887–1953), an old friend of Chink, had commanded the Western Desert Force from April to September 1941.
14. Lt. General Sir Alan Cunningham (1887–1983) commanded the newly formed Eighth Army from August to November 1941.
15. Lt. General Sir Alfred Godwin-Austen (1889–1963) led the Western Desert Force from April 1941 to February 1942

16. Lt General William Henry 'Strafer' Gott (1897–1942), King's Royal Rifle Corps (KRRC), had commanded a mixed force in Operations Brevity and Battleaxe, and been promoted to command of 7th Armoured Division for Operation Crusader. Gott had recently been given command of XIII Corps.

17. General Henry Maitland 'Jumbo' Wilson (1881–1964), the Rifle Brigade, commanded the Ninth Army in Syria and Palestine, and was honorary *Aide-de-Camp* General to the King.

18. General Sir Neil Ritchie (1887–1983), the Black Watch, now commanded the Eighth Army, as Cunningham's successor. In Chink's view, Ritchie was unqualified ('an Amateur') and, as such, the rival whose promotion he resented most.

19. Auchinleck.

20. Ritchie.

21. 4th and 10th Indian Infantry Divisions.

22. South African troops included the 1st and 2nd Infantry Divisions in North Africa.

23. The local hot, dry and dusty wind that blows from the south or south east.

24. Posh and snobbish.

25. Major-General John 'Jock' Campbell (1894 -1942) Royal Horse Artillery, had just been given command of 7th Armoured Brigade following Gott's promotion to lead the XIII Corps. He had won a VC 'for his magnificent example and utter disregard of personal danger' during Crusader the previous November. In 1940 those same characteristics at the head of his mobile, all-arms flying column had led to the nickname 'Jock Columns'.

26. Air Vice Marshal Arthur Coningham (1895 -1948) commanded Air Headquarters Western Desert. (Air Chief Marshal Sir Arthur Tedder was head of RAF Middle East Command.)

27. Lt General Charles Willoughby Norrie (1893–1977), 11th Hussars, commanded 30th Corps.

28. Lt General Thomas 'Tom' Corbett (1888–1981) had joined the 9th Hodson's Horse in the Indian Army after Sandhurst. Until this diary entry, he had been commander of IV Corps in Iraq.

29. Director of Military Training.

30. Chief of the General Staff.

31. Deputy Chief of the General Staff.

32. General Headquarters, Middle East, in Cairo.

33. Condemning the practice of armoured divisions operating under separate command, his proposals included composite battle groups on the German model. Lighter vehicles to be weeded out from tanks for a mobile Light Armoured Division. Unsound static lines, fortified by isolated therefore vulnerable defensive boxes, tied up brigade groups which fragmented their artillery. Concentration of artillery was essential, and control of artillery had to be centralised. 'Here I am with a head running wild with ideas for reshaping this army because we have got to reshuffle it and its weapons before we find the right tactical combination.'

34. Sir James Grigg (1890–1964), whom both Auchinleck and Chink had known and liked in New Delhi during Grigg's Civil Service posting as Finance Member with the Government of India. He had been recalled to London in 1939.

35. South African Major Gen Frank Theron, General Officer Administration, Union Defence Force (UDF), Middle East Forces.

36. General Sir Harold Alexander (1891–1969) was categorised by Chink, after previous meetings in their small world, as an amateur and a snob. Before Burma, Alexander had commanded the 1st Division in France from 1939. On France's surrender in May 1940, as commander of the I Corps, he had directed the evacuation of British and French troops from Dunkirk.

37. Germany, Italy and Japan.

38. Chink's youngest brother.

39. During the surrender of Hong Kong, Japanese soldiers had burst into St Stephen's College, a temporary front-line British hospital, killing doctors who tried to stop them and

bayoneting patients. A second wave had killed survivors. and gang-raped and mutilated the nurses. Over a hundred, Canadians in particular, had died in the massacre.

40. Eden's full speech as Secretary of State for Foreign Affairs on 10 March 1942 is on the Hansard website.
41. Chink's plan for a Higher Command Staff School in South Africa (the Union).
42. Lt General Sir Alexander 'Sandy' Galloway (1895–1977), the Cameronians, was Auchinleck's Deputy Chief General Staff (DCGS) at GHQ Mid East.
43. His brother, 14 months younger than Chink.
44. Bir Hacheim, an old Ottoman fortress and oasis in the Libyan desert, 80 km south of Tobruk.
45. Gazala was on the coast, 48 km west of Tobruk. The Gazala Line was a series of defensive boxes laid out across the desert behind minefields and wire, each holding (roughly) a brigade.
46. Prince Henry, Duke of Gloucester, King George V's third son, had been sent on a four-month military and diplomatic mission to the Middle East, India and East Africa.
47. Tactically important high inland desert track.
48. A valley in the north-western Nile delta of Egypt
49. Australian Richard Casey, Lord Casey (1890–1976) was Churchill's recent appointment as Minister Resident in the Middle East.
50. The needs of a military force excluding manpower – supplies, equipment, weapons.
51. Mena, on the outskirts of Cairo overlooking the pyramids of Giza, was a tented and hutted military town. Conditions bore no relation to life in an active service unit in the desert.
52. Cairo's leading hotel. Popular with the military, its bar was known as 'the long bar' for being so crowded that impatient waiting was compulsory. 'Wait until Rommel gets to Shepheards' ran a 1942 joke. 'That will hold him up!'
53. Miles Lampson (1880–1964), the British Ambassador to Egypt, and his wife Jacqueline (1910–2015).
54. The Western Desert.
55. Considered to be the Diplomatic Club in Cairo, the Mohammad Ali offered fine dining but was less fashionable and popular in military circles than the Turf Club.
56. The exclusive Gezira Sporting Club on Cairo's island lived up to its name, offering a wide range of activities for those wanting to keep fit, from swimming and tennis to polo. There was even the occasional wartime race meeting.
57. Major-General H.B Klopper, who had been appointed commander of the 2nd South African Division holding Tobruk as recently as 14 May that year.
58. German intelligence signals, forwarded by code-breaking Bletchley Park.

The First Battle of El Alamein: June–July 1942

1. Brigadier John 'Jock' Whiteley (1896–1970), Royal Engineers, currently Auchinleck's choice inherited from Wavell, as Deputy Chief General Staff/ at GHQ ME in Cairo.
2. Marshal Rodolfo Graziani (1882–1955), loyal to Mussolini. He had commanded the Italian 10th Army until defeat by O'Connor in Operation Compass in 1940, and been replaced in March 1941.
3. What would become known as the Battle of Mersa Matruh had been lost that day, after four days of heavy fighting. Although six thousand troops had fallen into Rommel's hands, as well as a great quantity of supplies and equipment, owing to the foresight of retreat, the Eighth Army had survived.
4. A reproof, punishment, or imminent penalty.
5. Beda Fomm was a small coastal town in Cyrenaica, Libya, between the larger port of Benghazi to its west and the large town of el Agheila to the south-west. It had been the final engagement of Operation Compass in 1940.
6. The Battle of Crécy, during the Hundred Years' War, took place on 26 August 1346 in north-east France between the French army, commanded by King Philip VI, and a greatly

outnumbered English army led by King Edward III, fighting a defensive battle but armed with deadly longbows. Victory went to England, with heavy loss of life for the French.

7. Slang for Egyptian.

8. The motion to censure Churchill, stemming from British losses in North Africa, took place on 2 July 1942. It was defeated by 476 votes to 25.

9. Brigadier Frederick Herman Kisch (1888–1943), commanding officer of the Royal Engineers in the North African campaign, was admired by Chink after working closely with him on the design of defence plans at First Alamein in June/July 1942. 'Here for the first time', wrote the BBC war correspondent Denis Johnson, 'we had the spectacle of an ace German general not only being outfought, but outfoxed.' Kisch was killed in action the following year.

10. During the run of British disasters at the Battle of Gazala.

11. An Appreciation is the setting down, under seven headings, of arguments and considerations before arriving at a military operation decision.

12. Prince Henry, Duke of Gloucester.

13. An oppressively hot, dry, dusty south or south-east wind from the Sahara Desert, blowing most frequently between April and June.

14. Freddie de Guingand.

15. Deputy Chief of the General Staff Operations, Plans and Intelligence.

16. In layman's terms, an Appreciation must consider: 1) The objective. 2) Positions and strength of your own forces. 3) Estimation of the enemy's forces. 4) Trial balance of your advantages, and of his advantages. 5) Things the enemy may do to increase his advantages and minimise yours. 6) Various things you can do to increase your chances and minimise his. 7) Definite proposal of how to achieve your objective. Exact wording of the long Appreciation which Chink wrote that day can be found in *The Desert Generals* (Barnett) and *Chink* (Greacen).

17. Chief of the Imperial General Staff.

18. To 'Levant' is to run for cover. From the French *faire voile en Levant*: set the sail with the Levant, an easterly wind that blows in the Mediterranean.

19. Delays by smart cavalry regiments (the armour), with their separate command structure.

20. Where Auchinleck was due to meet Brooke on arrival.

21. Churchill.

22. Wavell.

23. Churchill, Brooke, and Wavell.

24. [Arthur Smith's successor at GHQ Cairo.]

25. Brooke, who was the CIGS.

26. He was well-aware of Brooke's dislike and distrust, with power (for or against) at his disposal.

27. Roughly, the proposal was to create mobile divisions of two motorized infantry brigades and one armoured brigade, mixing armour, trucked infantry and guns, which in fact Brooke was firmly against. (It would become the standard NATO pattern.)

28. Major General John Harding (1896–1989) of the Somerset Light Infantry, was at that time Deputy Director of Military Training, Middle East Command. Chink and he knew each other well.

29. A military transport plane.

30. Brooke.

31. [Churchill]

32. Gott had already been chosen as Army Commander, although that was not yet public knowledge.

33. The potentially overstretched Iraq-Persia Command.

34. Dismissal.

35. On June 24 1942 Rommel's forces had been poised to drive deep into Egypt, confident that the battle was as good as won. 'We are at this moment,' Churchill told the House of Commons, 'in the presence of a recession from hopes ... unequalled since the fall of France.'

36. An inter-service committee to study the strategically flawed proposal of splitting the Middle East command: 1) Egypt, Palestine and Syria. 2) Iraq and Persia.

37. A remote town in the mountainous region of northern India, up near the Khyber Pass.

38. The inter-service committee examining the division of the Middle East command had not yet produced its report.

39. Brooke.

40. Rommel, and his Afrika Korps.

41. At the news of Gott's death, Bernard Law Montgomery (1887–1976), whom Chink had detested at the Staff College when 'Monty' was the senior lecturer there, was Brooke's next choice for Commander of the Eighth Army.

42. Freddie de Guingand had been appointed on Chink's recommendation in February 1942 to the vacant post of Director of Military Intelligence, Middle East. He would be working for Montgomery now, if retained.

43. Richard Casey, Minister of State for the Middle East.

44. Baghdad. The trip did not take place.

45. General Sir Oliver Leese (1894–1978) the Coldstream Guards, had been a friend at the Staff College, when Alexander (P3) had also known them both. In suspecting imminent 'Guards' fraternity, Chink was partially right, but it was Monty who gave Leese command of the Eighth Army's XXX Corps in September, 1942.

46. Lieutenant General Sir Richard McCreery (1898–1967), 12th Lancers, had been rated 'a dolt' by Chink when both were fellow students at the Staff College. Chink now saw him as Brooke's protégé, and had witnessed his very recent dismissal by Auchinleck for refusing to merge cavalry and infantry divisions. The antagonism was mutual.

47. Alexander was an old friend of Chink's wife, Estelle, and had worked with Reggie lately when C-in-C of Land Forces in Burma.

48. The British contingent for Churchill's meeting with Stalin in Moscow.

49. Wavell had lost an eye when wounded in action in 1915, on the same day and in the same area that Chink had won his MC.

50. Alfred Duff Cooper (1890–1954), Conservative Party politician and diplomat. Reggie had arranged a private meeting in 1935, when Cooper was Secretary of State for War, for Chink to warn him of the danger of a German *blitzkrieg* across the Low Countries.

51. Leslie Hore Belisha (1893–1947), Liberal MP and Secretary of State for War until 1940. Chamberlain had removed him for ruffling senior army feathers.

52. [Royal Automobile Club]

53. When DMT in India, Chink and Estelle had often dined with Grigg and his intelligent wife as a foursome, enjoying their congenial company.

54. He had met Amery in 1925 on a Swiss mountaineering holiday, but had not talked to him since. Staying in the same hotel then, they had climbed regularly together, joined on the last day by Amery's young son, governess in tow. Chink had called the summit 'Pitz Julian' in the boy's honour. Ironically on 3 July 1942 it was the now-adult Julian Amery's urging, on leave from serving in the ME, for a 'personal visit' by the PM that caused Churchill to join Brooke's trip.

55. Chink had worked well under 'Jumbo' Wilson, then Commander of British Troops in Egypt, during Dick O'Connor's Libyan triumphs almost two years earlier. But since then Wilson had been on the powerful Selection Board, and, although they socialised, Chink concluded that Wilson had deliberately ignored him for promotion.

England: October 1942–March 1944

1. Ironically used, because the cenotaph in Whitehall commemorates the dead in British wars.

2. Lieutenant General Frederick 'Boy' Browning (1896–1965), Grenadier Guards, whose wife was the novelist Daphne du Maurier (1907–89). Both were old friends. Chink had admired

'Boy' since teaching with him at Sandhurst in 1924, and had encouraged Daphne to marry him. Since November 1941 Browning had been GOC of the new 1st Airborne Division.

3. Brigade Major at Welsh 160th Brigade HQ in the Tudor house Provender, outside Sittingbourne in Kent. Rex Cohen was a civilized wartime soldier who hoped to return to his world of high finance. He would prove to be understanding and a good friend to Chink.

4. Lord Woolton (1883–1964), managing director of Lewis's department store in Liverpool, was Minister of Food, despite being non-political. Rationing was in full force, and his popular 'Woolton Pie' used carrots, parsnips, potatoes and turnips with a pastry or potato crust, served with brown gravy.

5. Where Liddell Hart – as under-utilised and unpopular with the army hierarchy as Chink – was living.

6. Sir Richard Stafford Cripps (1889–1952), Labour MP until expelled pre-war for his left-wing views, had been appointed Ambassador to the Soviet Union by Churchill in 1940. After entry of the USSR into the war he had returned to London, and had recently been appointed Minister of Aircraft Production.

7. At last the Allies had gained the advantage in the Mediterranean, evidenced that month by a British convoy reaching Malta undisturbed.

8. Permanent Under-Secretary, who would turn out to be a colleague from India, Archie Rowlands.

9. Gen Montague 'Monty' Stopford (1892–1971) the Rifle Brigade, was GOC of the 56th (London) Infantry Division. A first-line Territorial Army formation serving in Kent, it was one of three divisions in XII Corps.

10. Sir Archibald Rowlands (1892–1953) was a senior civil servant. Seconded to the Indian government in the 1930s as an advisor on Military Finance, he had been taught about Army needs by Chink, and was now Permanent Secretary to the Minister of Air Production.

11. Yet in his India days as DMT he and Estelle had socialised with Rowlands and his wife.

12. 'And yet it moves.' Attributed to Galileo after being forced to recant his claim that the earth moves around the sun.

13. Churchill.

14. Francois Darlan (1881–1942), who had been Chief of Staff of the French Navy in 1939 and joined the pro-German Vichy regime in 1940. The highest-ranking Vichy officer in North Africa in 1942, Darlan had made a deal for control of French forces there in exchange for joining the Allies. Within two months he had been assassinated.

15. In fact, the drink may have loosened Chink's tongue, heightened by his personal dislike of Churchill, so much so that in one opinion his arrogance towards the end of lunch almost amounted to disloyalty to Auchinleck.

16. Lines of Communication.

17. George Goschen, 2nd Viscount Goschen (1866–1952), had been Governor of Madras in the 1920s, standing in as Viceroy at one point. Back in England, he had been president of a group in favour of Indian federation, known as the Union of Britain and India.

18. 30 January 1943, the tenth anniversary of the Nazi regime taking power, should have been a celebration. Instead, it was marked by long-range daytime RAF bombing raids on Berlin, and looming defeat at Stalingrad could no longer be ignored. Goering's speech in a radio broadcast that evening, included: 'Every German will one day speak of a Stalingrad of sacrifice … as the rule of honour and the conduct of war have ordained that we must do, for Germany's sake. It may sound harsh to say that a soldier has to lay down his life at Stalingrad, in the deserts of Africa … but if we soldiers are not prepared to risk our lives, then we would do better to get ourselves to a monastery.'

19. Major General J.F.C. 'Boney' Fuller (1878–1966), Oxfordshire Light Infantry, who had been Chief Instructor when Chink was at the Staff College. Fuller's experience in the 1914–18

war had shown him the need for mechanisation and modernisation. He had retired in 1933 to convey that urgent message, often writing in collaboration with Liddell Hart.

20. Exercise Spartan was the try-out for the invasion of the Continent. It turned out to be limited, as Chink had predicted, by its format of a bridgehead.

21. Monty's Eighth Army had outflanked Axis defences in late March on the Mareth Defensive Line in Tunisia, forcing them into further retreat.

22. The Eighth Army was now in a diversionary role in an operation intended to end the Tunisian campaign. But being on the right of the US main assault by US II Corps, they were making only limited progress at a heavy cost.

23. In bullfighting, a bull may stake out his *querencia*, where he feels strong and safe.

24. Wavell deployed two monocles: one to read through, and the other to camouflage the eye lost in action in World War I. Incidentally, the Military Secretary who personally phoned Chink about the meeting was General Sir Henry Colville Wemyss.

25. Field Marshal.

26. Auxiliary Territorial Service, the womens branch of the Army.

27. A result of his general reorganisation suggestions, forwarded by Rowlands.

28. Railway carriage of the train to London.

29. The Soviet counter-offensive earlier that year on the Eastern Front (24 January – 17 February) had recaptured the city of Voronezh, taken by the German 4th Panzer Army the previous summer.

30. WASP was an acronym for 'White Anglo-Saxon Protestant,' originally described as being from the elite families.

31. Women's Auxiliary Air Force.

32. Herbert Stanley Morrison (1888–1965), a senior Labour MP, was Home Secretary in the wartime coalition.

33. Colonel 'Boney' Fuller, nicknamed Boney because he looked like Napoleon.

34. Major Gen. Robert Ross (1893–1951) was GOC of the 53rd Division, and described by Chink as 'a sound, unimaginative, regimental soldier'.

35. Ritchie, who had been replaced as Eighth Army commander by Auchinleck on 24 June 1942, was due to take over command of 12th Corps on 27 November 1943.

36. General Sir Bernard Paget (1887–1961) was GOC of GHQ Home Forces, and usually respected by Chink for 'constructive criticism' having been taught by him at the Staff College. Paget had instilled his regular habit of analysing military problems through an Appreciation. 'He showed me how to arrange my thoughts into meaning.' He would continue to praise him, once he had simmered down.

37. General Sir Henry Colville Wemyss (1891–1959), Royal Engineers and Royal Corps of Signals. As Military Secretary to the Secretary of State for War, Wemyss was responsible for policy direction on personnel management.

38. The report could only be answered in a single word: either Yes or No.

39. A soldier mentioned in the Bible as having been sent into battle by his king to be conveniently killed.

Anzio: April–August 1944

1. Harold Nicolson (1886–1968) diplomat, diarist and Labour MP, had been Parliamentary Secretary and official Censor at the Ministry of Information in Churchill's 1940 government for a year. He was now a respected backbencher on foreign policy issues, and a governor of the BBC.

2. Wilson had succeeded Eisenhower as Supreme Allied Commander in the Mediterranean in January, and was based at Allied Forces HQ in Algiers.

3. Général Henri Honoré Giraud (1879–1949) had cooperated with the Allies after his adventurous escape from German capture in 1942.

4. Major General Ronald Penney (1896–1964), Royal Corps of Signals, had been Chink's contemporary at the Staff College in 1927/28. Recently he had been Signal Officer-in-Chief in the Middle East under Wavell, Auchinleck and Alexander. Appointed GOC of 1st Infantry Division in late 1943, Penney had been wounded by shellfire at Anzio, and was awaiting sick leave.

5. Combined Military Force.

6. The King's Shropshire Light Infantry.

7. Once pinned down, the 1st Division had drawn the worst of the fighting, and Chink's brigade had been driven up into the German cordon with horrific casualties.

8. When DMT in India at the outset of war, Chink had held 'sharpish sessions' with Careless, then Cassell's loyal *aide-de-camp*.

9. Chink's stepdaughter, from Estelle's previous marriage.

10. Chink did not confide that he'd heard from Estelle's letters that the marriage was in deep trouble.

11. Flak, or *Fliegerabwehrkanone*, enabled artillery class cannons to be used for Anti-Air duties. Timed fuses in the shells detonated bombs in the air, igniting the high explosive compound, and the shells contained ball bearings, fragment shards, phosphorous or even worse.

12. Stalin wanted Churchill and Roosevelt to open a second front of an Anglo-American landing in France, Belgium or Holland, which would force Germany to withdraw troops from the Eastern front. The Allies had differed about priorities, but finally agreed on a major offensive in spring 1944.

13. After their two years of training for the Allied invasion of France, 160th Brigade (in 53rd Division), reached Normandy in June as part of Operation Overlord, and soon faced fierce fighting around Caen. Montgomery was the Commander of 21st Army Group, which comprised all land forces.

14. Lieutenant General Sir John Hawkesworth (1893–1945) was GOC of 46th Infantry Division, a Territorial Army formation which had fought at Monte Cassino, sustaining heavy losses. He was taking temporary command of the 1st Infantry Division at Anzio in Penney's absence.

15. The Normandy Landings, which began on D-Day 6 June 1944.

16. Still formidably held by Axis forces.

17. Wynford Vaughan-Thomas (1908–87), a proud Welshman with a distinctive radio voice, was one of the BBC's most distinguished war correspondents.

18. Uriah the Hittite, according to the biblical Old Testament, had been sent into battle to be killed because King David lusted after his wife, Bathsheba.

19. High Explosive.

20. Major Henry Leask (1913–2004). Royal Scots Fusiliers. As brigade major with the 3rd Infantry Brigade, Leask's conspicuous ability at Anzio and coolness in action after the breakout would win him an MBE.

21. *Generalfeldmarschall* Albert Kesselring (1885–1960) was Commander-in-Chief, South (a joint command) and also of Army Group C (an army command), covering all German forces in Italy. Hitler respected Kesselring's political idealism and military optimism.

22. The Gustav Line, built by Kesselring, was a defensive line drawn across central Italy just south of Rome. It had to be broken by Allied forces before Rome could be taken. But by attacking it, the Allies also hoped to force Germany to commit more German troops to Italy, easing conditions in France for the expected D-Day landings.

23. Scrubland vegetation of the Mediterranean region.

24. The river Liri valley was a long flat corridor through miles of rugged mountains, with Monte Cassino at one end and Rome at the other. The last German defences in the Liri valley would be penetrated before the end of May, leading to the collapse of the Gustav Line.

25. Weakness of command, which would result in a negative report and his demotion.

26. Montgomery's Eighth Army pursuit of the enemy to the Sangro river in November 1943 had been hindered by bad weather. Vehicles and tanks were bogged down in their descent of the valley.

27. Chink's Staff College contemporary 'Dreary' McCreery was now in command of X Corp, part of the US Fifth Army in the Italian campaign under General Mark Clark.

28. Radio Transmission

29. Codename.

30. The DWR and Foresters on the right of the attack bore the brunt of casualties, not the KSLI on the left as stated in the biography, Chink. Subsequently a rumour spread that an overnight deserter accounted for the greater DWR and Forester losses.

31. Canadian Brigadier Charles Loewen (1900–86), Royal Field Artillery, had recently been awarded a CBE in the New Year's Honours List. He would be appointed GOC of the 1st Infantry Division in July 1944.

32. Unnamed. It could be Careless, Webb-Carter or Hackett.

33. Cisterna's capture was crucial now that Axis forces were retreating north, and had been fiercely defended by the Germans in January. On 25 May press reports confirmed that a US armoured column had smashed through the German flank on the Rome side of Cisterna. Its fall, at last, was 'imminent'.

34. The town of Littoria had been built in 1932 during the Fascist government's drive to reclaim the unhealthy Pontine Marshes by drainage and the clearance of vegetation.

35. Operation Overlord, postponed in May, would begin on D-Day, 6 June 1944.

36. General Mark W. Clark (1896–1984), the controversial commander of the US Fifth Army.

37. The enemy were being forced to retire north to avoid being trapped in the Liri valley.

38. The BBC war correspondent, whom Chink had met before the breakout.

39. Part of the chain of the Alban hills, between Anzio and Rome. It had caught the imaginations of Byron, Goethe and Stendhal, among a host of writers and artists.

40. Brigade Major.

41. During his convalescent role as Commandant of the British troops in Milan, where he and Hemingway had met on Armistice Day 1918. At the end of January 1919 Chink had been posted farther from Rome, to Taranto.

42. The infamous London prison near the Old Bailey, demolished in 1904. By then 1,169 prisoners had been executed there.

43. After the Japanese attack on Pearl Harbor his 18 year old godson Bumby, known to friends now as Jack Hemingway and scarcely seen by Chink since his parents' divorce, had enlisted in the US Army. He wore the uniform that so offended Chink, and was serving in France that year. (Captured in October, he subsequently became a POW in Germany.)

44. Distinguished Service Order.

45. About the Normandy landings, begun four days earlier.

46. From the 1882 comic novel *Vice Versa* by F. Anstey (1856–1934), a book Chink had loved as a boy. The name explains the plot and Chink's comparison. When City businessman Mr Bultitude's young son Dick cringes at his return to a cruel boarding school, Bultitude snaps that schooldays are the best years of life, and says he wishes that he were the one going. Thanks to a magic stone, this happens, and simultaneously Dick swaps roles with his father. Both are restored to their own bodies in the end, newly aware of the problems of each other's world.

47. Vittorio, in the north-east of Italy, had in November 1918 been the setting for the famous battle now known as Vittorio Veneto. At that time, Chink had just been discharged from hospital in Genoa to an easy billet in Milan as Commandant of the British Troops.

48. A poignant comment, in retrospect.

49. Observation Post.

50. Castle.

51. Count Vaselli.
52. General Sir Colville Wemyss (1891–1959).
53. After a quiet year following his unhappy experiences in Cairo, Auchinleck had been appointed to his old role of C in C of the Indian Army in June 1943. (This was possible because Wavell had since been appointed Viceroy.) Prosecution of the war with Japan moved to a newly created South East Asia Command under Mountbatten in November 1943. Auchinleck's responsibility included internal security, North-West Frontier defence, and reorganisation and training of the Indian Army. His wife having remarried, he invited Chink's stepdaugher Elaine to be his official hostess after she was widowed, taking her and her two children under his wing.
54. Catholics, like Chink.
55. Lieutenant General Sir Charles (Allfrey 1895–1964), Royal Artillery. Unpopular with Alexander and Montgomery, kindly Allfrey would be severely criticised himself as an amateur by Brooke.
56. Ostia is at the mouth of the river Tiber, near Rome. Being on the coast, it is cooler in summer and warmer in winter.
57. Josip Broz Tito (1882–1980), leader of the partisans in Yugoslavia. As the likelihood of an Allied invasion in the Balkans built, Axis attacks on Tito's partisans increased, leading to official recognition by Churchill and Stalin, plus Tito's help behind Axis lines.
58. Dick O'Connor was no longer a prisoner of war. He had escaped (succeeding on his third attempt) in September 1943, and on his return to England had been knighted, backdated to 1941, and promoted. Montgomery had wanted O'Connor to succeed him as Eighth Army commander, but O'Connor had been given VIII Corps instead, now part of the Normandy Landings.
59. A four-monthly report was necessary for command of an operational brigade before the next step of command of a division could be decided.
60. Ethnic Germans from Poland.
61. Part of the 185th Paratroopers Division – 'folgore' means lightning. So tough was their reputation for fighting to the end that survivors did not have to raise their hands or show a white flag in surrender.
62. Hundredweight.
63. The puns were too good to resist. After intense house-to-house fighting which left the old city razed to the ground, Brest ultimately surrendered to the US VIII Corps on 19 September 1944.
64. From *The Lays of Ancient Rome* by Thomas Babington Macaulay (1800–59), the stirring narrative poems so popular with public schoolboys of Chink's vintage that he still knew many by heart.
65. V-1 Flying bombs, better known as 'doodlebugs' or 'buzz' bombs. Of the 10,000 fired at Britain during the war, 2,419 hit London, killing 6,184 and injuring 17,981.
66. The failed assassination attempt by Claus von Stauffenberg and fellow officers, including Rommel, to kill Hitler had happened the previous day at the Wolf's Lair field HQ in East Prussia. Chink's 'Party v *Wehrmacht*' speculation focused on Hitler's prompt broadcast threat: 'No military installation, no commander of a unit, no soldier is to obey any order by these usurpers. [They] should be shot on the spot.'
67. On the downfall of Mussolini in July 1943 Marshal Pietro Bardoglio had became Prime Minister of Italy. His armistice with the Allies had led to unconditional surrender, and to Italy declaring war on Germany. Badoglio had recently resigned and left politics.
68. Allied Military Government for Occupied Territories.
69. Among those executed for the plot would be 3 field marshals, 19 generals, and 26 colonels. Lieutenant General Paul von Hase was in the first eight hanged with piano wire suspended from meat hooks, with his agonising death filmed for Hitler. Rommel's known popularity

with the public meant that on Hitler's orders he was to commit suicide by poison, which took place on 14 October 1944.

70. The senior civil servant friend from Chink's time in India whom he had contacted in London, Rowlands was now Permanent Secretary to the Minister of Air Production.

71. Winston Spencer Churchill.

72. German soldiers.

73. Brest Litovsk, where the historic peace treaty ending Russia's participation in World War I had been signed with Germany, Austria-Hungary, Bulgaria and the Ottoman Empire, was taken by Soviet troops on 28 July 1944.

74. A biblical canticle taken from Luke II, verses 29–32. Translated from the Latin: 'Lord, now lettest thou thy servant depart in peace.'

75. Heinrich Himmler (1900–45) whose authority and powers increased after the 20 July plot. He succeeded the implicated General Fromm as C in C of the two-million-strong Reserve Army, and began merging army officer recruitment with the *Waffen-SS*. As Minister of the Interior and *Generalbevollmächtigter für die Verwaltung*, he was also in charge of prisoners of war and the police service, and controlled the development of *Wehrmacht* armaments.

76. Joseph Goebbels (1897–1945) had been appointed Plenipotentiary for Total War (his brainchild) on 23 July 1944, and was charged with maximising both the *Wehrmacht* and the German armaments industry manpower. His efforts gained half a million more men for military service.

77. Receiving the news coldly of his son-in-law's death, Chink gave Careless the impression, as the latter would testify in due course, that 'he couldn't care less'. Chink chose not to reveal his knowledge from Estelle's latest letter that Bols had left Elaine for someone else, leaving the young family with his huge gambling debts. He loathed Bols as a result.

78. Charles Loewen.

79. Lieutenant General Sir Sidney Kirkman (1895–1982), Royal Artillery, was GOC XIII Corps. He had been specifically requested for that command by Chink's erstwhile Staff College contemporary Oliver Leese, now GOC Eighth Army.

80. Lieut. General Sir Oliver Leese, 3rd baronet (1894–1978), Coldstream Guards, had succeeded Montgomery at the head of Eighth Army in late December 1943. Chink had liked Leese personally, despite rating him an amateur, during their two years spent in close proximity almost 20 years earlier, and he banked on getting a fair hearing.

81. Originally used to qualify those horses whose bloodlines had never been mixed with other species, but in this case to mock pure 'blue' blood.

82. *Journey's End* by R.C, Sherriff is a play first performed in 1928. Set in the claustrophobic Officers' quarters of the British trenches of St Quentin, France, in 1918, it features s group of young infantry officers, one of whom is already mentally disintegrating. A major attack from the German trenches less than a hundred yards away is expected, and tension builds when an order comes for a raid on the enemy. The men know it will risk their lives, and is futile.

83. A wry comparison of Anzio with the 1565 landing on the west coast of America (now Florida) by a band of Spanish settlers to found a mission to St. Augustine. Subsequent converts were most numerous in a local village known as Nombre de dios.

84. Tom Wintringham, (1898–1949), military historian, Marxist, politician and author, had fought in World War I and the Spanish Civil War. Rejected for an officer's commission in World War II, he set up a private Home Guard training school near London, only to be sidelined when it was taken over by the army. Vocal and energetic, he founded the left-wing Commonwealth Party in 1943 and got 48 per cent of the vote in a previously safe Tory seat.

Civvy Street: 1945–1958

1. Founded in 1772, the *Morning Post* had merged with the *Daily Telegraph* in 1937. A conservative newspaper, Chink's hopes of its public support were optimistic, if not naive.
2. From the deep, out of the depths.
3. Éamon de Valera.
4. Clann na Poblachta was a small, radical, republican political party, founded by Seán MacBride in 1946 to challenge Éamon de Valera's entrenched Fianna Fáil. It had grown rapidly in 1947 and in February 1948 joined the new coalition for the First Inter-Party Government which ousted de Valera. MacBride, son of Maud Gonne and a former Chief of Staff of the Irish Republican Army, was appointed Minister for External Affairs.
5. *Across the River and Into the Trees.*
6. At Sidi Barrani in November 1940, which led to O'Connor's successful follow up.
7. Flavius Belisarius (c. 500-565) was a famous military commander under Emperor Justinian in the Byzantine Empire. He is now considered outstanding for his innovative battle tactics and unconventional strategy.
8. In *Across the River and Into the Trees.*
9. In the case he was now thinking about taking against Churchill over *The Hinge of Fate*, the fourth volume of his *History of the Second World War*. The implication there was that the removal of Auchinleck and Chink from the Eighth Army had been fully justified.
10. Hugh Morrow (1915–76) editor of the *Saturday Evening Post*, had served with the US Navy during the war. Chink had begun an affair with Elise Morrow, Hugh's wife and fellow journalist, on his US tour the previous year. He corresponded with them together and separately.
11. Sir Hartley Shawcross (1902–2003), barrister and Labour politician, was well-known for being the leading British prosecutor at the postwar Nuremberg War Crimes Trial. He had been president of the Board of Trade until Labour's defeat in the 1951 general election, but was tiring of party politics. Admiration for Churchill's staunch public leadership during the war enabled him to be legal advisor to the leader of the Conservative Party. Aneurin Bevan wittily named him Sir Shortly Floorcross.
12. The intriguing case is the subject of a monograph being prepared by barrister Charles Lysaght, author of the acclaimed biography of Brendan Bracken (London: Allen Lane, 1979 and Kindle edition 2022).
13. Reggie had been elected in 1939, and membership of the Privy Council is conferred for life.
14. Malaria was spread by mosquitoes from the nearby overgrown Pontine Marshes.
15. Montgomery's wartime practice of colouring reports to entirely favour himself was by now wellknown. Churchill had drawn on these for *The Hinge of Fate*, the fourth volume of his popular History of the Second World War, over which Chink had threatened proceedings for libel in 1953 on behalf of Auchinleck, as well as himself.
16. Churchill is always right.
17. Gaelic spelling for Chink's son Christopher (born 1946) and daughter Rionagh (b. 1948).
18. In the early 1920s, while drunkenly walking back to their Swiss *pension* along the little Chamby to Aigle road, they had argued about the best description of the chestnut blossoms overhead. 'Waxen candelabra' had finally been agreed.
19. In 1954 Hemingway had been awarded the Nobel Prize in Literature for his 'mastery of the art of narrative [and] for the influence that he has exerted on contemporary style.'
20. A Swiss winter resort, developed for skiing since the 19th century.
21. The Chief, a Memoir of Fathers and Sons, by Lance Morrow, Random House 1984. The description of Morrow's visit to Chink is in Chapter 9 Breaking Away, pp 144-150. Chink's love letters to Elise Morrow are in the possession of the Morrow family.
22. Bradley (1893-1981) was a 5 star General of the US Armed Forces who served in both World Wars. Following a popular news story in 1944, he was known as 'the GI's general'.

Fighting Back: 1958–1969

1. Correlli Barnett (1927–2022) had graduated from Exeter College, Oxford in Modern History, his special subject being Military History and the Theory of War. He rated Clausewitz's *On War* as the most influential book on his course; coincidentally, Chink had discovered Clausewitz at much the same age. Barnett's National Service experience had been with the Army Intelligence Corps in Palestine.
2. Gatehouse and Galloway were erstwhile contemporaries of Chink, and like-minded colleagues when working with him on O'Connor's team in 1940.
3. *The Memoirs of Field Marshal the Viscount Montgomery of Alamein KG*, published by Collins that autumn in time for the Christmas market.
4. A wry dig at his old friend Brian Horrocks (1895–1985), known since their schooldays at Uppingham together, who had reached the rank of Lieutenant General, been knighted, and secured the ultimate Establishment approval. Since 1949 Horrocks had been Gentleman Usher of the Black Rod in Parliament's House of Lords.
5. Auchinleck, increasingly disbelieving and angry, had just finished reading Montgomery's *Memoirs*, to which Freddie de Guingand had contributed. No longer fatalistic, he would obtain a settlement.
6. If Auchinleck decided to sue Montgomery over the slur in his memoirs that Auchinleck was intending to retreat to the Delta in the summer of 1942.
7. Colonel Ian Jacob (1899–1993), Royal Engineers, had been Military Assistant Secretary to the war cabinet at the time of First Alamein, accompanying Churchill and Brooke on the flight to Cairo. He considered Chink vain, and had disliked him strongly since the '1000 out of 1000' reputation at the Staff College. In a change of role, since 1952 Jacob had been Director General of the BBC.
8. Robert Boothby MP (1900–86) had been made a life peer (Baron Boothby) the previous year, triggering a by-election. His interest in war was personal but eclectic, having been commissioned into the Brigade of Guards in 1918, too young to see active service, and, after a short term in Churchill's war cabinet, joining the RAF Volunteer Reserve. He had served as a junior staff officer with Bomber Command, and then as a liaison officer with the Free French Forces.
9. John Connell, the military historian, had published a biography of Auchinleck in 1959.
10. South African Frank Theron had become quite friendly with Chink in Cairo during the eventful summer of 1942.
11. Colonel Ian Jacob, who had accompanied Churchill and Brooke to Cairo in August 1942.
12. Respectfully promoted as being by Lieutenant General Sir Brian Horrocks, *A Full Life* was published by Collins in 1960.
13. Two bestselling scholarly books – promptly banned in Ireland – from the American Kinsey Institute for Research in Sex, Gender and Reproduction, founded by zoologist Alfred Kinsey. *Sexual Behaviour in the Human Male* (1948) was followed by *Sexual Behaviour in the Human Female* (1953) and they were jointly referred to on both sides of the Atlantic as The Kinsey Report.
14. *The Memoirs of Lord Ismay*, published by Heinemann in 1960.
15. John Profumo (1916–2006) had served with the Royal Armoured Corps during the war, including a stint in North Africa with the acting rank of major, where he was mentioned in despatches. A well-connected Conservative MP with a good war record, Profumo was new to the role of Secretary of State for War, having been appointed in July, only three months earlier.
16. After Anzio, following Loewen's tip-off about Penney.
17. In 1894 Captain Alfred Dreyfus, a French artillery officer of Jewish descent, was convicted of treason for – allegedly – communicating French military secrets to the German Embassy

in Paris, and was sentenced to life imprisonment on remote Devil's Island. The scandal, known as The Dreyfus Affair, dominated the press when, within a year, evidence came to light of the real culprit and was suppressed by the Army. Accusations of antisemitism and rightwing nationalism abounded. In 1898 the case was taken up by the leading French novelist Émile Zola in a powerful front page open letter to the Président, entitled 'J'Accuse', which stiffened public opinion. Dreyfus was eventually exonerated in 1906.

18. General Sir Hubert Gough (1870–1963), 16th Lancers, Commander of the Fifth Army 1916–18, had faced public disgrace based on wrong information about retreat, after the war. Widespread press indignation had led to his military rehabilitation and a letter from the Secretary of State for War, Winston Churchill, praising the Fifth Army's gallant fight and promising that Gough would be considered for command 'appropriate to [his] rank'. Gough's anger had been more for the men of Fifth Army than for himself.

19. Richard 'Dick' Crossman (1907–74), left-wing Labour MP, party spokesman on Education, member of the National Executive Committee and party chairman. Crossman contributed regularly to the *New Statesman*, having been assistant editor there until 1955.

20. Desmond Young (1892–1966), author of the popular biography *Rommel, the Desert Fox* (1950), with forewords by Auchinleck and Wavell. Young had served in North Africa, once being taken prisoner by Rommel's troops, and his biography drew on interviews with Rommel's widow, son and friends, with the support of Liddell Hart. In time the book would be criticised as hagiography.

21. The '8th Army' bar to the Africa Star medal was only awarded for army service between 23 October 1942 (dating from the opening of Montgomery's battle of Alamein) and 12 May 1943. Chink believed that it should also encompass the earlier campaigns.

22. Panzerarmee Afrika, consisting of 90th Light Division and the Afrika Korps.

23. Professor Carlos Baker (1909–87) of Princeton University had been appointed Hemingway's authorised biographer. His earlier book, *Hemingway: The Writer as Artist*, published in 1952, had been acclaimed as a definitive portrait of the writer and his generation, and included a critique of Hemingway's novels.

24. Lieutenant Colonel Gustavo Durán Martinez (1906–69), composer and soldier, was a commander in the army of the Spanish Republic throughout the civil war, where his courage influenced Hemingway. He is praised in Chapter 30 *For Whom the Bell Tolls*.

25. Major General Charles 'Buck' Lanham (1902–78), short story writer and professional soldier, was commanding the US 22nd Infantry Regiment in Normandy in 1944 when Hemingway covered the war there for *Collier's*. Decorated for his role in the Huertgen Forest fighting, he contributes to the character of Cantwell in *Across the River and Into the Trees*.

26. Of the start of World War I, already being planned for August 1964.

27. Ernest Wood, Senior Counsel (1909–91) was a distinguished Dublin barrister who specialized in defamation cases.

28. William 'Bill' Finlay, Senior Counsel (1921–2010), cultivated and debonair, was also Dean of Law at University College, Dublin. He and his wife Verette would become personal friends.

29. Ulick O'Connor (1928–2019) combined his zest for the law with his interest in sport, especially boxing, and his literary leanings. His biography *The Times I've Seen: Oliver St John Gogarty* had just been published in New York.

30. Freddie de Guingand.

31. 'There is, in my opinion, no doubt that an unnecessary and vicious vendetta was waged against him by the Army hierarchy.... I feel sure that it would have been wiser and more profitable to have placed him in a position better suited to his particular gifts.' [*Generals at War*, Hodder and Stoughton 1964, p. 185.]

32. Mesopotamia.

33. Paul Mowrer (1887–1971), American journalist awarded the first Pulitzer Prize for Correspondence in 1929 for his coverage of international affairs for the *Chicago Daily News*. He had met Hadley in Paris two years earlier, shortly after her divorce from Hemingway, and they had married in 1933.

34. Carlos Baker.

35. M.R.D. Foot (1919–2012), Professor of Modern History at Manchester University, had first-hand experience of warfare, having joined the SAS as an intelligence officer and been parachuted into France after D-Day. While a prisoner of war he had been injured in an escape attempt, and postwar retained his interest in European resistance movements. Foot was the author of *SOE in France*, the official history.

36. Fritz Bayerlein (1899-1970) had headed the Afrika Korps in Rommel's time, commanded 3rd Panzer Division during the breakout on the Eastern Front, and led the Panzer Lehr Division in Normandy and the Ardennes in 1944.

37. Despite Manchester University's enthusiasm, the idea did not, eventually, grow into regular joint academic work.

38. The Indian Medical Service, which looked after the British military stationed there during both wars. Officers were both British and Indian, and many served in civilian hospitals.

39. Chink had hoped to meet the writer Compton Mackenzie twelve years earlier when debating at the Wexford Festival, but illness had prevented Mackenzie from attending. This time the correspondence was due to Mackenzie wishing to meet him, on a flying visit to Dublin..

40. An Italian musical term, meaning 'repeat from the beginning'.

41. Ematris Saint John (E) Parish Church, within three miles of Cootehill in County Monaghan.

Index